Lecture Notes in Physics

New Series m: Monographs

The Editorial Policy for Monographs

The series Lecture Notes in Physics reports new developments in physical research and teaching - quickly, informally, and at a high level. The type of material considered for publication in the New Series m includes monographs presenting original research or new angles in a classical field. The timeliness of a manuscript is more important than its form, which may be preliminary or tentative. Manuscripts should be reasonably self-contained. They will often present not only results of the author(s) but also related work by other people and will provide sufficient motivation, examples, and applications.
The manuscripts or a detailed description thereof should be submitted either to one of the series editors or to the managing editor. The proposal is then carefully refereed. A final decision concerning publication can often only be made on the basis of the complete manuscript, but otherwise the editors will try to make a preliminary decision as definite as they can on the basis of the available information.
Manuscripts should be no less than 100 and preferably no more than 400 pages in length. Final manuscripts should preferably be in English, or possibly in French or German. They should include a table of contents and an informative introduction accessible also to readers not particularly familiar with the topic treated. Authors are free to use the material in other publications. However, if extensive use is made elsewhere, the publisher should be informed.
Authors receive jointly 50 complimentary copies of their book. They are entitled to purchase further copies of their book at a reduced rate. As a rule no reprints of individual contributions can be supplied. No royalty is paid on Lecture Notes in Physics volumes. Commitment to publish is made by letter of interest rather than by signing a formal contract. Springer-Verlag secures the copyright for each volume.

The Production Process

The books are hardbound, and quality paper appropriate to the needs of the author(s) is used. Publication time is about ten weeks. More than twenty years of experience guarantee authors the best possible service. To reach the goal of rapid publication at a low price the technique of photographic reproduction from a camera-ready manuscript was chosen. This process shifts the main responsibility for the technical quality considerably from the publisher to the author. We therefore urge all authors to observe very carefully our guidelines for the preparation of camera-ready manuscripts, which we will supply on request. This applies especially to the quality of figures and halftones submitted for publication. Figures should be submitted as originals or glossy prints, as very often Xerox copies are not suitable for reproduction. In addition, it might be useful to look at some of the volumes already published or, especially if some atypical text is planned, to write to the Physics Editorial Department of Springer-Verlag direct. This avoids mistakes and time-consuming correspondence during the production period.
As a special service, we offer free of charge LaTeX and TeX macro packages to format the text according to Springer-Verlag's quality requirements. We strongly recommend authors to make use of this offer, as the result will be a book of considerably improved technical quality. The typescript will be reduced in size (75% of the original). Therefore, for example, any writing within figures should not be smaller than 2.5 mm.
Manuscripts not meeting the technical standard of the series will have to be returned for improvement.
For further information please contact Springer-Verlag, Physics Editorial Department II, Tiergartenstrasse 17, W-6900 Heidelberg, FRG.

P. Christe M. Henkel

Introduction to Conformal Invariance and Its Applications to Critical Phenomena

Springer-Verlag Berlin Heidelberg GmbH

Authors

P. Christe
Institute for Theoretical Physics
University of Berne, Sidlerstrasse 5
CH-3012 Berne, Switzerland

M. Henkel
Department of Theoretical Physics
University of Geneva, 24 quai Ernest Ansermet
CH-1211 Geneva, Switzerland

ISBN 978-3-662-13922-6 ISBN 978-3-540-47575-0 (eBook)
DOI 10.1007/978-3-540-47575-0

Originally published by Springer-Verlag Berlin Heidelberg New York in 1993
Softcover reprint of the hardcover 1st edition 1993

Typesetting: Camera ready by authors
58/3140-543210 - Printed on acid-free paper

Preface

The history of critical phenomena goes back to the year 1869 when Andrews discovered the critical point of carbon dioxide, located at about 31 °C and 73 atmospheres pressure. In the neighborhood of this point the carbon dioxide was observed to become opalescent, that is, light is strongly scattered. This is nowadays interpreted as coming from the strong fluctuations of the system close to the critical point.

Subsequently, a wide variety of physical systems were realized to display critical points as well. Of particular importance was the observation of a critical point in ferromagnetic iron by Curie. Further examples include multicomponent fluids and alloys, superfluids, superconductors, polymers and may even extend to the quark-gluon plasma and the early universe as a whole. Early theoretical investigations tried to reduce the problem to a very small number of degrees of freedom, such as the van der Waals equation and mean field approximations and culminating in Landau's general theory of critical phenomena. In a dramatic development, Onsager's exact solution of the two-dimensional Ising model made clear the important role of the critical fluctuations. Their role was taken into account in the subsequent developments leading to the scaling theories of critical phenomena and the renormalization group. These developements have achieved a precise description of the close neighborhood of the critical point and results are often in good agreement with experiments. In contrast to the general understanding a century ago, the presence of fluctuations on all length scales at a critical point is today emphasized. This can be briefly summarized by saying that at a critical point a system is *scale invariant.*

Conformal invariance has been known for almost a century in connection with scale invariance. For example, Maxwell's equations in the vacuum are scale invariant as well as conformal invariant. This feature arises generally in quantum field theories with a local energy-momentum tensor. However, the first application of conformal invariance to critical phenomena was made only in 1970 by Polyakov. At that time, the consequences of conformal invariance for an arbitrary number of space dimensions were considered to be far from spectacular. This is completely different in two dimensions as was pointed out in the seminal work by Belavin, Polyakov and A. Zamolodchikov in 1984. This is because, for two dimensions, the conformal group is infinite-dimensional and much stronger constraints on the multipoint correlation functions of the system are obtained.

In a sense, conformal invariance is the logical extension of scale invariance. Scale invariance requires the invariance of the system under a *uniform* length rescaling. Conformal invariance also permits a non-uniform, local, rescaling and only requires that angles are kept unchanged. This extension is in fact very natural since it can be shown that for any system which is invariant under translations and rotations, at least in the continuum limit, is scale invariant and has short-ranged interactions, conformal invariance follows automatically.

For two-dimensional systems, conformal invariance has considerably extended our knowledge of the nature of a critical point. While scale invariance alone is capable of casting systems into universality classes only dependent on a few selected properties like the global symmetry, the dimension of the space and of the number of components of the order parameter, two-dimensional conformal invariance yields a classification of the critical point partition functions and thereby furnishes exact values of the critical exponents. Furthermore, the critical multipoint correlation functions of the local variables of the system can be determined exactly.

These prospects should place conformal invariance on the center stage of investigations on critical phenomena. However, the technical tools required for its understanding are quite elaborate. One of the central notions, the *central charge*, requires some profound background on anomalies which is not necessarily in the toolkit of a condensed-matter physicist. Rather, many of the basic concepts and techniques were developed in the context of string theory and most of the existing reviews on the subject assume quite some knowledge on quantum field theory on the side of the reader. On the other hand, we believe that the time is ready for conformal invariance techniques to enter condensed matter applications on a wide front. In writing this introduction, we have tried to meet the needs of a reader with some exposure to scaling and the renormalization group without being an expert in quantum field theory. To do so, we give a joint presentation of both the field theory techniques required as well as their explicit application in lattice systems. Numerical techniques in connection with finite-lattice systems will be emphasized, having in mind a reader with a good physical understanding of the lattice model who is curious about what conformal invariance techniques can reveal about its behaviour. At the same time, we have laid some accent on immediate applications to lattice systems and have tried to illustrate the phenomenological consequences of conformal invariance as explicitly as possible. Finite-size effects, rather than being a nuisance, will appear at a central position throughout.

This book has grown out of a joint two-trimester course held at the University of Fribourg in the winter 1991/1992. We hope it may serve as a first glance into the field and will prepare the reader for the study of more advanced presentations, some of which are given in the general references. In selecting the material to be presented in a first introduction to conformal invariance, we had to restrict ourselves to the basic foundations of the theory and many of the more advanced applications had regretably to be left out completely. Although string theory stands at the origin of conformal invariance, no mention is made of it here. While conformal invariance provides one of the building blocks of strings in higher dimensions, it has also led to profound studies of field theories on a fluctuating metric. These studies include pure two-dimensional quantum gravity as well as spin systems on lattices with random connectivity, rewritten in the form of matrix models. We skip completely the fascinating subject of exactly integrable systems with the exceptions only of the two-dimensional Ising model and a brief sketch of A. Zamolodchikov's theory of two-dimensional systems perturbed away from their critical point. The most spectacular result of this has been the proof that the two-dimensional Ising model at its critical temperature, but in a magnetic field, is integrable. These developments have also stimulated further investigations in mathematics, uncovering for example very deep and interesting relationships to the theory of knots and links. We merely mention here that polynomial invariants of knots have reappeared in connection with the partition functions of two-dimensional integrable systems. Conformal invariance has also added to the general understanding of the flow under the renormalisation group, referred to as the c-theorem. We restrict ourselves here exclusively to its occurrence in two dimensions and do not go into the existing generalisations in four dimensions. Since conformal invariance acts primarily as a dynamical symmetry which allows one to write the spectrum of the transfer matrix in terms of the irreducible representations of the conformal algebra, it is worth looking for conformal symmetries more general than a Lie group. Of the extensions of the conformal group we only briefly mention $N = 1$ superconformal invariance and skip higher superconformal algebras as well as W algebras, braid groups and quantum groups. The only Kac-Moody algebra treated here in any detail is the $U(1)$, including its shifted representations. Temperley-Lieb algebras are merely mentioned although they provide but the first example of a whole class of new symmetry structures present in conformally invariant and integrable systems. On the side of the more advanced applications, we do not cover weakly disordered systems, polymers and random walks, the quantum Hall effect or the Kondo effect. It would be tempting to use conformal invariance for a better understanding of high T_c superconductivity. Conformal invariance techniques have also been useful in calculating exactly the fractal dimensions of clusters as defined for example by the order parameter of spin systems. For an introduction to these active reasearch topics, see the general references in the bibliography or the references to review articles in the text.

In chapter 1, we recall some well-known notions of scaling relevant in connection with their subsequent generalisation to conformal invariance. We also repeat the correspondence between quantum field theory and classical statistical equilibrium mechanics. This is formulated via the *transfer matrix*, which will play a major role in what follows. In chapters 2 to 7, we then give the field theory point of view of conformal invariance. We shall show how this can be used to calculate explicitly the critical multipoint correlation functions and shall go through the example of the Ising model in full detail. Following our two-pronged approach, we present at an early stage (chapter 3) the theory of finite-size scaling and the important contributions to it from conformal invariance. Conversely, the new conformal results provide a simple and efficient means for the calculation of critical exponents and the central charge from a given lattice system. We then turn to lattice systems. A large part of the original work done on conformal invariance uses as a simplifying technical device an extremely anisotropic limit of the transfer matrix, which has the virtue of turning a fully populated matrix into a sparse one. This is quite useful for numerical calculations and also sheds more light on universality, as detailed in chapter 8. We describe the numerical techniques needed with an accent laid on the subleties of finite-lattice extrapolation in chapter 9. The Ising model example is then studied again in chapter 10 to show how a lattice system can be treated from the start from the point of view of conformal invariance.

The subsequent chapters deal with more advanced applications. Modular invariance is treated in chapter 11 and is shown to lead to a classification of the critical point partition functions. Examples beyond the Ising model, along with additional concepts, are presented in chapter 12. In chapters 13 and 14, we consider the effects of both relevant and irrelevant perturbing operators, culminating in the beautiful developments involing S-matrix theory which led to the recognition of the integrability of the two-dimensional Ising model in a magnetic field. Surface critical phenomena are treated in chapter 15 and we close with an outlook towards possible applications in critical dynamics.

To keep track of every new piece of work in this rapidly expanding field has been beyond our capabilites. We compiled in the bibliography the references we used in writing this text and added some more intended as suggestions for further reading. The first half of the general references gives introductory texts on critical phenomena, followed by earlier reviews on conformal invariances and a few suggestions for catching up with the current lines of research. The selection is certainly incomplete and we sincerely apologize to any author who might feel that we did not give proper credit to his contribution to the field. Inevitably the bibiography reflects our interests and/or our lack of knowledge of some directions under the conformal umbrella. Citations in the text are often in historical order.

Finally, we have the pleasant task of thanking all those who have contributed to making this work possible. We thank Prof. D. Baeriswyl for his kind invitation to give the lectures from which this book has grown. We are deeply indebted to R. Flume and V. Rittenberg for introducing us to the subject and for continuous support and encouragement. One or other or both of us have received advise or support from and/or had the pleasure to collaborate with J.L. Cardy, E. Domany, M. Droz, L. Frachebourg, G.v. Gehlen, H.J. Herrmann, A.W.W. Ludwig, M.J. Martins, G. Mussardo, A. Patkós, R. Peschanski, V. Privman, F. Ravanini, H. Saleur, G. Schütz, N.M. Švrakić, R.A. Weston and J.-B. Zuber. Our warmest thanks go to all of them.

We thank the Institut für Theoretische Physik of the Universität Bern and the Département de Physique Théorique of the Université de Genève for support. This work was supported in part by the Swiss National Science Foundation.

Bern and Genève
January 1993

Philippe Christe
Malte Henkel

Contents

List of Tables

List of Figures

1 Critical Phenomena: a Reminder

We begin by recalling some basic facts about critical phenomena. This is not intended to provide a complete and exhaustive introduction to the field. We rather put emphasis on those standard concepts which will form the basis for the extension to conformal invariance in later chapters. Besides this, we establish our notation.

After briefly discussing the phenomenology of second-order phase transitions, the conventional bulk critical exponents are defined. Scale invariance, as implemented via the renomalization group, will be used to derive the well-known scaling relations between the critical exponents. Some of the most conventional model systems used in the study of critical phenomena will be introduced. Finally, we recall the analogy between statistical mechanics and quantum theory.

The material of this chapter is fairly standard and further information can be found in the general overviews quoted.

1.1 Phase diagrams and critical exponents

Consider the simple example of a classical uniaxial ferromagnet. It is described by the Ising (1925) model [200]. The Hamiltonian is

$$\mathcal{H}[\sigma] = -\frac{1}{2}\sum_{i,i'} J(i,i')\sigma_i\sigma_{i'} - B\sum_i \sigma_i \tag{1.1}$$

where B is the external magnetic field and the coupling J decays with the distance and is positive to ensure ferromagnetism. The interaction is supposed to be short-ranged (usually just with nearest neighbours). The classical spin variable σ_i located at site i takes the values ± 1. We consider a lattice with $\mathcal{N}$ sites. We can vary both the temperature T and the external magnetic field B. The phase diagram is simple:

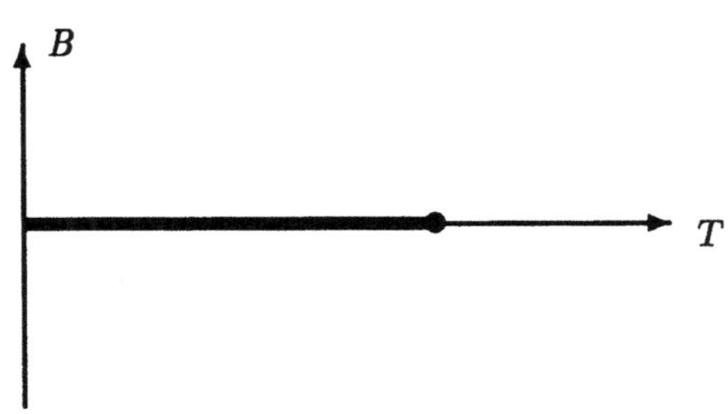

Fig. 1. Typical phase diagram of a ferromagnet. The transition is first order on the thick line, second order at the black dot.

Thermodynamic quantities like the specific heat or susceptibility are analytic functions of T and B except across the line $B = 0$, $T < T_c$. The magnetization M as a function of B depends on the temperature (see Figure 2). At low temperature, M is discontinuous

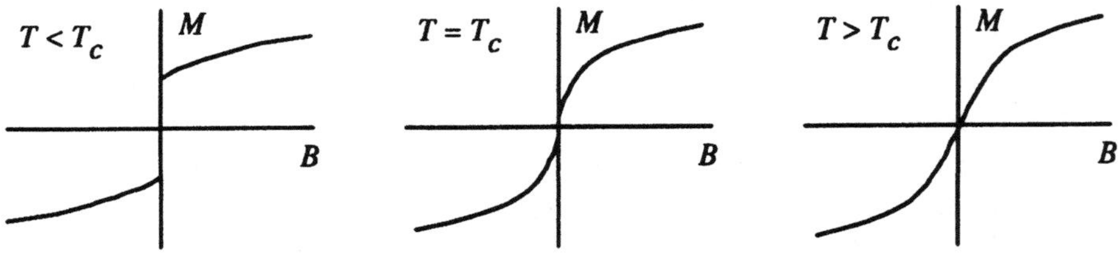

Fig. 2. Magnetization as function of B and T.

across the line $B = 0$ and the model has a first-order transition there. In the high temperature phase, large clusters of spins of the same orientation occur. The size of these clusters defines the *correlation length* ξ. Roughly speaking, if the distance r between two spins is small compared to ξ, the spins are highly correlated, while if $r \gg \xi$, the correlation decreases exponentially with r. As T decreases, at $B = 0$, the size of the clusters grows and so does ξ. At the critical point $T = T_c$, $B = 0$, the correlation length becomes infinite. This divergence characterizes a second order critical point. Fluctuations on all length scales become important. The fact that one has fluctuations on **all** scales is nicely illustrated by Kadanoff's droplet picture [209] (see Figure 3) which shows configurations of nested droplets with spin "up" inside droplets with spin "down" and so on. At the

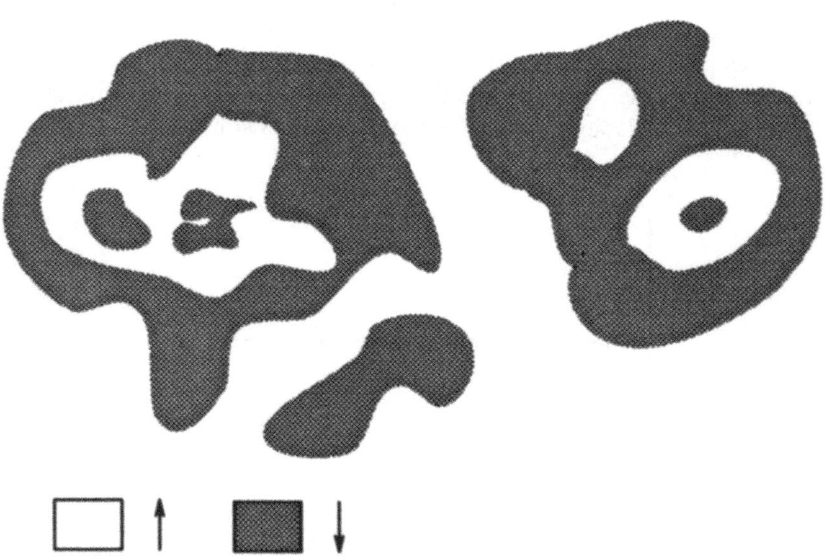

Fig. 3. Droplet picture (schematic).

same time, microscopic informations such as the lattice structure become irrelevant and one expects the critical behaviour, as described by the critical exponents, to be independent of them. For example, if we had nearest neighbour and next-to-nearest neighbour interactions (both of them ferromagnetic), the critical behaviour would be independent of the ratio of the two couplings. This is an example of ***universality***. Critical points can be cast into ***universality classes*** characterized by the values of the critical exponents. We shall see later that for two-dimensional systems, ***conformal invariance*** will lead to a partial classification of the possible universality classes. Critical behaviour is described by the deviation from the critical point which is given by (in the context of the Ising model example considered) the ***reduced temperature*** t and the ***reduced magnetic field*** h where

$$t = \frac{T - T_c}{T_c} \qquad \text{and} \qquad h = \frac{B}{k_B T_c} \tag{1.2}$$

From now on we use units such that the Boltzmann constant $k_B = 1$.

We recall the calculation of the fundamental thermodynamic quantities. The ***partition function*** is

$$\mathcal{Z} = \mathbf{Tr}_\sigma \, e^{-\frac{1}{T}\mathcal{H}[\sigma]} \tag{1.3}$$

Tr means here the sum over all possible spin configurations. The free energy is

$$F = -T \ln \mathcal{Z} \tag{1.4}$$

and we calculate the thermodynamic quantities in the standard fashion

$$\begin{aligned} C &= -\frac{T}{\mathcal{N}} \frac{\partial^2 F}{\partial T^2} \\ M &= -\frac{1}{\mathcal{N}} \frac{\partial F}{\partial B} \\ \chi &= \frac{\partial M}{\partial B} \end{aligned} \tag{1.5}$$

where C is the specific heat, M is the magnetization per site and χ is the susceptibility per site and $\mathcal{N}$ is the number of sites. Furthermore, the ***two point correlation function*** $G(i,j)$ of the spins σ_i and σ_j at sites i and j is defined as (r is the distance between the sites i and j)

$$G(r) = G(i,j) = <\sigma_i \sigma_j> - <\sigma_i><\sigma_j> \tag{1.6}$$

where the ***average value*** $< X[\sigma] >$ for any functional $X[\sigma]$ is

$$< X[\sigma] >= \frac{1}{\mathcal{Z}} \mathbf{Tr}_\sigma \left(X[\sigma] e^{-\frac{1}{T}\mathcal{H}[\sigma]} \right) \tag{1.7}$$

For example, the magnetization per site M can be expressed

$$M =< \sigma_i >= \frac{1}{\mathcal{N}\mathcal{Z}} \frac{\partial \mathcal{Z}}{\partial B} \tag{1.8}$$

Away from criticality and at large distances $r \gg 1$, $G(r)$ decays exponentially

$$G(r) \approx e^{-|r|/\xi}/|r|^\tau \ , \qquad T \neq T_c \, , \, B = 0 \tag{1.9}$$

which defines the ***correlation length*** ξ and d is the dimension of the lattice. The exponent τ depends on whether T is larger or smaller than T_c, in the two-dimensional Ising model one has for example $\tau = 1/2$ for $T > T_c$ and $\tau = 2$ for $T < T_c$ [G10].

We write $X(t) \sim |t|^{x_\pm}$ for some thermodynamic quantity X if

$$\lim_{t \to \pm 0} \frac{\ln X(t)}{\ln |t|} =: x_\pm \tag{1.10}$$

exists. This defines the critical exponents $x_\pm$. In most cases, the exponent obtained is the same for $t < 0$ and $t > 0$, that is $x_+ = x_-$. We shall henceforth assume this to be the case.

The conventional *critical exponents* α, β, γ, and δ are now defined as follows

$$\begin{aligned} C_{(h=0)} &\sim |t|^{-\alpha} \\ M_{(h=0)} &\sim (-t)^{\beta} \quad , \quad t < 0 \\ \chi_{(h=0)} &\sim |t|^{-\gamma} \\ M_{(t=0)} &\sim h^{1/\delta} \end{aligned} \tag{1.11}$$

However, the associated *scaling amplitudes* (not detailed in equation (1.11)) do depend on the sign of t. We do not introduce them here but refer to the excellent review [283] for further information. The correlation length diverges as we approach the critical point along the line $B = 0$ like

$$\xi \sim |t|^{-\nu} \tag{1.12}$$

defining the exponent ν. The large distance behaviour of $G(r)$ at criticality defines the critical exponent η

$$G(r) \sim |r|^{-d+2-\eta}, \qquad T = T_c \, , \, B = 0 \tag{1.13}$$

This completes the definition of the conventional bulk critical exponents. One of the main problems in critical phenomena is their explicit computation. As we shall see, conformal invariance permits to obtain exact results in two dimensions.

1.2 Scaling relations

Not all of the bulk critical exponents defined are independent. In fact, scale invariance can be used to derive a number of relations between them, as we now show.

A convenient way to implement scale invariance is via a real-space renormalization group procedure. The system is characterized by its number of degrees of freedom and the values of the parameters t and h, as specified by the Hamiltonian $\mathcal{H}[\sigma; t, h]$. Now consider a blocking of spins. One prescribes a rule to assign a value for the *block spin* of a block of size $b \times b \times \ldots$ as a function of all the spins contained in that block, where $b > 1$ is some constant. For example, one might choose the *majority rule* which gives the value of the block spin as $\mathrm{sgn}\left(\sum_{i \in \text{block}} \sigma_i\right)$. The Hamiltonian $\mathcal{H}'[\sigma'; t', h']$ describes the system written in terms of the block spins σ'. It only contains one degree of freedom for each block spin σ' and the degrees of freedom of the individual spins σ within a block have been integrated out. This transformation must be implemented such that the partition function remains invariant, up to a rescaling factor. The new Hamiltonian $\mathcal{H}'$ should have the same symmetries as the old Hamiltonian $\mathcal{H}$. This procedure defines a mapping from the old parameters (t, h) to the new ones (t', h'). For simplicity, we discard here the apperance of new scaling fields in the transformed Hamiltonian $\mathcal{H}'$. While in principle these extra fields must be kept, they are often irrelevant (in a precise sense defined below) for the description of the leading critical singularity and their inclusion does not change the results of the scaling analysis of this section. Close to the critical point the new parameters are in first approximation a linear combination of the old ones and we can find functions of the parameters, called *scaling fields* $\{u_t, u_h\}$ which transform according to

$$u'_{t'}(t', h') = b^{y_t} u_t(t, h) \quad , \quad u'_{h'}(t', h') = b^{y_h} u_h(t, h) \tag{1.14}$$

The numbers y_t and y_h are called the *renormalization group eigenvalues*. We will relate them below to the critical exponents defined above. If t, h, are small enough, we have in first approximation

$$u_t \sim t, \qquad u_h \sim h \tag{1.15}$$

The scaling fields u can be classified as follows. If $y > 0$, u is called *relevant* and repeated renormalization group transformations drive the system away from its critical point if initially $u \neq 0$. If $y < 0$, u is called *irrelevant* and if initially u is small enough, the system will converge towards the fixed point. Finally, if $y = 0$, the scaling field u is called *marginal.*

While for the Ising model, one has just the two relevant fields corresponding to t and h, for more general models one can have additional relevant scaling fields. We shall encounter examples of these as we go along.

The free energy $F = F^{reg} + F^{sin}$ generically contains a regular and a singular part. While the regular part F^{reg} remains analytic at the critical point and, acting essentially as a background term, is of little interest here, the singular part F^{sin} does contain the terms which produce, for example, the singular behaviour of the specific heat. We shall exclusively concentrate on studying F^{sin}. The singular part of the *free energy density* $f = F/\mathcal{N}$ rescales as follows

$$f^{sin}(u_t, u_h) = b^{-d} f^{sin}(b^{y_t} u_t, b^{y_h} u_h) \tag{1.16}$$

The factor b^{-d} results from the fact that the number of sites $\mathcal{N}$ which appears in the definition of f is reduced by a factor b^{-d}. After n repeated transformations

$$f^{sin}(u_t, u_h) = b^{-dn} f^{sin}(b^{ny_t} u_t, b^{ny_h} u_h) \tag{1.17}$$

We cannot let n grow too large because u_t and u_h become eventually too large and the linear approximation involved in this approach would be invalid. But we can iterate until we reach a specific point $|b^{ny_t} u_t| = K$, K small. Solving for n we find

$$f^{sin}(u_t, u_h) = K^{-d/y_t} u_t^{d/y_t} f^{sin}(K, K^{y_h/y_t} u_t^{-y_h/y_t} u_h) \tag{1.18}$$

This leads to the scaling functions $W_\pm$ which enter into the free energy

$$f^{sin}(t, h) = \left|\frac{t}{t_0}\right|^{d/y_t} W_\pm \left(\left(\frac{h}{h_0}\right) \left(\frac{t}{t_0}\right)^{-y_h/y_t} \right) \tag{1.19}$$

where the index $\pm$ refers to $t > 0$ or $t < 0$, respectively. The scaling functions $W_\pm$ are universal, the information described by all the irrelevant scaling fields being condensed into t_0 and h_0.

It is now easy to obtain relations between the different critical exponents cited in the previous section. We obtain α from the specific heat which is given by

$$C_{(h=0)} = \partial^2 f/\partial t^2|_{(h=0)} \sim |t|^{d/y_t-2} \Longrightarrow \alpha = 2 - d/y_t \tag{1.20}$$

The other exponents β, γ and δ are computed from $M_{(h=0)} = -\partial f/\partial h|_{(h=0)}$ and $\chi_{(h=0)} = -\partial^2 f/\partial h^2$. The result is

$$\begin{aligned} \alpha &= 2 - d/y_t \\ \beta &= (d - y_h)/y_t \\ \gamma &= (2y_h - d)/y_t \\ \delta &= y_h/(d - y_h) \end{aligned} \tag{1.21}$$

Taking advantage of the dependence in y_t and y_h, we obtain the celebrated *scaling relations*

$$\boxed{\begin{aligned} \alpha + 2\beta + \gamma &= 2 \\ \alpha + \beta(1+\delta) &= 2 \end{aligned}} \tag{1.22}$$

For the other exponents, consider the correlation function. We expect the scaling behaviour

$$G(r; t, h) = b^{-2x} G(r/b; tb^{y_t}, hb^{y_h}), \qquad r = |i - j| \tag{1.23}$$

for some x. For simplicity, we put $h = 0$. As before, after n iterations so that $tb^{y_t} = K$, we get

$$G(r; t) \sim t^{2x/y_t} \hat{G}\left(r/(K/t)^{1/y_t}; K\right) \tag{1.24}$$

Comparing this with the exponential decay, $\hat{G}$ must depend on r/ξ, and we conclude

$$\xi \sim t^{-1/y_t} \Longrightarrow \nu = 1/y_t \tag{1.25}$$

We now want to have a finite limit for G as $t \to 0$. This means that at criticality $\hat{G} \sim r^{-2x}$. The susceptibility may be expressed as (fluctuation-dissipation theorem)

$$\chi = T^{-1} \sum_r G(r) \tag{1.26}$$

Replacing the sum by an integral and by suitably changing the variable, we find

$$\chi \sim t^{(2x-d)/y_t} \Longrightarrow x = d - y_h \tag{1.27}$$

and $\eta = d + 2 - 2y_h$. This leads to

$$\boxed{\gamma = \nu(2 - \eta)} \tag{1.28}$$

Finally, we discuss hyperscaling. From the scaling behaviour of the singular free energy density (1.16), we read off

$$\boxed{\alpha = 2 - d\nu} \tag{1.29}$$

This is only valid provided that the scaling functions $W_\pm$ remain finite at the critical point. One can show that no singularities occur in the scaling functions below the upper critical dimension d^*. For the Ising model, $d^* = 4$. For $d \geq d^*$, on the other hand, singularities of the free energy scaling function occur generically and are due to so-called *dangerous irrelevant* scaling fields (Fisher [126]). While dangerous irrelevant variables are absent for $d < d^*$, the breaking of hyperscaling is still possible if the scaling functions vanish precisely at the critical point.

One might think that the scaling eq. (1.23) of the correlation function is independent of the scaling eq. (1.16) of the free energy. That this is not so is seen as follows. From (1.26), we have $\chi \sim \int drG(r)$. Integrating (1.23), we get for the scaling of the susceptibility

$$\chi(t,h) = b^{d-2x}\chi(tb^{y_t}, hb^{y_h}) \tag{1.30}$$

Recall that $\chi = -\partial^2 f/\partial h^2$. Therefore, using (1.27), we recover indeed (1.16). The importance of this remark comes from the fact that exactly at criticality $t = h = 0$ and eq. (1.23) expresses the *covariance* of the correlation function under scale transformations. The generalisation towards the covariance of the critical point correlation functions under more general transformations will be the starting point for conformal invariance in chapter 2.

It is customary to introduce the following notation. We call

$$x_\phi = d - y_\phi \tag{1.31}$$

the *scaling dimension* of the *scaling operator* ϕ. The scaling operators are canonically conjugated to the scaling fields. In particular, y_t corresponds to the scaling field t and y_h corresponds to the scaling field h. The conjugated lattice operators are in first approximation the energy density ε for t

$$\varepsilon_i = \sum_{i'} J(i,i')\sigma_i\sigma_{i'} \tag{1.32}$$

and the magnetization density σ for h. In the example of the $2D$ Ising model, one has for the relevant scaling operators the following numerical values of the scaling dimensions

$$x_\varepsilon := x_t = \frac{1-\alpha}{\nu} = 1 \ , \ x_\sigma := x_h = \frac{\beta}{\nu} = 1/8 \tag{1.33}$$

We have thus seen that scale invariance, as implemented by the renormalization group, yields relations between the critical exponents (or equivalently the relevant scaling dimensions), but without fixing their numerical value. In later chapters we shall use conformal invariance techniques to calculate exactly the scaling dimensions of *all* scaling operators (that is, relevant, marginal and irrelevant) for a large variety of systems. This amounts to give the full *operator content* of a spin system.

1.3 Some simple spin systems

We now give a list of some well-known models, which we shall study later in much detail. We have already defined the Ising model and have given the scaling dimensions of the relevant scaling operators. The models presented here are arranged such that they show a steady evolution from the simple Ising to more complex structures. We only discuss isotropic models on square lattices in this section. In later chapters, we will often consider anisotropic models where the couplings in the two different directions of the square lattice are different. However, the universal aspects of the critical behaviour of such anisotropic models are the same as those of the corresponding isotropic ones and conformal invariance still applies.

The Yang-Lee edge singularity. We begin with a particular aspect of the Ising model which Hamiltonian is

$$\mathcal{H} = -J \sum_{<i,i'>} \sigma_i \sigma_{i'} - B \sum_i \sigma_i \tag{1.34}$$

and $< i, i' >$ indicates that the sum is taken over the nearest neighbours only. Yang and Lee [332] first discussed the analytic behaviour of the magnetization per site M in the Ising model. This can be deduced from the asymptotic distribution of the zeroes of the partition function $\mathcal{Z}$ as a function of a complex magnetic field $B = \mathrm{Re}B + i\mathrm{Im}B$. The Yang-Lee theorem [332] states that the zeroes are exclusively located on the imaginary axis $\mathrm{Re}B = 0$. Below the critical temperature, they accumulate at $\mathrm{Im}B = 0$ and there is a first-order phase transition with a jump in M when the line $\mathrm{Re}B = 0$ is crossed. However, above the critical temperature, there is a T-dependent lower bound $|B_c|$ for the absolute values of the imaginary parts of the zeroes of the partition function. The situation is

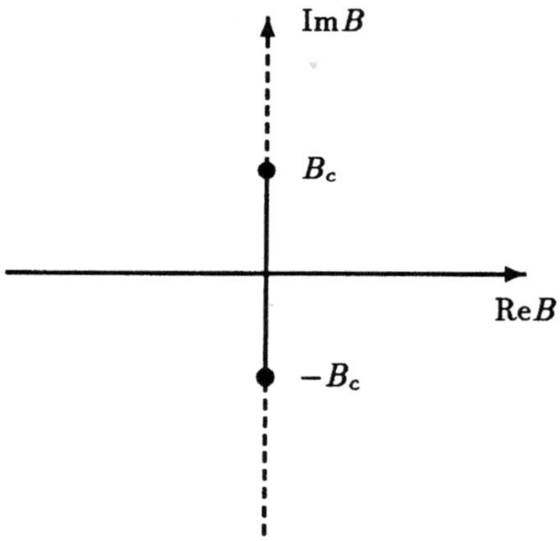

Fig. 4. Zeroes of the Ising model partition function at $T > T_c$. The zeros are located on the dashed lines and accumulate at the branch points $\pm B_c$ (thick dots).

depicted in Figure 4. At these branch points, called *Yang-Lee edge singularities*, the sets of zeroes are dense. The magnetization per spin scales as

$$M \sim (\mathrm{Im}B - B_c)^{1/\delta_{YL}} \tag{1.35}$$

for $|\mathrm{Im}B| \gtrsim |B_c|$. Fisher [125] showed that the points $B = \pm iB_c$ are new critical points of the theory. We wrote δ_{YL} to avoid confusion with the Ising value at the usual critical point $T = T_c$ (the exponent $\sigma = 1/\delta_{YL}$ is often referred to as Fisher exponent). The critical theories at $B = \pm iB_c$ are conformal. We shall see later that in two dimensions $\delta_{YL} = -6$ and that the corresponding scaling operator has scaling dimension $x = -2/5$. We note for later reference that this critical model is obtained in higher dimensions as the infrared limit of the non-unitary scalar field theory with the Landau-Ginzburg-Wilson action [125]

$$S = \int d^d x \left(\frac{1}{2}(\partial\phi)^2 + i(\mathrm{Im}B - B_c)\phi + \frac{1}{3} i\lambda\phi^3 \right) \tag{1.36}$$

Landau-Ginzburg-Wilson actions will be discussed in more detail in chapter 8. From this action, the critical exponents can be calculated in $6-\epsilon$ dimensions with the result [125]

$$\delta_{YL} = 2 - \frac{\epsilon}{3} + \mathcal{O}(\epsilon^2) \ , \ \eta_{YL} = -\frac{\epsilon}{9}\left(1 + \frac{43}{81}\epsilon\right) + \mathcal{O}(\epsilon^3) \tag{1.37}$$

If this is extrapolated down to two dimensions, one finds indeed to good accuracy $\delta_{YL} \simeq -6$, in agreement with the result of conformal invariance.

The tricritical Ising model. The tricritical Ising model (for reviews, see [237, 224]) combines the classical Ising spins $\sigma_i = \pm 1$ with annealed vacancies $v_i = 0, 1$ and the classical Hamiltonian is

$$\mathcal{H} = -J \sum_{<i,i'>} \sigma_i \sigma_{i'} v_i v_{i'} - \mu \sum_i v_i - B \sum_i \sigma_i v_i \tag{1.38}$$

The first sum is made ove the nearest neighbour pairs $< i, i' >$. The lattice is only partially filled with Ising spins (where $v = 1$). The local effective spin which couples to the the magnetic field B is therefore $v_i \sigma_i$. The coupling constant J is positive and μ is the chemical potential because $\sum_i v_i$ counts the number of occupied sites. The model has a global symmetry group Z_2, if $B = 0$.

Let us discuss qualitatively the phase diagram. In the limit $\mu \to \infty$, the ground state is obtained with $v = 1$ everywhere and we are back to the ordinary classical Ising model. When $\mu \to -\infty$, all sites are empty ($v = 0$) and no magnetic properties are observable at any temperature. Suppose now that $B = 0$. In the zero temperature limit $T = 0$ of the partition function $\mathcal{Z} = \exp(-\mathcal{H}/T)$, there will be two possible ground-states, one of the Ising model type and the other being non-magnetic. Between these two regimes, the system (1.38) has therefore a first-order phase transition for some specific value of μ. From this point begins a line of first-order phase transition as we increase the temperature. On the other side, we know that a second-order critical line must begin at the critical Ising point when $\mu = \infty$. Along this line, the system is in the ordinary Ising model universality class. Both lines meet in the $(T, e^{-\mu})$ plane at a point $(T_t, e^{-\mu_t})$. This is the *tricritical point* of the tricritical Ising model. The tricriticality comes from the fact that if we introduce the magnetic field B, two critical wings attached to the first-order transition line will symmetrically develop on both sides of the plane $(T, e^{-\mu})$ in the three-dimensional space of parameters $(T, e^{-\mu}, B)$. In other words, at a tricritical point, **three** phases become indistinguishable, as opposed to just two at an ordinary critical point. For a finite non-vanishing value of B, a first order line of Figure 5 becomes a first order surface in the enlarged space of coupling constants, starting at zero temperature and ending at a second-order critical line. These two surfaces for $B > 0$ and $B < 0$ meet the first order $B = 0$ surface at a line of *triple points*, which ends at the tricritical point $(T_t, e^{-\mu_t}, 0)$.

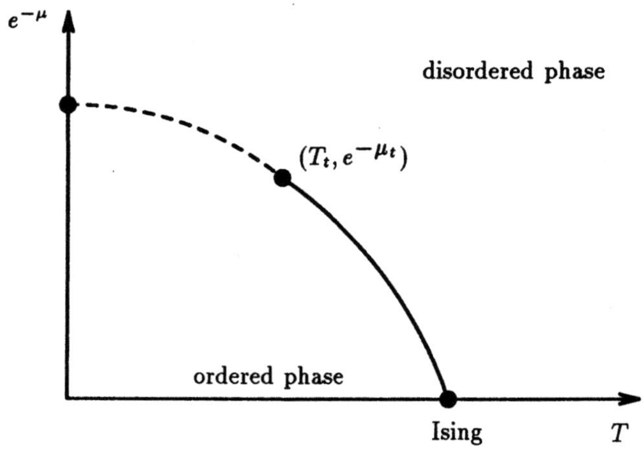

Fig. 5. Phase diagram of the tricritical Ising model. The dashed line is a first-order transition line and the continuous line is of second order.

The Ising tricritical point $(T_t, e^{-\mu_t})$ is characterized by several relevant scaling operators (see chapter 12). In two dimensions, the energy-density operator ε has scaling dimension $x_\varepsilon = 1/5$. The vacancy-density operator $v = \varepsilon'$ has $x_v = 6/5$. There is a magnetic operator σ with scaling dimension $x_\sigma = 3/40$, and a sub-leading magnetic operator σ' with scaling dimension $x_{\sigma'} = 7/8$.

The *p*-state Potts model. The p-state Potts [276] model (1951) extends the Ising model to the case where the spin variable Θ can take p different values, for example $\Theta = 0, 1, \ldots, p-1$. For reviews, see [329, 247]. The coupling between nearest neighbours is given by the Hamiltonian

$$\mathcal{H} = -J \sum_{<i,i'>} \delta(\Theta_i - \Theta_{i'}) \tag{1.39}$$

We make use sometimes of the unusual notation $\delta(\Theta)$ for the Kronecker delta matrix $\delta_{\Theta,0}$. The global symmetry of the model is the permutation group S_p of p objects. In the particular case of the Ising model $p = 2$, we have the simple correspondence $\delta(\Theta_i - \Theta_{i'}) = (\sigma_i \sigma_{i'} + 1)/2$ with $\sigma = 2(\Theta - \frac{1}{2})$. As in the Ising model, there is a disordered high-temperature phase and an ordered low-temperature phase. The transition between these two phases is of second order for $p = 2, 3, 4$ and of first order for $p > 4$. We will be of course interested in the critical cases $p \leq 4$. The operator contents of the 3-state and 4-state Potts model are more complicated and they will be discussed in chapter 12.

For an experimental example, a convenient test system for studying two-dimensional critical behaviour is provided by adsorbed monolayers on single-crystal surfaces. For the system O/Ru(0001) at 1/4 monolayer, the following exponent values were found [271]

$$\alpha = 0.60 \pm 0.04 \;\; , \;\; \beta = 0.085\,^{+0.015}_{-0.005} \;\; , \;\; \gamma = 1.08 \pm 0.07 \;\; , \;\; \nu = 0.68 \pm 0.03 \tag{1.40}$$

and agree to within 3 – 10% with the values found for the four-states Potts model from conformal invariance (see Table 1 below). The precise relationship of the spin systems as defined here and the experementally realized lattice gases is quite intricate and will not be detailed here. For experiments on the critical behaviour of two-dimensional magnets, see [E2, E3].

The ordered low-temperature phase and the disordered high-temperature phase are related by a *duality transformation*. Here, we shall give the conventional derivation (see e.g. [329]), which relates the high-temperature and low-temperature expansions of the model. Alternative techniques to derive this result will be discussed in chapters 8 and 9. For simplicity, we absorb the temperature T into the coupling constant J and consider the partition function with periodic boundary conditions

$$\mathcal{Z} = \sum_{\{\Theta\}} e^{J \sum_{<i,i'>} \delta(\Theta_i - \Theta_{i'})} \tag{1.41}$$

We suppose that we have a square lattice with $\mathcal{N}$ sites and therefore $2\mathcal{N}$ links connecting them.

a) *J* large (*T* small): In this case, we factorize a major contribution from the sum by writing $\delta = 1 + (\delta - 1)$

$$\mathcal{Z} = e^{2\mathcal{N}J} \sum_{\text{closed graphs}} \left(e^{-J}\right)^{L} \nu(\text{graph}) \tag{1.42}$$

At low temperatures, the ground state of the model shows large ordered patches where Θ_i is constant. The only contributions to the sum (1.42) come from the *domain walls* which separate two values of Θ for a given configuration. These walls form closed graphs as depicted in Figure 6. The walls connect points which are at the center of the square spin

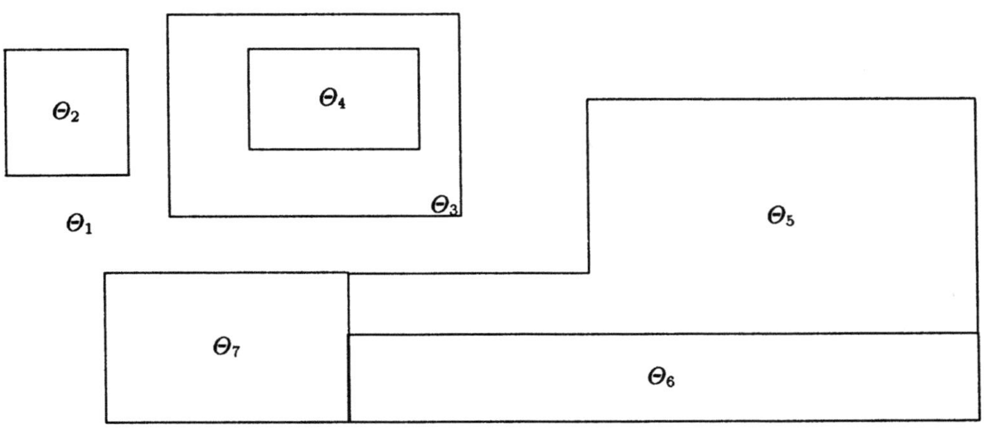

Fig. 6. Example of a closed graph in the p-state Potts model.

plaquettes. The graphs are therefore defined on the *dual lattice*. By definition, the dual lattice is obtained from a given lattice by placing a *dual site* on each bond of the original lattice. The multiplicity ν(graph) of a given graph depends on the number of possible configurations of spin variables for a given graph, and L is the total perimeter of that graph. The low-temperature expansion of the free energy per site $f = -\lim_{\mathcal{N}\to\infty} \mathcal{N}^{-1} \ln \mathcal{Z}$ can be written

$$f(J) = g\left(e^{-J}\right) + 2J \tag{1.43}$$

The function g represents the result from the contributions to the summation.

b) J small (T large): We replace the Kronecker delta term by the sum

$$p\delta(\Theta_i - \Theta_{i'}) = \sum_{k=0}^{p-1} \left(\Gamma_{\Theta_i}\Gamma^*_{\Theta_{i'}}\right)^k =: 1 + A_{\Theta_i,\Theta_{i'}} \tag{1.44}$$

where

$$\Gamma_{\Theta_j} = \exp\left(i\frac{2\pi}{p}\Theta_j\right) \tag{1.45}$$

and the matrix $A_{\Theta,\Theta'}$ is defined by

$$A_{\Theta,\Theta'} = \sum_{k=1}^{p-1} \left(\Gamma_\Theta \Gamma^*_{\Theta'}\right)^k = \begin{cases} p-1 \text{ if} & \Theta = \Theta' \\ -1 \text{ if} & \Theta \neq \Theta' \end{cases} \tag{1.46}$$

We can now work out the summand of the partition function and write

$$\begin{aligned} e^{J\delta(\Theta_i - \Theta_{i'})} &= 1 + \delta(\Theta_i - \Theta_{i'})\left(e^J - 1\right) \\ &= 1 + \frac{1}{p}\left(e^J - 1\right) + \frac{1}{p}\left(e^J - 1\right) A_{\Theta_i,\Theta_{i'}} \\ &= \frac{p + e^J - 1}{p}\left(1 + \frac{e^J - 1}{p + e^J - 1} A_{\Theta_i,\Theta_{i'}}\right) \end{aligned} \tag{1.47}$$

The partition function is now expressed as the infinite product

$$\mathcal{Z} = \left(\frac{p + e^J - 1}{p}\right)^{2\mathcal{N}} \sum_{\{\Theta\}} \prod_{<i,i'>} \left(1 + \frac{e^J - 1}{p + e^J - 1} A_{\Theta_i,\Theta_{i'}}\right) \tag{1.48}$$

We factorized out the term which is independent of the local value of the spin. There is one of those factors for each link of the lattice. We can now expand the product as a power series in $(e^J - 1)/(p + e^J - 1)$. We can graphically represents this by drawing a line between the pair of sites i and i' each time a term $A_{\Theta_i,\Theta_{i'}}$ contributes. This creates graphs on the lattice itself. We will now show that the only non-zero graphs are the ones equivalent to those depicted in Figure 6. This means that if a line terminates, the sum over the possible spin values of the last site cancels the whole graph. This is true because we can write independently of the site i' and of the details of a given graph

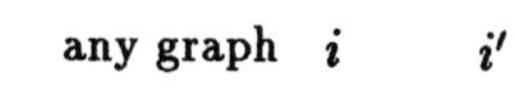

$$\sim \sum_{\Theta_{i'}=0}^{p-1} A_{\Theta_i,\Theta_{i'}} = \sum_{k=1}^{p-1} \Gamma^k_{\Theta_i} \sum_{\Theta_{i'}=0}^{p-1} \exp\left(-i\frac{2\pi}{p}k\Theta_{i'}\right)$$

$$= \sum_{k=1}^{p-1} \Gamma^k_{\Theta_i}\left(p\,\delta_{k,0}\right) = 0 \tag{1.49}$$

Therefore the non-trivial graphs, though residing on the lattice itself, are equivalent to the dual graphs of the large-J expansion. Moreover the sites which do not belong to any walls of a given graph give a factor p to partition function. This corresponds to the number of possible configurations at the given site. The sites of the graphs can be seen as n-point vertices, $n = 2$, 3 or 4, labeled by the site i where we want to sum over the p configurations. The neighbours on the graph are labeled by $(s) = (1), \ldots, (n)$ and we can chose arbitrarily $A_{\Theta_{(s)},\Theta_i}$ or $A_{\Theta_i,\Theta_{(s)}}$ for each link. The sum over all spin values at such a site is given by the sum

$$\sum_{\Theta_i=0}^{p-1} A_{\Theta_i,\Theta_{(1)}} \cdots A_{\Theta_{(s)},\Theta_i} \cdots A_{\Theta_{(n)},\Theta_i} \sim \sum_{\Theta_i=0}^{p-1} e^{-i\frac{2\pi}{p}(\pm k_{(1)} \pm \cdots \pm k_{(n)})\Theta_i}$$
$$= p\,\delta_{\pm k_{(1)} \pm \cdots \pm k_{(n)}, 0 \bmod (p)} \tag{1.50}$$

The sign $\pm$ depends of course of the particular choice $A_{\Theta_{(s)},\Theta_i}$ or $A_{\Theta_i,\Theta_{(s)}}$ and therefore the sum of the indices $\pm k_{(s)}$ must be zero *modulo* p (actually 0, p, $2p$ or $3p$). There is anyway a factor p for each site and we can write the partition function

$$\mathcal{Z} = \left(\frac{p + e^J - 1}{p}\right)^{2\mathcal{N}} p^{\mathcal{N}} \sum_{\text{closed graphs}} \left(\frac{e^J - 1}{p + e^J - 1}\right)^L \tilde{\nu}(\text{graph}) \tag{1.51}$$

This expression is very similar to eq. (1.42) for the large J regime. The perimeter of the graph is again L and the factor $\tilde{\nu}$ counts the multiplicity of the number of compatible combinations of the parameters k to satisfy the constraint given by the delta function on the r.h.s. of eq. (1.50). It is now a simple exercice to check that the combinatoric problems in both eqs. (1.51) and (1.42) are in fact equivalent and that we have

$$p\,\tilde{\nu}(\text{graph}) = \nu(\text{graph}) \tag{1.52}$$

The free energy can therefore be written with the same g function as in eq. (1.43) and we finally obtain

$$f(J) = g\left(\frac{e^J - 1}{p + e^J - 1}\right) + \ln\left(\frac{(p + e^J - 1)^2}{p}\right) \tag{1.53}$$

Notice that in this expansion, we never had to assume J to be small. The free energy is therefore valid for any J and the function g must satisfy

$$g\left(e^{-J}\right) = g\left(\frac{e^J - 1}{p + e^J - 1}\right) + 2\ln\left(\frac{p + e^J - 1}{e^J p^{1/2}}\right) \tag{1.54}$$

The last equation is invariant under the duality transformation which relates large and low values of J

$$e^J \longrightarrow 1 + \frac{p}{e^J - 1} \tag{1.55}$$

If the critical point is unique, which is the case for this model, we can conclude that it must coincide with the fixed point of the *duality transformation*. Therefore the critical value for the parameter J is given by

$$J_c = \ln(p^{1/2} + 1) \tag{1.56}$$

This result will be confirmed later by other methods.

The vector Potts model. There is a variant of the p-state Potts model, sometimes called *clock model*, but which is equivalent to the p-state Potts model only for $p = 2, 3$. The site variables are defined by $\Theta = \frac{2\pi}{p} r$, $r = 0, \ldots, p-1$ and the classical Hamiltonian is

$$\mathcal{H} = -J \sum_{<i,i'>} \cos(\Theta_i - \Theta_{i'}) \tag{1.57}$$

and the model is self-dual. The global symmetry group is Z_p. The structure of the phase diagram is quite different from the one of the ordinary Potts model, however [117]. While there is a single second-order transition for $p \leq 4$, for $p > 4$, the model has **several** distinct critical points. These separate a high-temperature disordered phase, an intermediate phase and a low-temperature vortex phase. While the transition between the first two of these phases is a conventional second-order transition (for a discussion of the operator content see [292]), the other transition is in the universality class of the XY model to be discussed below.

We shall see later that conformal invariance yields a classification of the partition functions of critical systems. It is therefore interesting to compare with previous attempts to get an overwiew on the possible universality classes in two dimensions. In the context of the lattice gas realizations of the discrete spin systems, Landau theory arguments were used to propose [106, 299] that only four universality classes were possible for the critical melting transitions of adatoms adsorbed to a surface. This is illustrated in Table 1 (after [299, E4]). The columns give the possible realizations together with the exactly known

Table 1. Possible melting transitions of adatoms.

		Ising	Potts-3	Potts-4	XY/cubic aniso.	Examples
Adsorption	α	0(log)	1/3	2/3		
Site	β	1/8	1/9	1/12	non-universal	
Symmetry	γ	7/4	13/9	7/6		
	ν	1	5/6	2/3		
Skew (p1)		(2×1)				Ni(110)
or		(1×2)				Au(110)
Rectangular (p2mm)		$c(2 \times 2)$				W(112)
Centered Rect. (c2mm)					(2×2)	Ni(100)
or		$c(2 \times 2)$			(1×2)	W(100)
Square (p4mm)					(2×1)	W(110)
Triangular (p6mm)			$(\sqrt{3} \times \sqrt{3})$	(2×2)		graphite
Honeycomb (p6mm)		(1×1)		(2×2)		
Honeycomb in crystal field (p3m1)			$(\sqrt{3} \times \sqrt{3})$	(2×2)		Ni(111)

critical exponents and some experimental examples. The non-universal critical behaviour of the XY model with cubic anisotropy will be seen later in the example of the Ashkin-Teller model [106] (see chapters 11,12). For the details of the classification scheme and the crystallographic notation used we refer to the literature. In fact, there are many more universality classes known in two dimensions. These can for example be defined in terms of restricted RSOS models which are introduced below.

The XY model. The classical XY model is defined by the Hamiltonian

$$\mathcal{H} = -J \sum_{<i,i'>} \cos(\Theta_i - \Theta_{i'}) \tag{1.58}$$

where the spin variables Θ_i are continuously varying in the interval $[0, 2\pi]$. It can therefore be considered as the $p \to \infty$ limit of the vector Potts model. The XY model has a global $U(1)$ symmetry. There is a huge variety of systems whose critical behaviour can be described in terms of the XY model, including superfluids, superconductors, ordering in liquid crystals or the roughening transition. In two dimensions, the Mermin-Wagner theorem [255] states that no **local** order parameter in a model with a continuos symmetry can take a non-zero value at a non-vanishing temperature. As a consequence, the transition from the disordered to the ordered phase is described by the formation of *vortices* (Kosterlitz and Thouless [229]). This unusual transition mechanism implies that most of the thermodynamic quantities do no longer show power-law singularities. Rather, their singular behaviour is described by essential singularities of the form

$$\xi \sim \exp\left(bt^{-1/2}\right) \quad , \quad \chi \sim \xi^{2-\eta} \tag{1.59}$$

(where b is a constant) while the specific heat is finite at T_c. The conventional critical exponents can no longer be defined. However, it is still possible to define scaling dimensions. At the critical point, the two-point correlation function has the long-distance behaviour

$$G(r) \sim r^{-1/4}(\ln r)^{1/8} \tag{1.60}$$

which implies $\eta = 1/4$ since logarithmic factors are suppressed in the critical exponent, as a consequence of eq. (1.10). Alternatively, the XY model can be defined in terms of two-dimensional spin vectors $\vec{S}_i$ of absolute value $\vec{S}_i^2 = 1$

$$\mathcal{H} = -J \sum_{<i,i'>} \vec{S}_i \cdot \vec{S}_{i'} \tag{1.61}$$

If one were to consider n-dimensional vector spins, one would obtain the $O(n)$ model. The $n \to \infty$ limit of these is the exactly solvable *spherical model* [45, 309] and for $n = 1$ one is back to the Ising model again. There is a very rich body of results on the critical exponents of these models in more than two dimensions. For example, in $4-\epsilon$ dimensions the critical exponents are known to $\mathcal{O}(\epsilon^5)$ [152, 156], but we merely list the lowest orders

$$\eta = \frac{n+2}{2(n+8)^2}\epsilon^2 + \mathcal{O}(\epsilon^3) \tag{1.62}$$

$$\gamma = 1 + \frac{n+2}{2(n+8)}\epsilon + \frac{(n+2)(n^2+22n+52)}{4(n+8)^3}\epsilon^2 + \mathcal{O}(\epsilon^3)$$

from which the other critical exponents can be found via the scaling relations. For further details, see [G7, G18]. However, the Mermin-Wagner theorem shows that these systems (for $n > 2$) do not have a phase transition at a non-vanishing temperature in two dimensions. They are thus of little interest here. On the other hand, if $-2 \leq n \leq 2$, a critical point exists and the operator content can be found (see chapter 11).

Restricted solid-on-solid models. This class of models is defined in a slightly different way. To each site i of the lattice, we attribute a *height* variable l_i bound to take only a restricted set of integer values $l_i = 1, \ldots, m$. Moreover, the heights of the nearest neighbours i, j are constrained to satisfy $l_i = l_j \pm 1$. The interactions depend on the heights of the four sites (i, j, k, l) of a given square plaquette $P(i, j, k, l)$

$$\mathcal{H} = - \sum_{P(i,j,k,l)} \epsilon(i, j, k, l) \tag{1.63}$$

By convention the site i is at the lower left corner of the plaquette and we label the other sites anticlockwise as in Figure 7. The partition function becomes the product of

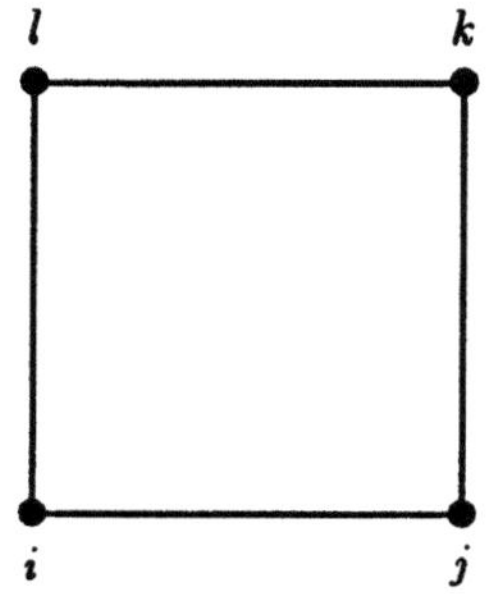

Fig. 7. Plaquette of a square lattice.

the *Boltzmann weights* $W(i, j, k, l) = \exp(\epsilon(i, j, k, l))$ associated to each plaquette

$$\mathcal{Z} = \sum \prod_{P(i,j,k,l)} W(i, j, k, l) \tag{1.64}$$

These models are often called *IRF models* (Interaction Round a Face) [G11, 19]. There are only six possible types of plaquettes due to the constraints on the nearest neighbours. We depict them in Figure 8. Such a model can be compared with the fluctuating surface of a liquid where the points on the surface move up and down. Neighbouring points must have heights which only slightly differ from each other. The local energy density ϵ is given by the surface tension. The phase structure is quite complicated, see [197, 19]. However, the model has a critical point which can be discussed in detail. The condition $l_i = l_j \pm 1$ can be diagrammatically represented as

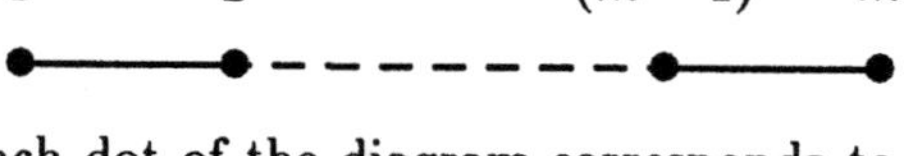

In this representation, each dot of the diagram corresponds to the m possible states of a lattice site. It is required that only those states can occur on neighboring sites on the lattice which are connected by a link in the above diagram. This diagram is called the *A_m Dynkin diagram*, which normally serves to classify the Lie algebras of $SU(m + 1)$, see [149]. The Dynkin diagram thus encodes the fact that if the height at a given lattice site has the value corresponding to a particular dot of the diagram, the heights of its four

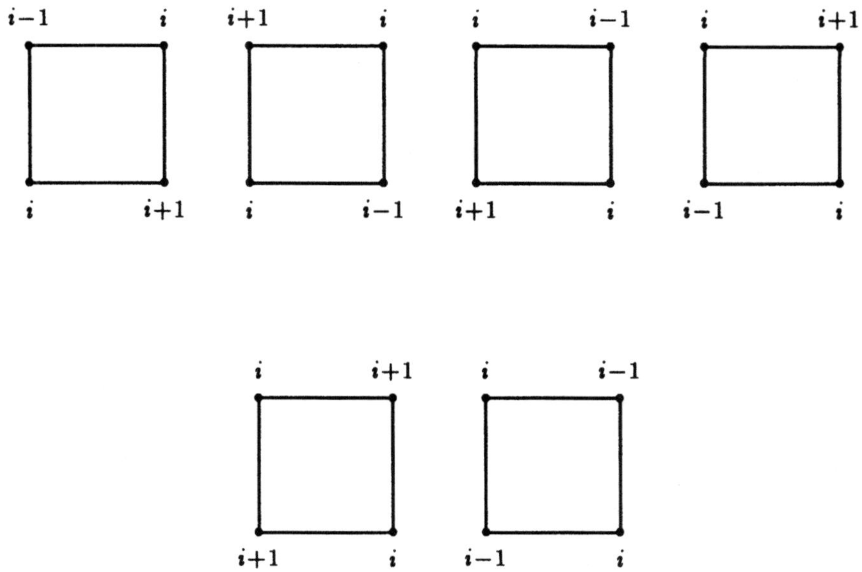

Fig. 8. The six possible plaquettes.

nearest neighbours on the lattice must take neighboring values in the Dynkin diagram. This model so defined is called the A_m-RSOS model.

The classification of simple Lie algebras plays a major role in the classification of critical models. Actually, it is possible to generalize the definition of the RSOS model to more complicated graphs, in particular any simply-laced Lie algebra (D_m, E_6, E_7, and E_8)[1)].

It is easy to show that some of the simplest RSOS models are equivalent to known models. The A_3 model is equivalent to the Ising model. We give the three heights of the A_3 model the values $l = -1, 0, +1$ instead of $1, 2, 3$. Suppose that at the site i, we have $l_i = \pm 1$. Automatically, the four nearest neighbours are forced to have $l_{i'} = 0$. This fact implies that any configuration on the square lattice splits into two distinct parts: 1) a regular square lattice with double lattice constant where all heights are zero and 2) another square lattice, dual to the previous one, where the heights are ± 1. Each plaquette of the full lattice has two opposite corners where the heights are zero while the remaining corners have the Ising spins ± 1 and the energy density ϵ implies a coupling between the two spins of the sublattice which causes the critical point of the system to fall in the Ising universality class.

A similar discussion would show that the A_4-RSOS model has the critical behaviour of the tricritical Ising model. The three-state Potts model corresponds to the D_4-RSOS model. The Lie algebra D_4 is related to the group $SO(8)$ and the Dynkin diagram with the corresponding heights is given by

---

[1)] And to their affine extensions as defined in chapter 7. The reader not familiar with Dynkin diagrams should for the moment simply accept them as a convenient means to label an algebra.

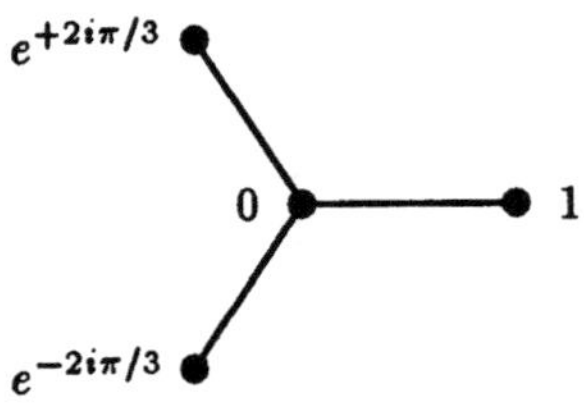

The model has a natural $\mathbf{Z}_3$ symmetry and the presence of the height 0 in the center implies again a decomposition of the lattice into two sublattices, one with constant configuration $l = 0$, while the other is $\mathbf{Z}_3$ symmetric with three possible states. This is a characteristic feature of the critical three-state Potts model. Notice finally that the four-state Potts model requires a diagram related to the so-called affine $\hat{D}_4$ Lie algebra obtained by adding a new dot (root) to the previous diagram. The Dynkin diagram is

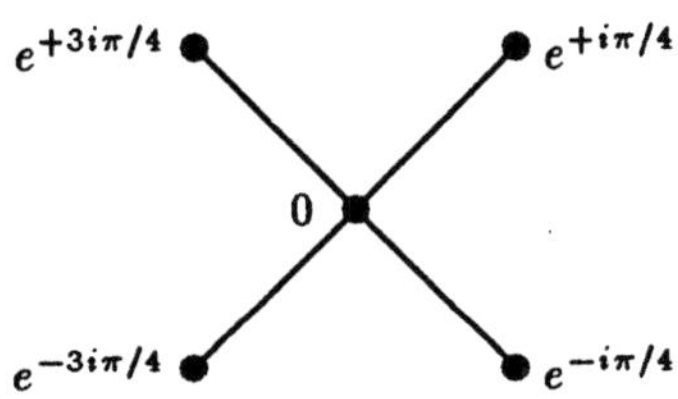

The relation to the critical behaviour of the four-state Potts model is as simple as in the previous examples.

We shall see later that the RSOS models just defined are generic models incorporating most of the other spin systems to be considered. For the relationship of RSOS models to the theory of knots and links see e.g. [330].

1.4 Correspondence between statistical systems and field theory

As a preparation for the field-theoretic considerations to follow in the next chapters, we now recall the formal correspondence between statistical systems and quantum field theory. We have already seen that the important aspect to be emphasized in a second-order transition is the role of the fluctuations, as illustrated by the droplet picture (see Figure 3). Due to universality, the presence of strong fluctuations is important, but for a description of the critical behaviour, many of their detailed properties are irrelevant in the renormalization group sense and can be neglected. This is the physical idea behind the formalism which establishes a formal correspondence between the thermal fluctuations in statistical mechanics and the quantum fluctuations of field theory.

Consider a quantum mechanical system. We shall use the non-relativistic notation for simplicity, though the results extend to the relativistic case. The time evolution is described by the time-evolution operator

$$\mathcal{U}(t,t') = \exp\left(-\frac{i}{\hbar}H\,(t-t')\right) \tag{1.65}$$

where H is the Hamiltonian, which we have assumed to be time-independent. $\mathcal{U}$ gives the time evolution of a wave function $\psi(t)$ via $\psi(t') = \mathcal{U}(t,t')\psi(t)$. Now take the time difference $t-t'$ to be infinitesimal. We can write the transition amplitude $\mathcal{Z}$ between two space-time points (t_a, x_a) and (t_b, x_b) via a Feynman path integral

$$\mathcal{Z} = \int \prod_i \mathcal{U}(t_{i+1}, t_i) dx_i = \int \mathcal{D}x \exp\left(-\frac{i}{\hbar} S\right) \tag{1.66}$$

where the integral represents the sums over all possible paths (see Figure 9) and $S(t_1, t_2) = \int_{t_1}^{t_2} dt \mathcal{L}$ is the classical action. We also note that a matrix element of $\mathcal{U}$ takes the form

$$\mathcal{U}(t_{i+1}, t_i) = \exp\left(-\frac{i}{\hbar} S(t_{i+1}, t_i)\right) \tag{1.67}$$

It is not necessary to describe how the functional integral can be precisely defined, since for our purposes it is sufficient to show a formal correspondence with analogous expressions arising in the context of statistical mechanics.

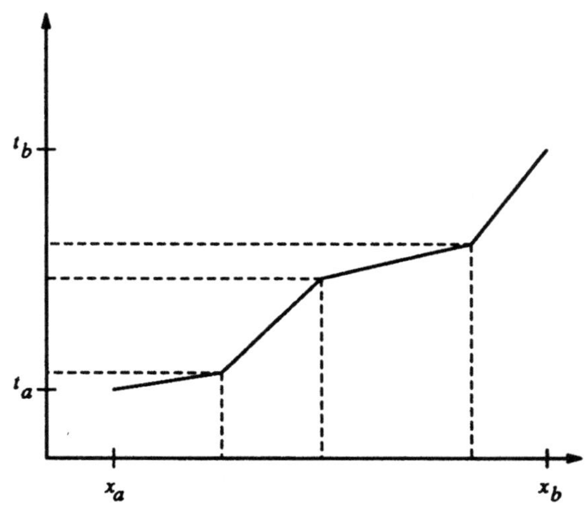

Fig. 9. Classical path.

Now, consider a statistical system defined on a hypercubic lattice (see Figure 10), for example an Ising model. Single out one of the d directions and call it *"time"* τ. The remaining $d-1$ other directions will be referred to as *"space"* $\vec{x}$. Consider two times τ_1, τ_2 and their τ-dependent configurations of (Ising) spins $\{\sigma\}(\tau_1)$, $\{\sigma\}(\tau_2)$. The *transfer matrix* T links these configurations and has the elements

$$\langle\{\sigma\}(\tau_1)|T|\{\sigma\}(\tau_2)\rangle = \exp\left(-\beta\tilde{\mathcal{H}}(\tau_1, \tau_2)\right) \tag{1.68}$$

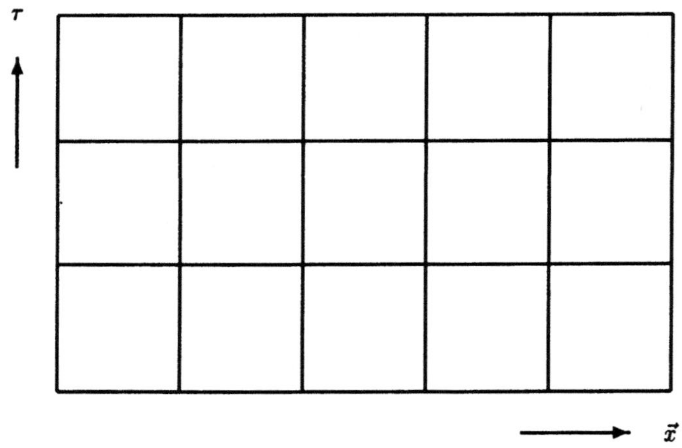

Fig. 10. Hypercubic lattice with "time" and "space" directions.

where $\mathcal{H} = \sum_\tau \tilde{\mathcal{H}}(\tau, \tau+1)$ is the classical Hamiltonian of the spin system and $\tilde{\mathcal{H}}$ includes the sum over the interacting pairs in the space directions. The partition function $\mathcal{Z}$ is

$$\begin{aligned}\mathcal{Z} &= \sum_{\{\sigma\}} e^{-\beta\mathcal{H}} \\ &= \sum \langle\{\sigma\}(\tau_a)|\mathcal{T}|\{\sigma\}(\tau_1)\rangle\langle\{\sigma\}(\tau_1)|\mathcal{T}|\{\sigma\}(\tau_2)\rangle \cdots \langle\{\sigma\}(\tau_T)|\mathcal{T}|\{\sigma\}(\tau_b)\rangle \\ &= \operatorname{tr}\mathcal{T}^{M+1}\end{aligned} \tag{1.69}$$

where M is the number of sites in the time direction.

Comparing these expressions with the ones for the quantum mechanical system, we note a few formal analogies, which are collected in Table 2. In particular, this suggests

Table 2. Analogies between statistical mechanics and quantum theory.

quantum mechanics		statistical mechanics	
real time	t	distance	$\tau = -it$
Planck's constant	$1/\hbar$	coupling	$\beta = 1/k_B T$
generating function	$\mathcal{Z}$	partition function	$\mathcal{Z}$
time evolution operator	$\mathcal{U}$	transfer matrix	$\mathcal{T}$

to interpret $\mathcal{T}$ as an Euclidean evolution operator. Then

$$\mathcal{T} = \exp -\tau H \tag{1.70}$$

defines the *quantum Hamiltonian* H. Therefore we obtain a map of a d-dimensional problem with classical, commuting variables onto a $(d-1)$-dimensional problem with quantum, non-commuting variables. In particular, we stress that $1/\hbar$ and $\beta = 1/k_B T$

parametrize the quantum and thermal fluctuations in a similar way. Absence of fluctuations corresponds to the classical ($\hbar \to 0$) or the zero-temperature ($T \to 0$) limits, respectively.

1.5 Correspondence of physical quantities

We now establish the explicit dictionary between quantities calculated in the context of statistical mechanics with those obtained within quantum theory. The results of this section are summarized in Table 3.

Table 3. Correspondence of physical quantities of statistical and quantum systems.

Statistical systems		Quantum systems	
classical Hamiltonian	$\beta\mathcal{H}$	classical (Euclidean) action	$\frac{1}{\hbar}S$
infinitesimal transfer matrix	$\mathcal{T}$	quantum Hamiltonian	H
Equilibrium state		ground state	
Ensemble averages		ground state expectation values	
Correlation functions		propagators	
free energy density	f	ground state energy density	ω_0
inverse correlation lengths	ξ_i^{-1}	energy gaps	$E_i - E_0$

Free energy density. Since the classical Hamiltonian is real, the transfer matrix is hermitian and has a real eigenvalue spectrum $e^{-E_i\tau}$ and an eigenvector basis $|i\rangle$. One can arrange the eigenvalues to be ordered, i.e. $E_0 \leq E_1 \leq E_2 \leq \ldots$. Then, using the spectral representation $\mathcal{T} = \sum_i |i\rangle e^{-E_i\tau}\langle i|$, one has[2)]

$$\mathcal{T}^N = \sum_i |i\rangle e^{-E_i N\tau}\langle i| \tag{1.71}$$

For a system with a finite number of degrees of freedom per site and a finite number of sites, $\mathcal{T}$ is a finite-dimensional matrix. If the Boltzmann weights $e^{-\beta\mathcal{H}}$ are positive, the Perron-Frobenius theorem (e.g. [G3]) states that the largest eigenvalue of $\mathcal{T}$, corresponding to E_0, is unique. Taking the limit $N \to \infty$, only the contribution from the eigenstate $|0\rangle$ survives, with exponentially small corrections

$$\mathcal{T}^N \to |0\rangle e^{-E_0 M}\langle 0| \quad , \quad \text{with } M = N\tau \tag{1.72}$$

and the partition function $\mathcal{Z} = \exp -E_0 M = \exp -F$, where F is the free energy (into which we have absorbed a factor $k_B T$). F is extensive in both T and the "spatial" volume V, while E_0 is extensive in V

$$F = fMV \quad , \quad E_0 = \omega_0 V \tag{1.73}$$

2) We have implicitly added a constant to make the eigenvalues of $\mathcal{T}$ positive. This corresponds to adding a smooth background term which does not influence the critical behaviour.

and we have for the free energy density $f = \omega_0$.

From eq. (1.72) we see that the lattice extent M in the temporal direction plays the role of an inverse temperature. This relationship is sometimes used in field theory calculations.

Correlation functions. The field theory propagator is defined by a vacuum expectation value

$$\Delta(t,\vec{x}) := \langle 0|\phi(t,\vec{x}),\phi(0,\vec{0})|0\rangle \tag{1.74}$$

The field operator $\phi(t,\vec{x})$, given in the Heisenberg picture, can be rewritten in the Schrödinger picture

$$\phi(t,\vec{x}) = e^{i/\hbar Ht}\phi(\vec{x})e^{-i/\hbar Ht} \tag{1.75}$$

$$\Delta(t,\vec{x}) = \langle 0|\phi(\vec{x})e^{-i/\hbar Ht}\phi(\vec{0})|0\rangle e^{i/\hbar E_0 t} \tag{1.76}$$

The statistical mechanics correlation function is defined by

$$\begin{aligned} G(n_0,\vec{n}) &= \frac{1}{\mathcal{Z}} \sum_{\phi(n_0',\vec{n}')} \phi(n_0,\vec{n})\phi(0,\vec{0})e^{-\beta\mathcal{H}} \\ &= \frac{\operatorname{tr}\left(\mathcal{T}^A\phi(\vec{n})\mathcal{T}^{n_0}\phi(\vec{0})\mathcal{T}^B\right)}{\operatorname{tr}\mathcal{T}^{M+1}} \end{aligned} \tag{1.77}$$

where n_0 measures the distance in time-direction and $\vec{n}$ in space direction. In writing this, we assume that $<\phi>=0$, which can be always achieved by redefining $\phi \to \phi- <\phi>$. Periodic boundary conditions in the τ-direction are implied and $A + n_0 + B = M + 1$. Taking the limit $A, B, n_0 \to \infty$, one has from the spectral representation (τ is the lattice constant in the time direction)

$$G(n_0,\vec{n}) \simeq \langle 0|\phi(\vec{n})e^{-n_0 H\tau}\phi(\vec{0})|0\rangle e^{n_0 E_0 \tau} \tag{1.78}$$

and comparison yields as the final result

$$G(n_0,\vec{n}) = \Delta\left(\frac{\hbar}{i}n_0\tau,\vec{n}\right) \tag{1.79}$$

which illustrates again the analytic continuation between the time in quantum mechanics and the Euclidean time direction of statistical mechanics, see Table 2.

Correlation lengths. Correlation lengths can be defined by the exponential fall-off of the two-point function

$$G(n_0,\vec{0}) \sim \exp\left(-|n_0|/\xi\right) \quad , \quad \text{if } |n_0| \gg \xi \tag{1.80}$$

On the other hand, we have from equation (1.78)

$$G(n_0,\vec{0}) = \sum_i \exp\left(-(E_i - E_0)n_0\tau\right) \left|\langle 0|\phi(\vec{0})|i\rangle\right|^2 \tag{1.81}$$

Consider all the (scaling) operators contained in a theory. These operators can be given a number i such that i corresponds to the first eigenstate $|i>$ for which $\langle|\phi(\vec{0})|i\rangle \neq 0$. One obtains an ordered sequence $\phi_0 < \phi_1 < \phi_2 \ldots$ of operators. If necessary, one has to redefine operators, e.g. $\phi_0 = 1, \phi_1$, $\phi_2' = \phi_2 - a_{2,1}\phi_1 - a_{2,0}\phi_0$, $\phi_3' = \phi_3 - a_{3,2}\phi_2 - a_{3,1}\phi_1 - a_{3,0}\phi_0, \ldots$ such that $\langle 0|\phi_1(\vec{0})|1\rangle \neq 0$, $\langle 0|\phi_2(\vec{0})|1\rangle = 0$, $\langle 0|\phi_2(\vec{0})|2\rangle \neq 0, \ldots$. For the Ising model, ϕ_1 would be the order parameter, ϕ_2 the energy density (then $a_{2,0} = 0$) and so on. Then $G_{\phi_i}(n_0, \vec{0}) \sim \exp(-(E_i - E_0)\, n_0 \tau)$ and we thus have for the correlation length ξ_i

$$\xi_i^{-1} = E_i - E_0 \tag{1.82}$$

where the lattice constant in time direction τ is put to unity.

In writing this, we have implicitly assumed **isotropic** critical behaviour. More precisely, we have obtained a time correlation length $\xi_{\parallel}$, which in principle could be different from a space correlation length $\xi_{\perp}$. Further, the choice of the lattice units is arbitrary. This leads to an ambiguous, non-universal prefactor for ξ. While this prefactor is not important for the definition of the critical exponents (see eq. (1.10)), it does become important for the calculation of critical amplitudes. We shall later encounter techniques to fix its value uniquely.

2 Conformal Invariance and the Stress-Energy Tensor

We discuss in this chapter the geometrical consequences of the scale invariance property which characterizes a second order critical point. For systems with short-ranged interactions, it is reasonable to suppose that global scale invariance ($\vec{r}' = b\vec{r}$) generalizes to local scale invariance ($\vec{r}' = b(\vec{r})\vec{r}$). However such transformations may change the global geometry of the system, therefore affecting the local covariant transformation properties of the scaling operators. This will be shown to lead to the *conformal anomaly*, parametrized by the *central charge* c, the existence of which describes the presence of long-ranged fluctuations. We define the transformation properties of physical scaling operators under smooth local transformations. This fixes the two-point and three-point correlation functions in d dimensions. The stress-energy tensor will be shown to act as a generating function for these local transformations.

2.1 Conformal group

In the infinite size limit, a critical system gains the invariance properties of a larger geometrical group. Dilatation invariance complements now the usual translational and rotational invariance. Such transformations are called conformal. By definition, conformal transformations preserve the angle of intersection of two arbitrary curves as illustrated in Figure 11. The assumption that the fundamental interaction is short-ranged suggests that

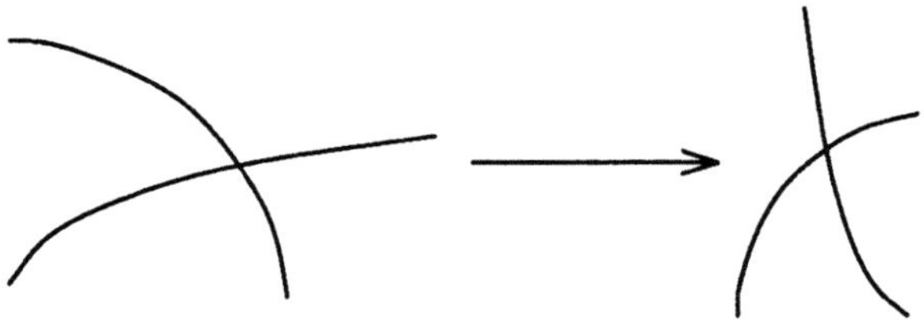

Fig. 11. Conformal transformation in d dimensions.

local smooth transformations do not affect the global behaviour of the critical system. This is verified for large classes of physical classical systems. This is not true any more if long distance interactions or strongly anisotropic effects occur.

Suppose that a system is invariant under local combinations of rotations, translations, and dilatations. In $d > 2$ dimensions, the conformal group is finite-dimensional. It is generated by global rotations $\vec{r} \to \vec{r}' = \Lambda\vec{r}$, global translations $\vec{r} \to \vec{r}' = \vec{r} + \vec{a}$, global dilatations $\vec{r} \to \vec{r}' = b\vec{r}$, and the so-called *special conformal transformation*

$$\vec{r} \to \vec{r}' = \frac{\vec{r} + \vec{a}r^2}{1 + 2\vec{a}\cdot\vec{r} + a^2r^2} \tag{2.1}$$

A geometric interpretation of the last transformation is apparent when we write

$$\frac{\vec{r}'}{r'^2} = \frac{\vec{r}}{r^2} + \vec{a} \tag{2.2}$$

which is the combination of an inversion $\vec{r} \longrightarrow \vec{r}' = \vec{r}/r^2$ followed by a translation $\vec{a}$ and again an inversion.

More mathematically, a line element in a space with metric $g_{\mu\nu}$ has the measure

$$ds^2 = g_{\mu\nu} dr^\mu dr^\nu \tag{2.3}$$

Under a local change of coordinates $\vec{r} \longrightarrow \vec{r}'$ the metric changes covariantly

$$g'_{\mu\nu}(r') = \frac{\partial r^\alpha}{\partial r'^\mu} \frac{\partial r^\beta}{\partial r'^\nu} g_{\alpha\beta}(r) \tag{2.4}$$

The conformal group is the subgroup of coordinate transformations which keeps the metric invariant up to a local scale factor

$$g'_{\mu\nu}(r') = \Omega(r) g_{\mu\nu}(r) \tag{2.5}$$

This preserves the angle $\cos\theta = \vec{r} \cdot \vec{r}'/(r^2 r'^2)^{1/2}$ between the two vectors $\vec{r}$ and $\vec{r}'$ with $\vec{r} \cdot \vec{r}' = g_{\mu\nu} r^\mu r'^\nu$.

Under infinitesimal transformations $r^\mu \longrightarrow r^\mu + \epsilon^\mu(r)$ the metric transforms as

$$ds^2 \longrightarrow ds^2 + (\partial_\mu \epsilon_\nu + \partial_\nu \epsilon_\mu) dr^\mu dr^\nu \tag{2.6}$$

The last term on the r.h.s. must be proportional to $g_{\mu\nu}$. Let $g^{\mu\nu}$ be the inverse matrix of $g_{\mu\nu}$ and take the trace $g^{\mu\nu}(\partial_\mu \epsilon_\nu + \partial_\nu \epsilon_\mu)$. We end up with the constraint

$$\partial_\mu \epsilon_\nu + \partial_\nu \epsilon_\mu = \frac{2}{d}(\partial \cdot \epsilon) g_{\mu\nu} \tag{2.7}$$

Translations are obviously solutions to this equation. Dilatations and rotations are polynomials of order one in $\{r^\mu\}$. Infinitesimal special transformations are quadratic solutions to (2.7). Higher order solutions do not exist for $d > 2$.

2.2 Conformal algebra in two dimensions

Equation (2.7) becomes the Cauchy-Riemann equations for $d = 2$ in a Euclidean metric

$$\partial_1 \epsilon_1 = \partial_2 \epsilon_2 \qquad \partial_1 \epsilon_2 = -\partial_2 \epsilon_1 \tag{2.8}$$

It is natural to introduce the complex coordinates

$$\begin{aligned} z &= r_1 + i r_2 \qquad & \bar{z} &= r_1 - i r_2 \\ \epsilon(z) &= \epsilon_1 + i\epsilon_2 \qquad & \bar{\epsilon}(\bar{z}) &= \epsilon_1 - i\epsilon_2 \end{aligned} \tag{2.9}$$

The group of conformal transformations in 2 dimensions is therefore isomorphic to the (infinite-dimensional) group of arbitrary analytic coordinate transformations $z \longrightarrow w(z)$ and $\bar{z} \longrightarrow \bar{w}(\bar{z})$. A line element transforms under such a transformation as

$$ds^2 \longrightarrow \left(\frac{\partial w}{\partial z}\right) \left(\frac{\partial \bar{w}}{\partial \bar{z}}\right) ds^2 \tag{2.10}$$

The infinitesimal transformations read

$$\epsilon_n = -z^{n+1} \qquad \bar{\epsilon}_n = -\bar{z}^{n+1} \tag{2.11}$$

They are generated by

$$l_n = -z^{n+1}\partial_z \qquad \bar{l}_n = -\bar{z}^{n+1}\partial_{\bar{z}} \tag{2.12}$$

which satisfy the following commutation relations

$$\begin{aligned} [l_n, l_m] &= (n-m)\, l_{m+n} \\ [\bar{l}_n, \bar{l}_m] &= (n-m)\, \bar{l}_{m+n} \\ [l_n, \bar{l}_m] &= 0 \end{aligned} \tag{2.13}$$

This algebra is called the *loop algebra*. The third commutator shows that this algebra splits into the direct sum of two algebras, one generated by $\{l\}$ and the other by $\{\bar{l}\}$. For that reason we shall often discard the $\bar{z}$ dependence in the calculations. As long as the global invariance of the metric is not explicitly required, we can treat z and $\bar{z}$ as separate independent coordinates.

The properties of $\{l_n\}$ ($\{\bar{l}_n\}$ resp.) deserve some comments. Obviously l_n is regular at $z = 0$ for $n \geq -1$. However if we perform the transformation $z = 1/v$, we found $l_n \longrightarrow -v^{-n+1}\partial_v$ singular at $v = 0$ (i.e. $z \to \infty$) for $n > 1$. Correspondingly l_n is singular at $z = 0$ for $n < -1$ but regular at infinity. Therefore only the subset $\{l_{-1}, l_0, l_1\}$ is defined everywhere and induces analytic transformations of the complex plane $\mathbf{C}$ into itself. The other transformations are therefore not analytic everywhere. They do not possess a global inverse. Their existence is unique to two dimensions.

The meaning of global conformal transformations is straightforward. Translations are generated by l_{-1} and $\bar{l}_{-1}$. A dilatation is a combination $l_0 + \bar{l}_0 = r_\mu\partial_\mu$, and $i(l_0 - \bar{l}_0) = r_\mu\partial_\nu - r_\nu\partial_\mu$ generates a rotation. The last generators l_1 and $\bar{l}_1$ correspond to the special conformal transformations. The set $\{l_{-1}, l_0, l_1\}$ generates the *projective conformal transformations* $SL(2,\mathbf{C})/\mathbf{Z}_2$ which can be written

$$z \longrightarrow w_p(z) = \frac{az+b}{cz+d} \tag{2.14}$$

with $a, b, c, d \in \mathbf{C}$ and $ad - bc = 1$ The relation to $SL(2,\mathbf{C})/\mathbf{Z}_2$ is apparent when we write the transformation in a matrix form

$$\begin{pmatrix} a & b \\ c & d \end{pmatrix} \tag{2.15}$$

Successive application of projective conformal transformations corresponds to matrix multiplication. Translations, rotations, dilatations, and special conformal transformations are expressed by

$$\begin{aligned} &\text{Translation} \Leftrightarrow \begin{pmatrix} 1 & b \\ 0 & 1 \end{pmatrix} \qquad &&\text{Rotation} \Leftrightarrow \begin{pmatrix} e^{i\theta/2} & 0 \\ 0 & e^{-i\theta/2} \end{pmatrix} \\ &\text{Dilatation} \Leftrightarrow \begin{pmatrix} \sqrt{a} & 0 \\ 0 & \sqrt{1/a} \end{pmatrix} \qquad &&\text{Special Transf.} \Leftrightarrow \begin{pmatrix} 1 & 0 \\ c & 1 \end{pmatrix} \end{aligned} \tag{2.16}$$

The subalgebra generated by $\{l_0, \bar{l}_0\}$ is the Cartan subalgebra. Physical operators will be eigenstates of this subalgebra with eigenvalues $\{\Delta, \overline{\Delta}\}$ called *conformal weights*. There is no need to have $\overline{\Delta}$ being equal to Δ. Since $l_0 + \bar{l}_0$ and $i(l_0 - \bar{l}_0)$ generate dilatations and rotations respectively, the *scaling dimension* x and the *spin* s are given by

$$\boxed{x = \Delta + \overline{\Delta}} \qquad \text{and} \qquad \boxed{s = \Delta - \overline{\Delta}} \tag{2.17}$$

The spin[3] s must not be confused with the spin operator encountered in the previous sections.

2.3 Conformal theory in two dimensions

Having defined the conformal group and written down its Lie algebra, we now have to implement the requirement of conformal invariance in a statistical system. We recall from chapter 1 that at the critical point, the correlation function transforms covariantly under global scale transformations. Conformal transformations generalize these to space-dependent rescaling factors. However, some care is required since not all possible scaling operators transform in the same way under conformal transformations. This makes it necessary to introduce a precise terminology [37], which distinguishes between so-called primary operators, an example of which is the order parameter density σ, which do transform covariantly, and more complex fields, such as $\partial\sigma$, which are called secondary and have a more complicated transformation behaviour. We now proceed to postulate those properties needed.

A conformal theory is described by the correlation functions of a large set of local scaling operators $A(z, \bar{z})$ (such as the order parameter density or the energy density) which satisfy some fundamental defining properties [37].

1. The set of local operators $\{A(z, \bar{z})\}$ is in general infinite. In particular, when it contains a certain operator, it also contains all derivatives of that operator.
2. There is a subset $\{\Phi(z, \bar{z})\} \subset \{A(z, \bar{z})\}$ of operator called *quasi-primary*, which transform covariantly under projective conformal transformations (sometimes also referred to as global) w_p of eq. (2.14)

$$\Phi(z, \bar{z}) \longrightarrow \left(\frac{\partial w_p}{\partial z}\right)^{\Delta} \left(\frac{\partial \bar{w}_p}{\partial \bar{z}}\right)^{\overline{\Delta}} \Phi(w_p(z), \bar{w}_p(\bar{z})) \tag{2.18}$$

3. The remaining local operators are linear combinations of the quasi-primary operators and their derivatives.
4. The *vacuum* is invariant under projective conformal transformations.

Property 2 means that correlation functions of quasi-primary operators transform covariantly under projective conformal transformations

$$\begin{aligned} < \Phi_1(z_1, \bar{z}_1) \cdots \Phi_n(z_n, \bar{z}_n) > = & \left(\frac{\partial w_p}{\partial z_1}\right)^{\Delta_1} \left(\frac{\partial \bar{w}_p}{\partial \bar{z}_1}\right)^{\overline{\Delta}_1} \cdots \left(\frac{\partial w_p}{\partial z_n}\right)^{\Delta_n} \left(\frac{\partial \bar{w}_p}{\partial \bar{z}_n}\right)^{\overline{\Delta}_n} \\ & \cdot < \Phi_1(w_p(z_1), \bar{w}_p(\bar{z}_1)) \cdots \Phi_n(w_p(z_n), \bar{w}_p(\bar{z}_n)) > \end{aligned} \tag{2.19}$$

[3] s is referred to as spin since it is the eigenvalue of the rotation operator $i(l_0 - \bar{l}_0)$. In distinction to the situation in higher dimensions, s is not restricted to integer or half-integer values.

It is easy to check that a derivative $\partial\Phi$ of a quasiprimary operator cannot be itself be quasiprimary, unless $\Phi = \mathbf{1}$. Property 4 will take a more precise meaning later in connection with the representation theory of the conformal algebra. Notice that finite holomorphic transformations affect in general the geometry of the system. For example, the logarithm maps the entire plane onto a strip. Finite transformations will therefore relate critical systems in different geometries. However projective transformations which do not change the geometry are supposed to keep the physical ground state invariant.

Quasi-primary operators are the most fundamental operators for $d > 2$. Unfortunately, in general there exists an infinite number of them. In two dimensions, we can take advantage of the special features of the conformal algebra and impose the much stronger condition of covariant transformation under ***arbitrary*** conformal transformations w, rather than just the projective ones w_p. A more fundamental subset of operators appears:

5. There is a subset $\{\phi(z,\bar{z})\} \subset \{\Phi(z,\bar{z})\}$ of the quasi-primary operators, called ***primary*** operators which transform under **any** conformal transformation $w(z)$ as

$$\phi(z,\bar{z}) \longrightarrow \left(\frac{\partial w}{\partial z}\right)^{\Delta} \left(\frac{\partial \bar{w}}{\partial \bar{z}}\right)^{\overline{\Delta}} \phi(w(z),\bar{w}(\bar{z})) \tag{2.20}$$

Correlation functions of primary operators transform similarly to eq. (2.19) but now for **any** conformal transformation w.

$$< \phi_1(z_1,\bar{z}_1)\cdots\phi_n(z_n,\bar{z}_n) >= \tag{2.21}$$
$$\left(\frac{\partial w}{\partial z_1}\right)^{\Delta_1} \left(\frac{\partial \bar{w}}{\partial \bar{z}_1}\right)^{\overline{\Delta}_1} \cdots \left(\frac{\partial w}{\partial z_n}\right)^{\Delta_n} \left(\frac{\partial \bar{w}}{\partial \bar{z}_n}\right)^{\overline{\Delta}_n} <\phi_1(w(z_1),\bar{w}(\bar{z}_1))\cdots\phi_n(w(z_n),\bar{w}(\bar{z}_n))>$$

The two $<>$ of this last equation have to be taken with caution. As mentioned above, they evaluate functionals of operators in the same geometry only if w is projective. Examples will be given below.

Obviously, every primary operator is also quasi-primary, but a quasi-primary operator may or may not be primary. Primary operators transform in a simpler manner than the remaining operators, called ***secondary***. A given secondary operator may or may not be quasi-primary. We shall see examples of both in the following sections. The important point of the distinction made between primary and secondary operators is, as we shall show, that all properties of any secondary operator can be entirely deduced from the knowledge of the primary operators. Moreover, many physical systems only have a finite number of primary operators. For example, the only two primary operators of the Ising model are just the order parameter density and the energy density.

Before developing the general theory further, we now show that some elementary correlation functions of quasiprimary operators can be completely determined.

2.4 Elementary correlation functions

The projective conformal group is sufficient to determine the shape of simple correlation functions. Under infinitesimal conformal transformations $z \longrightarrow z + \epsilon(z)$, $\bar{z} \longrightarrow \bar{z} + \bar{\epsilon}(\bar{z})$, primary operators transform as

$$\delta_\epsilon \phi(z,\bar{z}) = \left((\Delta \partial_z \epsilon + \epsilon \partial_z) + (\overline{\Delta} \partial_{\bar{z}} \bar{\epsilon} + \bar{\epsilon} \partial_{\bar{z}})\right) \phi(z,\bar{z}) \tag{2.22}$$

The same do quasi-primary operators Φ if infinitesimal projective transformations (written ϵ_p) are considered. Such a transformation do not change a correlation function of primary (or quasi-primary) operators. We have therefore

$$\sum_{i=1}^{n} \left((\Delta_i \partial_{z_i} \epsilon_p(z_i) + \epsilon_p(z_i)\partial_{z_i}) + (\overline{\Delta}_i \partial_{\bar{z}_i} \bar{\epsilon}_p(\bar{z}_i) + \bar{\epsilon}_p(\bar{z}_i)\partial_{\bar{z}_i})\right) < \phi_1 \cdots \phi_n >$$
$$= \delta_{\epsilon_p} < \phi_1 \cdots \phi_n > = 0 \tag{2.23}$$

Let us investigate the consequences for the two-point function [275]. We first consider the translation $\epsilon_p = 1, \bar{\epsilon}_p = 0$. Equation (2.23) becomes the differential equation

$$(\partial_{z_1} + \partial_{z_2}) < \phi_1 \phi_2 > = 0 \tag{2.24}$$

which proves that the two-point function only depend on $z_{12} = z_1 - z_2$, and $\bar{z}_{12} = \bar{z}_1 - \bar{z}_2$. We take now $\epsilon_p = z$, $\bar{\epsilon}_p = 0$.

$$(\Delta_1 + \Delta_2 + z_1 \partial_{z_1} + z_2 \partial_{z_2}) < \phi_1 \phi_2 > = 0 \tag{2.25}$$

Taking translation invariance into account, we conclude that

$$< \phi_1 \phi_2 > \sim 1/(z_{12}^{\Delta_1+\Delta_2} \bar{z}_{12}^{\overline{\Delta}_1+\overline{\Delta}_2}) \tag{2.26}$$

Finally, with $\epsilon_p = z^2$, $\bar{\epsilon}_p = \bar{z}^2$, we have to solve

$$(2z_1\Delta_1 + 2z_2\Delta_2 + z_1^2 \partial_{z_1} + z_2^2 \partial_{z_2}) \frac{1}{z_{12}^{\Delta_1+\Delta_2}} = 0 \tag{2.27}$$

This constrains the exponents to be equal $\Delta_1 = \Delta_2 = \Delta$ and $\overline{\Delta}_1 = \overline{\Delta}_2 = \overline{\Delta}$. Any two-point function of primary or quasi-primary operators must take the form

$$< \phi_1 \phi_2 > = C_{12}\, z_{12}^{-2\Delta} \bar{z}_{12}^{-2\overline{\Delta}} \tag{2.28}$$

The scaling dimension is given by $x = \Delta + \overline{\Delta}$ and the spin $s = \Delta - \overline{\Delta}$. The coefficient C_{12} can be arbitrarily chosen. Actually, we can always find a basis of primary operators, such that $C_{12} = 0$ if $\phi_1 \neq \phi_2$ for real operators or $\phi_1 \neq {\phi^*}_2$ for complex operators. Therefore, C_{12} is not a free parameter left undetermined by conformal invariance but a normalization factor.

We can repeat these calculations for any n-point function. The result is known under the name of *projective Ward identities*, written here for the z-dependence only. These express the covariance of the theory under translations, dilatations and special conformal transformations, respectively.

$$\begin{aligned} \textstyle\sum_i \partial_{z_i} < \phi_1 \cdots \phi_n > &= 0 \\ \textstyle\sum_i (z_i \partial_{z_i} + \Delta_i) < \phi_1 \cdots \phi_n > &= 0 \\ \textstyle\sum_i (z_i^2 \partial_{z_i} + 2\Delta_i z_i) < \phi_1 \cdots \phi_n > &= 0 \end{aligned} \tag{2.29}$$

They are obviously not sufficient to solve the correlation functions. However they reduced the number of degrees of freedom. For example the three-point function $< \phi_1\phi_2\phi_3 >$ must have the form [275]

$$< \phi_1\phi_2\phi_3 >= \frac{C_{123}}{(z_{12}^{\Delta_1+\Delta_2-\Delta_3} z_{23}^{\Delta_2+\Delta_3-\Delta_1} z_{13}^{\Delta_1+\Delta_3-\Delta_2})(\bar{z}_{12}^{\overline{\Delta}_1+\overline{\Delta}_2-\overline{\Delta}_3} \bar{z}_{23}^{\overline{\Delta}_2+\overline{\Delta}_3-\overline{\Delta}_1} \bar{z}_{13}^{\overline{\Delta}_1+\overline{\Delta}_3-\overline{\Delta}_2})} \tag{2.30}$$

This time, the coefficient C_{123} cannot be arbitrarily chosen. It remains a fundamental problem to compute such coefficients, to which we shall return later.

The four point functions are even less determined. Looking at the projective transformation (2.14), we can fix a, b, c, d so that the four points (z_1, z_2, z_3, z_4) are transformed into $(\eta, 0, 1, \infty)$ respectively with

$$\eta = \frac{z_{12}z_{34}}{z_{32}z_{14}} \tag{2.31}$$

The parameter η is invariant under projective transformations. The projective Ward identities tell us that we can understand the behaviour of a correlation function up to projectively invariant terms. The means that we can write

$$< \phi_1\phi_2\phi_3\phi_4 >= \prod_{i<j} z_{ij}^{-\gamma_{ij}} \bar{z}_{ij}^{-\bar{\gamma}_{ij}} F(\eta, \bar{\eta}) \tag{2.32}$$

with $\gamma_{ij}, \bar{\gamma}_{ij}$ are parameters which satisfy the conditions

$$\sum_i \gamma_{ij} = 2\Delta_j, \qquad \sum_i \bar{\gamma}_{ij} = 2\overline{\Delta}_j \tag{2.33}$$

and F is an unknown function of the projectively invariant parameters $\{\eta, \bar{\eta}\}$. We learn from those correlators that, due to global conformal invariance, the undetermined parts of the correlation functions only depend on projectively invariant terms. For a n-point functions, we can build $n-3$ independent invariants (twice as much if we count separately the $\bar{\eta}$) because there is always a projective transformation which maps three points onto $0, 1, \infty$.

The constraint of unitarity (and more generally minimality) will reduce the possibilities to choose the exponents $\Delta, \overline{\Delta}$, and will result in differential equations for the correlation functions. The unknown function F will be calculated. The first step towards this involves an analysis of the stress-energy tensor to which we turn now.

2.5 The stress-energy tensor

In the previous section, we investigated the consequences of projective transformations on correlation functions. Stronger results can only be obtained by considering the more restricted set of primary operators. However, primary operators transform covariantly under any conformal transformation (eq. (2.21)). We recall that according to Liouville's theorem [7], an analytic function cannot be bounded in the entire complex plane, unless it is a constant. This implies that non-trivial analytic transformations (eq. (2.11)) must have singularities and they cannot be small everywhere. To cope with this problem, we consider a local coordinate transformation $r' = r + \hat{\epsilon}(r)$ with $\hat{\epsilon}(r)$ small everywhere, which is analytic inside the region D_1 bounded by the curve C, and is a decreasing differentiable function in the external region D_2 and at the boundary C. We suppose that all points of the correlation function are in the internal region D_1 as in Figure 12. Conformal invariance is broken because $\hat{\epsilon}$ is not analytic in the external region. Therefore,

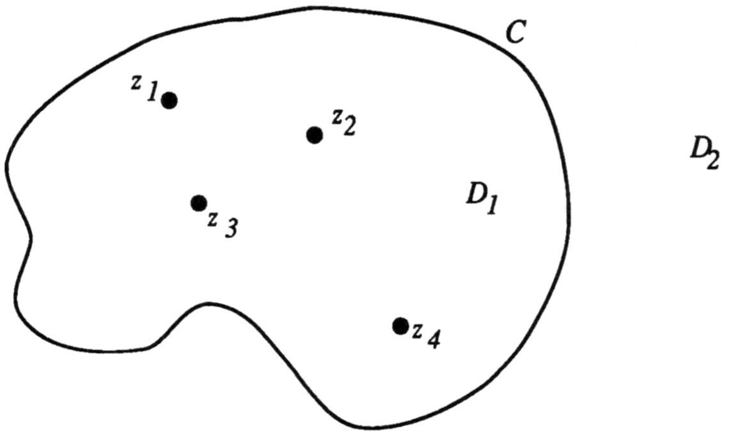

Fig. 12. Bounded coordinate transformation.

such a coordinate transformation drives the system away from criticality. The fixed point Hamiltonian is modified by a term which couples to the small perturbing parameter $\hat{\epsilon}$. In fact, it couples to derivatives of $\hat{\epsilon}$ because the particular case $\hat{\epsilon}(r) = 1$ corresponds to a translation. The scaling operator which couples to the lowest order in $\partial\hat{\epsilon}$ defines the *stress-energy tensor* $T_{\mu\nu}$.

$$\delta_{\hat{\epsilon}}\mathcal{H} = -\frac{1}{2\pi}\int_{D_2} d^2r\,\partial^\mu\hat{\epsilon}^\nu\, T_{\mu\nu}(r) \tag{2.34}$$

Invariance under rotations $\hat{\epsilon}^1 = r^2$, $\hat{\epsilon}^2 = -r^1$ implies that $T_{\mu\nu} = T_{\nu\mu}$ is symmetric. Global scale invariance $\hat{\epsilon}^\mu = r^\mu$ implies that $T_{\mu\mu} = 0$. We also note that the invariance under special conformal transformations does not add any further condition to the stress-energy tensor. We thus have the important result that translation, rotation and scale invariance automatically imply conformal invariance if the interactions are local.

The consequences of a infinitesimal transformation $\hat{\epsilon}$ on the correlation function $< \phi_1\cdots\phi_n >$ of primary operators can be perturbatively computed with the correlation function at criticality. Equation (2.23) generalized to an n-point function now reads

$$\begin{aligned}\textstyle\sum_i \left((\Delta_i\partial_{z_i}\hat{\epsilon} + \hat{\epsilon}\partial_{z_i}) + (\overline{\Delta}_i\partial_{\bar{z}_i}\bar{\hat{\epsilon}} + \bar{\hat{\epsilon}}\partial_{\bar{z}_i})\right) < \phi_1\cdots\phi_n >= \\ -\tfrac{1}{2\pi}\textstyle\int_{D_2} d^2r\,\partial^\mu\hat{\epsilon}^\nu(r) < T_{\mu\nu}(r)\phi_1\cdots\phi_n >\end{aligned} \tag{2.35}$$

The l.h.s. is justified because $\hat{\epsilon}$ is conformal within D_1. After integrating by part, the r.h.s. becomes

$$+\frac{1}{2\pi}\int_{D_2} d^2r\,\hat{\epsilon}^\nu(r) < \partial^\mu T_{\mu\nu}(r)\phi_1\cdots\phi_n > -\frac{1}{2\pi}\int_C d\Sigma\, n^\mu\hat{\epsilon}^\nu(r) < T_{\mu\nu}(r)\phi_1\cdots\phi_n > \tag{2.36}$$

where n^μ is orthonormal to the surface C. The fact that $\hat{\epsilon}$ can be arbitrarily chosen in D_2 implies that the first term of eq. (2.36) must vanish for any $\hat{\epsilon}$. This yields the *conservation law*:

$$\partial^\mu T_{\mu\nu} = 0 \tag{2.37}$$

T being traceless and satisfying eq. (2.37), we can define the two independent components of the stress-energy tensor as

$$\begin{aligned} T(z) &= T_{11} - T_{22} + 2iT_{12} \\ \overline{T}(\bar{z}) &= T_{11} - T_{22} - 2iT_{12} \end{aligned} \tag{2.38}$$

which satisfy

$$\partial_{\bar{z}}T(z) = \partial_z\overline{T}(\bar{z}) = 0 \tag{2.39}$$

We can now write equations in which the split between the z and $\bar{z}$-dependence becomes apparent. This can be achieved with the use of Cauchy theorem for the l.h.s. of eq. (2.35) and we obtain

$$\begin{aligned} &\frac{1}{2\pi i}\int_C dz\hat{\epsilon}(z) < T(z)\phi_1\cdots\phi_n > -\frac{1}{2\pi i}\int_C d\bar{z}\overline{\hat{\epsilon}(z)} < \overline{T}(\bar{z})\phi_1\cdots\phi_n >= \\ &\qquad \frac{1}{2\pi i}\int_C dz\hat{\epsilon}(z)\sum_i\left(\frac{\Delta_i}{(z-z_i)^2}+\frac{1}{z-z_i}\partial_{z_i}\right) < \phi_1\cdots\phi_n > \\ &\qquad -\frac{1}{2\pi i}\int_C d\bar{z}\overline{\hat{\epsilon}(z)}\sum_i\left(\frac{\overline{\Delta}_i}{(\bar{z}-\bar{z}_i)^2}+\frac{1}{\bar{z}-\bar{z}_i}\partial_{\bar{z}_i}\right) < \phi_1\cdots\phi_n > \end{aligned} \tag{2.40}$$

We know that the r.h.s. will vanish for any infinitesimal projective transformation, i.e. for $\hat{\epsilon} = 1, z, z^2$. A consistency condition for the vanishing of the l.h.s. is that $T(z)$ behaves for large z as z^{-4}. Moreover, being valid for any $\hat{\epsilon}$, we conclude that we can write this in a local form called *conformal Ward identities*

$$\begin{aligned} &< T(z)\phi_1(z_1,\bar{z}_1)\cdots\phi_n(z_n,\bar{z}_n) >= \\ &\qquad \sum_i\left(\frac{\Delta_i}{(z-z_i)^2}+\frac{1}{z-z_i}\partial_{z_i}\right) < \phi_1(z_1,\bar{z}_1)\cdots\phi_n(z_n,\bar{z}_n) > \end{aligned} \tag{2.41}$$

This is nothing but the formal expression of the covariance of the primary fields ϕ_i under an arbitrary analytic transformation. A similar relation holds for $\overline{T}$. An insertion of the holomorphic part T of the stress-energy tensor into a correlation function of primary operators amounts to apply a differential operator of dimension 2 to the correlation function without T. The scaling dimension of T is therefore canonically 2. Its spin however is $+2$ for T and -2 for $\overline{T}$.

The stress-energy tensor is more than the generator of conformal transformations. It is the Noether current which generates infinitesimal coordinate transformations in the sense that for any operator $A(z_1, \bar{z}_1)$ and any $\hat{\epsilon}$ (conformal or not)

$$\delta_{\hat{\epsilon}} A(z_1, \bar{z}_1) = \frac{1}{2\pi i} \oint_{C_1} d\Sigma\, n^\mu \hat{\epsilon}^\nu(r)\, T_{\mu\nu}(r)\, A(z_1, \bar{z}_1) \tag{2.42}$$

where the contour is chosen to surround the point z_1. This definition is true even if $\hat{\epsilon}$ is a general coordinate transformation because the definition of T does not say anything on the size of D_1 or the correlation function we want to study. We have to choose D_1 to contain the physical points only to derive the Ward identities eq. (2.41) for specific primary operators. More generally, for any functional $X(z_1, \ldots, z_n)$ (we skip in the notation the $\bar{z}$-dependence which is implicit)

$$\delta_{\hat{\epsilon}} < X(z_1, \ldots, z_n) >= \frac{1}{2\pi i} \oint_{C_1 \cup \cdots \cup C_n} d\Sigma\, n^\mu \hat{\epsilon}^\nu(r) < T_{\mu\nu}(r) X(z_1, \ldots, z_n) > \tag{2.43}$$

Integrating over a large curve surrounding all points amounts to sum over smaller curves surrounding each point. We assume that all points are different.

To extract the algebraic information hidden in the structure of T, we have to consider $\delta_\epsilon \delta_\epsilon < X >$ where T will appear twice. First notice that $< T >= 0$ by translation invariance. However, the two-point function has no reason to vanish, but due to scale invariance must have the form

$$< T(z_1) T(z_2) >= \frac{c/2}{(z_{12})^4} \tag{2.44}$$

This defines the *central charge* c, which plays a major role in the following. Below, we shall see that it describes the analogue of the Casimir effect and in chapter 3, we shall relate it to the finite-size scaling of the free energy density at the critical point. Of course holomorphic and antiholomorphic parts of $T_{\mu\nu}$ do not couple and $< T\overline{T} >= 0$. A consequence of the non-vanishing two-point function is that for ϵ analytic

$$\delta_\epsilon < T(z) >=< \delta_\epsilon T(z) >= \frac{1}{2\pi i} \oint_{C_z} dz'\, \epsilon(z') < T(z') T(z) >= \frac{c}{12} \partial_z^3 \epsilon(z) \tag{2.45}$$

$< \delta T >$ is zero only if the third derivative is zero, i.e. for the infinitesimal generators of the projective group. This is normal because general conformal transformation will change the geometry of the plane. It is well known that renormalization group transformations, here conformal coordinate transformations, do not change the canonical dimension of a Noether current. T must essentially transform like $(\partial_z)^2$ (i.e. like a rank 2 tensor) and δT (of dimension 3) must be expressed in terms of conserved operators (i.e. $\partial_{\bar{z}} A(z) = 0$). We only dispose of T, $\partial_z T$ (higher derivatives are ruled out for dimensional reasons), and of the trivial identity operator. Therefore, under a infinitesimal conformal transformation, T will be modified like

$$\delta_\epsilon T(z) = (2\partial_z \epsilon + \epsilon \partial_z) T(z) + \frac{c}{12} \partial_z{}^3 \epsilon \tag{2.46}$$

The first two terms mean that T is a quasi-primary operator and the factor 2 is fixed by the dimension of T. Dimensional arguments do not permit other terms and the transformation rule for T is the most general we can write. If we compute now $\delta < T\phi_1 \cdots \phi_n >$ with eqs. (2.43), (2.46) and (2.22), and if we proceed along the same line as to derive the Ward identities (making use of Cauchy theorem to write everything as contour integral), we obtain the local identity

$$
\begin{aligned}
&< T(z)T(z')\phi_1 \cdots \phi_n >= \\
&\left(\frac{2}{(z-z')^2} + \frac{1}{z-z'}\partial_{z'} + \sum_i \left(\frac{\Delta_i}{(z-z_i)^2} + \frac{1}{z-z_i}\partial_{z_i} \right) \right) < T(z')\phi_1 \cdots \phi_n > \\
&+ \frac{c/2}{(z-z')^4} < \phi_1 \cdots \phi_n >
\end{aligned}
\tag{2.47}
$$

T behaves almost like a primary operator, up to a c term which picks up the two-point singularity of T. We have seen that the last term is present as a consequence of the global properties of the conformal transformations. It signals the presence of an *anomaly* due to the fluctuations on **all** length scales for critical statistical systems. It numerical value will therefore contain important information on the nature of the fluctuating critical system and it will be one of our main tasks to decode this in terms of universality classes.

Ward identities can be derived for any operators once we know how they transform. We have seen that T acts onto other operators through singular terms $1/(z-z_i)^k$ for some k. k is smaller than or equal to 2 for primary operators and has the value $k = 4$ for T. Higher values of k are possible for general operators. A secondary operator like $\partial\phi$ will transform like

$$
\delta_\epsilon(\partial_z\phi) = ((\Delta+1)\partial_z\epsilon + \epsilon\partial_z)\partial_z\phi + \Delta(\partial_z^2\epsilon)\phi
\tag{2.48}
$$

Ward identities for such an operator contains a pole of order $k = 3$ due to the second derivative of ϵ once replaced by a contour integral. A derivative of order n of a primary operator gives rise to a pole of order $k = n+2$. It makes therefore sense to be interested in the Laurent coefficients of this development. This defines the *Virasoro generators* acting on any operator $A(z_1, \bar{z}_1)$.

$$
L_n A(z_1, \bar{z}_1) = \frac{1}{2\pi i} \oint_{C_{z_1}} dz\,(z-z_1)^{n+1}\, T(z)A(z_1, \bar{z}_1)
\tag{2.49}
$$

In order to avoid the delicate problem to define this in an operator sense, we always restrict to consider the well-defined correlations functions of the above. To apply such an operator many times, we need a convention on the choice of the contours. This is given from the situation where two Virasoro generators are applied to $A(z_1, \bar{z}_1)$.

$$
L_n L_m A(z_1, \bar{z}_1) = \left(\frac{1}{2\pi i}\right)^2 \oint_{C'_{z_1,z}} dz' \oint_{C_{z_1}} dz\,(z'-z_1)^{n+1}(z-z_1)^{m+1} T(z')T(z)A(z_1, \bar{z}_1)
\tag{2.50}
$$

The contour C_{z_1} surrounds z_1 and is surrounded by $C'_{z_1,z}$ (see Figure 13(a)).
If we apply p times L to the left of A, i.e. $L_{n_p} \cdots L_{n_1} A(z_1, \bar{z}_1)$, we have to choose p non-intersecting contours surrounding z_1, with C^p corresponding to the upmost left operator

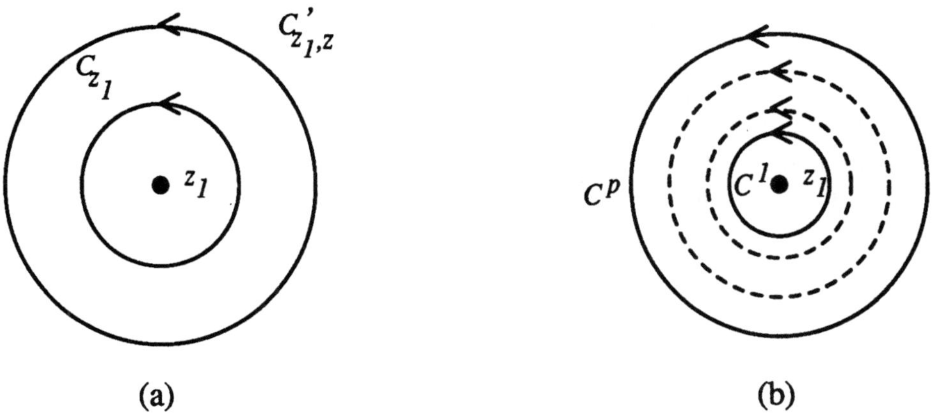

Fig. 13. Convention for the contours in the definition of the Virasoro generators.

L_{n_p} surrounding all others, and C^1 corresponding to L_{n_1} placed inside all the others (see Figure 13(b)). Permuting the order of two L amounts to interchange contours. For example in the case of eq. (2.50), commuting L_n and L_m means that the contour $C'_{z_1,z}$ must move inside C_{z_1}. The difference, i.e. the commutator $[L_n, L_m]$, is given by the residue of the integration over z' at the pole $z' \to z$.

$$[L_n, L_m]A(z_1, \bar{z}_1) = \left(\frac{1}{2\pi i}\right)^2 \oint_{C_z} dz' \oint_{C_{z_1}} dz\, (z' - z_1)^{n+1}(z - z_1)^{m+1} T(z')T(z)A(z_1, \bar{z}_1) \tag{2.51}$$

The contour C_z surrounds the point z only as illustrated in Figure 14. We know that the

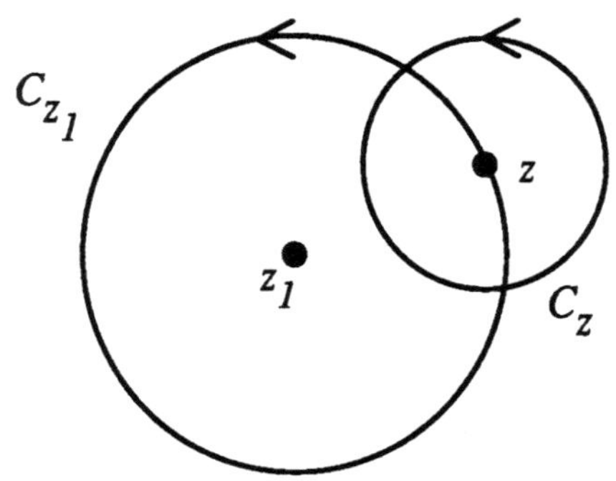

Fig. 14. Commutation contours for the Virasoro algebra.

insertion of $T(z')$ induces an infinitesimal transformation. Its action on $T(z)$ is given by eq. (2.46) which can be written in an integral form and gives rise to the pole structure

$$T(z')T(z) = \frac{c/2}{(z'-z)^4} + \frac{2T(z)}{(z'-z)^2} + \frac{\partial_z T(z)}{z'-z} + \text{regular terms} \tag{2.52}$$

This is similar to the derivation of eq. (2.47). The "regular terms" stand for the infinitesimal transformation of A and all other operators which may be in a general correlation function. For general A operators, the regular terms will be much more complicated then in eq. (2.47) but they do not contribute to the integration in the variable z' along C_z. We compute now the three contributions to the commutator $[L_n, L_m]$ coming from the three poles in $(z'-z)$ in (2.52). The first term is given by

$$\begin{aligned}
\frac{c}{2}\left(\frac{1}{2\pi i}\right)^2 \oint_{C_z} dz' \oint_{C_{z_1}} dz\, \frac{(z'-z_1)^{n+1}(z-z_1)^{m+1}}{(z'-z)^4} A(z_1,\bar{z}_1) &= \\
= \frac{cn(n^2-1)}{12}\left(\frac{1}{2\pi i}\right) \oint_{C_{z_1}} dz\,(z-z_1)^{n+m-1} A(z_1,\bar{z}_1) & \\
= \frac{cn(n^2-1)}{12}\delta_{n+m,0}\, A(z_1,\bar{z}_1) & \qquad (2.53)
\end{aligned}$$

This term gives the central extension of the loop algebra (2.13). The second and third contributions will couple to L_{n+m}. The second gives

$$\begin{aligned}
2\left(\frac{1}{2\pi i}\right)^2 \oint_{C_z} dz' \oint_{C_{z_1}} dz\, \frac{(z'-z_1)^{n+1}(z-z_1)^{m+1}}{(z'-z)^2} T(z) A(z_1,\bar{z}_1) &= \\
= 2\,(n+1)\left(\frac{1}{2\pi i}\right) \oint_{C_{z_1}} dz\,(z-z_1)^{n+m+1} T(z) A(z_1,\bar{z}_1) & \\
= 2\,(n+1)\, L_{n+m} A(z_1,\bar{z}_1) & \qquad (2.54)
\end{aligned}$$

and the last one is

$$\begin{aligned}
\left(\frac{1}{2\pi i}\right)^2 \oint_{C_z} dz' \oint_{C_{z_1}} dz\, \frac{(z'-z_1)^{n+1}(z-z_1)^{m+1}}{(z'-z)} \partial_z T(z) A(z_1,\bar{z}_1) &= \\
= \left(\frac{1}{2\pi i}\right) \oint_{C_{z_1}} dz\,(z-z_1)^{n+m+2} \partial_z T(z) A(z_1,\bar{z}_1) & \\
= -\left(\frac{1}{2\pi i}\right) \oint_{C_{z_1}} dz\,\partial_z (z-z_1)^{n+m+2} T(z) A(z_1,\bar{z}_1) & \\
= -(n+m+2)\, L_{n+m} A(z_1,\bar{z}_1) & \qquad (2.55)
\end{aligned}$$

We now sum up the three contributions and repeat the same operations for the conjugate algebra $\{\bar{L}_n\}$ generated by $\overline{T}$. The result is the celebrated double *Virasoro algebra* [322].

$$\boxed{\begin{aligned}
&[L_n, L_m] = (n-m)\, L_{n+m} + \tfrac{1}{12}\, c\, n(n^2-1)\, \delta_{n+m,0} \\
&[\bar{L}_n, \bar{L}_m] = (n-m)\, \bar{L}_{n+m} + \tfrac{1}{12}\, c\, n(n^2-1)\, \delta_{n+m,0} \\
&[L_n, \bar{L}_m] = 0
\end{aligned}} \qquad (2.56)$$

The *central extension* proportional to c disappears for $n = -1, 0, 1$ where we recover the loop algebra (2.13) for the projective transformations. The two algebras only differ by this central term. Notice how we have reduced a complicated field theoretic quantity like the stress-energy tensor to an infinite dimensional algebra, which representation theory will bring dramatic new results.

Let us finish this chapter with a last comment on T. We have only shown how it transforms under infinitesimal transformations. To complete our understanding, we need to write how it transforms under a general conformal transformation. This is given by

$$T_{\text{plane}}(z) = T_{\text{new}}(z')\,(\partial_z z')^2 + \frac{c}{12}\{z', z\} \tag{2.57}$$

where we introduce the subscripts "plane" and "new" to emphasize that the geometry is not the same. The last term introduces the *Schwarzian derivative* defined by

$$\{w, z\} = \frac{\partial_z^3 w}{\partial_z w} - \frac{3}{2}\left(\frac{\partial_z^2 w}{\partial_z w}\right)^2 \tag{2.58}$$

Schwarz introduced such a derivative while working on the problem of mapping circular polygons conformally [188, 189]. Such a derivative appears in the study of second order differential equations $y'' + Q(z)y = 0$. It obeys several important properties. Given a projective transformation, we have

$$\left\{\frac{aw+b}{cw+d}, z\right\} = \{w, z\} \qquad \text{and} \qquad \left\{w, \frac{az+b}{cz+d}\right\} = \{w, z\}\,(cz+d)^4 \tag{2.59}$$

Under a sequence of conformal maps $z \to w \to \tilde{w}$, we have the equality

$$\{\tilde{w}, z\} = \{\tilde{w}, w\}\,(\partial_z w)^2 + \{w, z\} \tag{2.60}$$

The two last equations reflect the consistency condition for eq. (2.57) to transform correctly under successive analytic transformations.

The particular case of the *strip geometry* will become very important in the sequel. It is defined by $z' = w(z) = (L/2\pi)\ln(z)$ which maps the entire plane onto a strip of width L. The negative part of the real axis corresponds to both edges of the strip, which has therefore the topology of a cylinder. Under this particular transformation, eq. (2.57) can be written

$$T_{\text{strip}}(z') = \left(\frac{2\pi}{L}\right)^2 \left(T_{\text{plane}}(z)z^2 - \frac{c}{24}\right) \tag{2.61}$$

The expectation value of T in the strip turns out to be different from zero

$$< T_{\text{strip}}(z') >_{\text{strip}} = -\frac{c}{24}\left(\frac{2\pi}{L}\right)^2 \tag{2.62}$$

even if $< T_{\text{plane}} >= 0$. This is the reason why the central charge is often called *conformal anomaly*. Recalling the definition of the stress tensor as the variation of the Hamiltonian under a coordinate transformation, we can see that the free energy is now modified

$$\delta_{\hat{\epsilon}} \ln \mathcal{Z} =< \delta_{\hat{\epsilon}}\mathcal{H} >= -\frac{1}{2\pi}\int d^2z'(< T_{\text{strip}}(z') >_{\text{strip}}\, \partial_{\bar{z}'}\hat{\epsilon} + < \overline{T}_{\text{strip}}(\bar{z}') >_{\text{strip}}\, \partial_{z'}\bar{\hat{\epsilon}}) \tag{2.63}$$

The parameter c measures therefore the variation of the free energy due to finite size effects. Recalling from chapter 1 that the free energy of two-dimensional systems corresponds to the ground state (or vacuum) energy of a one-dimensional quantum system, we realize the analogy with the *Casimir effect* in quantum electrodynamics (with c instead of $\hbar$). The equations (2.62) and (2.63) will be used to discuss the finite-size scaling of the free energy in the next chapter.

3 Finite Size Scaling

In the last chapter, conformal invariance was used to constrain the multipoint correlation functions of isotropic critical two-dimensional systems. Before following the field-theoretic developments further, we shall describe important applications to the study of finite-size effects. Besides being of interest in their own right, these results provide highly efficient computational tools for the practical calculations of central charges and scaling dimensions.

Finite-size effects are conventionally described by *finite-size scaling.* In this chapter, we shall give a presentation of the most basic facts useful for the later discussion in connection with conformal invariance. From a phenomenological description of the behaviour of observables of a statistical model like a specific heat, a magnetic susceptibility or a correlation length due to the influences of the presence of a finite geometry, we shall pass to recall the notions of finite-size scaling variables and finite-size scaling functions. We describe the phenomenological renormalization and discuss universality. Although usually presented here for the two-dimensional case, the results are in fact valid for all dimensions d up to the upper critical dimension, $d < d^*$. These considerations will set the frame into which the stronger results following from two-dimensional conformal invariance will be inserted.

For detailed reviews of finite-size scaling, we refer to [27, 283] and [G8].

3.1 Statistical systems in finite geometries

Consider a statistical system confined to a finite domain of volume V, surface A, etc. The free energy $F(T, V, \mathcal{N})$ of the system can be written as

$$F(T, V, \mathcal{N}) = V f_b(T, \rho) + A f_s(T, \rho) + \ldots \tag{3.1}$$

where T is the temperature, $\mathcal{N}$ the number of particles and we have suppressed other thermodynamic variables. Writing this, a limit $V \to \infty$, $\mathcal{N} \to \infty$ such that $\rho = \mathcal{N}/V$ is kept fixed is implied, that is, the ***bulk free energy density*** $f_b(T, \rho)$ is defined by

$$f_b(T, \rho) = \lim_{\mathcal{N}\to\infty, V\to\infty} F(T, V, \mathcal{N})/V \quad \text{with } \rho = \mathcal{N}/V \text{ fixed} \tag{3.2}$$

A similar limit is implied for the definition of the ***surface free energy density*** $f_s(T, \rho)$. Further finite-size terms can (and will) be defined if necessary.

We shall mainly use the following boundary conditions. Let $\Phi(r)$ be a space-dependent observable with r constrained to a finite volume, e.g. $0 \leq r \leq L$. We shall use *periodic* boundary conditions $\Phi(r + L) = \Phi(r)$ and also *free* boundary conditions $\Phi(r) = 0$ if $r < 0$ or $r > L$. For an example consider the $1D$ Ising model with Hamiltonian

$$\mathcal{H} = -\sum_{i=1}^{N} J_i \sigma_i \sigma_{i+1} \tag{3.3}$$

Periodic boundary conditions correspond to $\sigma_1 = \sigma_{N+1}$ and free boundary conditions correspond to $J_N = 0$.

The bulk free energy density f_b is independent of the boundary conditions, while the surface free energy density f_s does depend on the boundary conditions. For periodic boundary conditions and for values of T far enough from the critical temperature T_c, the finite-size corrections to f_b are in fact exponentially small.

In what follows we shall restrict our attention to the case where $\bar{d}$ dimensions of the lattice are infinite and in d' dimensions we take a hypercubic lattice of linear extent L. We shall for brevity refer to L as the *size of the system*. Then $d = \bar{d} + d'$ and we have for the free energy density

$$f(T,\rho) = f_b(T,\rho) + \frac{1}{L} f_s(T,\rho) + \frac{1}{L^2} f_2(T,\rho) + \dots \qquad (3.4)$$

In the following sections, we shall mainly consider the case $d = 2$, $\bar{d} = d' = 1$. The quantity f_2 will be shown to play an important role in the following.

3.2 Finite-size scaling hypothesis

The decomposition of the free energy density as written in eq. (3.4) is only valid when the temperature T is far enough from the critical point T_c. Close to T_c, singularities will appear which may invalidate the simple relationship written so far.

As an example, consider the specific heat of some lattice system as shown in Figure 15 for two lattices of finite sizes $L_1 > L_2$. Although the specific heat of the **infinite** system shows a divergence $C_\infty(T) \sim (T - T_c)^{-\alpha}$, this singularity is rounded when the system size L is finite. We now define a few quantities which are useful to describe these new finite-size features, following phenomenological ideas first proposed by Fisher and Barber [123].

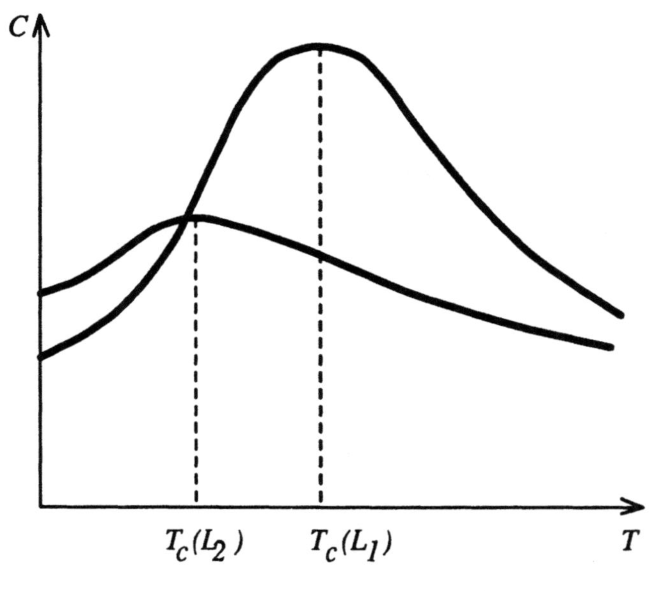

Fig. 15. Specific heat as a function of temperature for two lattices of sizes $L_1 > L_2$.

The curves show a maximum at some value of T which is shifted away from T_c. One can take the temperature of the finite-size maximum to be a definition of a finite-size estimate $T_c(L)$ of the true critical temperature $T_c(\infty) = T_c$. Then the finite-size shift in T_c is

$$\left(T_c(L) - T_c(\infty)\right)/T_c(\infty) \sim L^{-\lambda} \text{ if } L \to \infty \tag{3.5}$$

This defines the ***shift exponent*** λ. Another useful quantity is the ***rounding temperature*** $T^*(L)$. This is defined such that if $|T - T_c|/T_c \geq |T^* - T_c|/T_c$, then the finite-size specific heat is close to the bulk value: $C_L(T) \simeq C_\infty(T)$. We use this to define a ***rounding exponent*** θ via

$$\left(T^*(L) - T_c(\infty)\right)/T_c(\infty) \sim L^{-\theta} \text{ if } L \to \infty \tag{3.6}$$

Asymptotically, the value of θ characterizes the size of the finite-size scaling region.

Note that we have implicitly assumed here that there is no phase transition in the system for L finite. It is straightforward to generalize to this case as well. For the later applications of conformal invariance, we shall always have $\bar{d} = 1$ and no phase transition for $L < \infty$ can occur.

Now we state the basic hypothesis of finite-size scaling, namely that there should be only one relevant length describing the rounding and shifting of the thermodynamic singularities. This amounts to saying that, for the correlation length

$$\xi(T^*(L)) \sim L \tag{3.7}$$

which, using $\xi_\infty \sim (T - T_c)^{-\nu}$ yields for the rounding exponent

$$\theta = \frac{1}{\nu} \tag{3.8}$$

As we shall see later, this result is confirmed by explicit calculation in a large variety of systems. For the shift exponent, one expects, if there is only one length describing the finite-size effects

$$\lambda = \frac{1}{\nu} \tag{3.9}$$

This result is also found to hold true for many systems, but exceptions occur in the spherical model and the ideal Bose gas, see [27]. Since these models do not have a phase transition at a non-vanishing temperature for $d = 2$, we shall use both of the relations eqs. (3.8) and (3.9) in the sequel.

We now proceed to state the finite-size scaling hypothesis for the scaling function. Introduce the finite-size scaling variable

$$\tilde{z} = \frac{L}{\xi_\infty(T)} \tag{3.10}$$

and distinguish three cases

1. $\tilde{z} \ll 1$: the correlation length is much larger than the system size. Finite-size and boundary effects will be very important.
2. $\tilde{z} \simeq 1$: this defines the rounding temperature $T^*(L)$.
3. $\tilde{z} \gg 1$: the correlation length is much smaller than the system size. One expects a cross-over to the bulk behaviour. Here we assume that the correlation length is still much larger than some microscopic scale.

We thus see that the *finite-size scaling region* is characterized by those values of T and L such that the finite-size scaling variable $\tilde{z}$ remains finite.

For $L \to \infty$, the specific heat behaves close to T_c like

$$C_\infty(T) \simeq \gamma^\pm (T - T_c)^{-\alpha} \tag{3.11}$$

where $\gamma^\pm$ refers to $T - T_c \gtrless 0$, respectively. We now postulate [123] that in the case of the *finite-size scaling limit*

$$T \to T_c, \quad L \to \infty, \quad \text{but} \quad \tilde{z} = \frac{L}{\xi(T)} \text{ fixed} \tag{3.12}$$

we can write C_L in the form

$$C_L(T) \sim L^\omega \tilde{Q}(\tilde{z}) \sim L^\omega Q(L^{1/\nu} t) \tag{3.13}$$

where the reduced temperature is

$$t := (T - T_c)/T_c \tag{3.14}$$

This is the basic finite-size scaling form of a thermodynamic quantity. Using $\theta = \lambda = 1/\nu$, we can replace $T_c = T_c(\infty)$ by some finite-size estimate $T_c(L)$. This does not affect the finite-size scaling form as specified.

In order to reproduce for $\tilde{z} \gg 1$ the bulk behaviour, one makes the ansatz

$$Q(\tilde{x}) \sim q_\infty^\pm \tilde{x}^{-\rho} \quad \text{if } \tilde{x} \to \pm\infty \tag{3.15}$$

and finds by matching expressions $\omega = \rho/\nu, \rho = \alpha$ and $q_\infty^\pm = \gamma^\pm$. So we can say

$$C_L(T^*) \sim L^{\alpha/\nu} \tilde{Q}(\tilde{z}) \tag{3.16}$$

Again, with $\theta = \lambda = 1/\nu$ and assuming that $Q(\tilde{z})$ is finite and non-vanishing between $0 \leq \tilde{z} \leq 1$, we have

$$C_L(T_c(L)) \sim L^{\alpha/\nu} \tag{3.17}$$

This result will be very important for a practical calculation of critical exponents from finite-size data.

To summarize, we have seen how to relate the finite-size rounding of the singularities of thermodynamic quantities close to a second order phase transition to the exponents of the bulk system. The discussion, presented here for the specific heat, immediately generalizes to any other quantity and to other thermodynamic variables than the temperature. In Table 4, we give a few examples. We also recall that the free energy density or the specific heat must be decomposed into a regular and a singular part: $f = f^{reg} + f^{sin}$ for example. While the regular part f^{reg} is an analytic function of T close to T_c, non-analyticities appear in the singular part f^{sin} and its derivatives. The background term f^{reg} (and C^{reg}) must therefore be subtracted before the critical behaviour of either f or C can be analyzed. Background terms are absent for ξ, M or χ.

We note that the finite-size scaling behaviour of both ξ and f^{sin} at $\tilde{z} = 0$ does not depend on any exponent. This is going to be very useful in the context of phenomenological renormalization and for finding the numerical value of $T_c(L)$.

Table 4. Bulk scaling and finite-size scaling close to T_c with $\tilde{z} = 0$.

		bulk	finite-size
specific heat	C^{sin}	$t^{-\alpha}$	$L^{\alpha/\nu}$
magnetic susceptibility	χ	$t^{-\gamma}$	$L^{\gamma/\nu}$
correlation length	ξ	$t^{-\nu}$	L
free energy density	f^{sin}	$t^{2-\alpha}$	L^{-d}
order parameter[(a)]	M	t^{β}	$L^{-\beta/\nu}$
latent heat[(a)]	ℓ_h	$t^{1-\alpha}$	$L^{(\alpha-1)/\nu}$

[(a)] Special care is needed to define non-vanishing lattice expressions for both M and ℓ_h.

3.3 Universality

The critical behaviour of a statistical system is determined by its renormalization group (RG) properties. In particular, it follows from the renormalization group that the critical exponents only depend on very few characteristics of the model, like lattice dimension, number of components of the order parameter and the global symmetry, but are independent of the details of the lattice structure or the precise form of the interactions. This is referred to by saying that critical exponents are *universal*. In fact, similar universality properties also hold for the scaling functions. The presentation given follows the original work of Privman and Fisher [278].

In the case of a simple ferromagnet, the singular bulk free energy density can be written

$$f_{\infty}^{sin} \sim A_1 |t|^{2-\alpha} W^{\pm}(A_2 h |t|^{-\beta-\gamma}) \tag{3.18}$$

where $W^{\pm}$ is a scaling function, and the index $\pm$ refers to $t \gtrless 0$. It is one of the results of the renormalization group that all information on the "details" of the system can be absorbed into the two non-universal *metric factors* A_1, A_2. Therefore, $W^{\pm}$ is a universal function. From this, similar expressions are easily obtained for the derivatives of f_{∞}^{sin}. In writing this, one has implicitly assumed that the scaling function has neither zeroes nor poles close to criticality. This requires the absence of so-called *dangerous irrelevant variables* (which is the case if $d < d^*$), and of marginal variables.

Equation (3.18) is readily generalized to include finite-size effects

$$f^{sin}(t,h;L) \sim A_1 |t|^{2-\alpha} \tilde{W}^{\pm}(A_2 h |t|^{-\beta-\gamma}; L/\xi_{\infty}) \tag{3.19}$$

where $\tilde{W}^{\pm}$ is universal and $\xi_{\infty} \sim \xi_0 t^{-\nu}$ (as $t \to +0$) is the bulk correlation length. Note that there is no extra metric factor associated with L/ξ_{∞}. This is true if the dimension d is below the upper critical dimension d^*, but holds no longer for $d \geq d^*$ due to the presence of dangerous irrelevant variables, as can be seen explicitly in spherical model calculations [57, 242]. On the other hand, the non-universal constant ξ_0 does appear implicitly. It must also be present in the correlation function

$$\begin{aligned} G(\vec{r};t,h) &:= < s_{\vec{0}} s_{\vec{r}} > - < s_{\vec{0}} >< s_{\vec{r}} > \\ &\sim D_1 r^{2-d-\eta} X^{\pm}(\vec{r}/\xi_{\infty}, D_2 h |t|^{-\beta-\gamma}) \end{aligned} \tag{3.20}$$

where $X^{\pm}$ is universal and D_1, D_2 are metric factors. The magnetic susceptibility is obtained as

$$\chi_{\infty}(t,h) = \int d\vec{r}\; G(\vec{r};t,h) \sim D_1 \xi_{\infty}^{2-\eta} \tilde{X}^{\pm}(D_2 h|t|^{-\beta-\gamma}) \tag{3.21}$$

The same scaling behaviour is also expected for the total correlation function

$$\begin{aligned} \Gamma(\vec{r};t,h) &:= < s_{\vec{0}} s_{\vec{r}} > \\ &\sim D_1 r^{2-d-\eta} Z^{\pm}(\vec{r}/\xi_{\infty}, D_2 h|t|^{-\beta-\gamma}) \end{aligned} \tag{3.22}$$

with a new universal scaling function $Z^{\pm}$. In the limit $r \to \infty$, $\Gamma(\vec{r}) \to m_{\infty}^2$, where m_{∞} is the magnetization per spin. Thus

$$m_{\infty}^2(t,h) \sim D_1 \xi_{\infty}^{2-d-\eta} \tilde{Z}^{\pm}(D_2 h|t|^{-\beta-\gamma}) \tag{3.23}$$

Now, from eq. (3.18), one has by taking derivatives

$$\begin{aligned} m_{\infty}(t,h) &\sim A_1 A_2 |t|^{\beta} W_1^{\pm}(A_2 h|t|^{-\beta-\gamma}) \\ \chi_{\infty}(t,h) &\sim A_1 A_2^2 |t|^{-\gamma} W_2^{\pm}(A_2 h|t|^{-\beta-\gamma}) \end{aligned} \tag{3.24}$$

where $W_1^{\pm}(x) = \partial_x W^{\pm}(x)$ and $W_2^{\pm}(x) = \partial_x W_1^{\pm}(x)$. Comparing coefficients, we find for the susceptibility

$$A_1 A_2^2 |t|^{-\gamma} W_2^{\pm}(A_2 h|t|^{-\beta-\gamma}) = D_1 \xi_0^{2-\eta} |t|^{-(2-\eta)\nu} \tilde{X}^{\pm}(D_2 h|t|^{-\beta-\gamma}) \tag{3.25}$$

which returns the well-known scaling relation $\gamma = (2-\eta)\nu$, and for the magnetization density

$$A_1^2 A_2^2 |t|^{2\beta} W_1^{\pm\,2}(A_2 h|t|^{-\beta-\gamma}) = D_1 \xi_0^{2-\eta-d} |t|^{-(2-\eta-d)\nu} \tilde{Z}^{\pm}(D_2 h|t|^{-\beta-\gamma}) \tag{3.26}$$

which implies the hyperscaling relation $2\beta = (d-2+\eta)\nu$. Furthermore, the metric factors become related and one identifies [278] the universal combinations

$$\begin{aligned} A_1 \xi_0^d &=: Q_1 \\ A_2 / D_2 &=: Q_2 \\ D_1 A_1^{-\psi} A_2^{-2} &=: Q_3 \quad \text{if } \psi = 1 + \gamma/(d\nu) \end{aligned} \tag{3.27}$$

whose universality follows from the universality of the scaling functions. In particular, we have only two non-universal metric factors which describe, to leading order in the scaling behaviour, the non-universal details of the system under consideration. This universality goes a little beyond the conventional universality of ratios of some scaling amplitudes like γ^+/γ^- (eq. (3.11)) since it does involve hyperscaling. In other words, we have obtained a *hyperuniversality relation*

$$\lim_{t \to \pm 0} f^{sin}(t) \xi^d(t) = u^{\pm} = \text{universal} \tag{3.28}$$

This is known as *two-scale factor universality* or *Privman-Fisher hypothesis* [278].

Consequently, we can rewrite the singular free energy density with only two metric factors C_1, C_2

$$f^{sin}(t,h) = L^{-d} Y(C_1 t L^{1/\nu},\, C_2 h L^{(\beta+\gamma)/\nu}) \tag{3.29}$$

where $Y(x_1, x_2)$ is a universal finite-size scaling function. Similarly, one writes for the (inverse) correlation length

$$\xi^{-1}(t,h) = L^{-1}S(C_1 t L^{1/\nu},\ C_2 h L^{(\beta+\gamma)/\nu}) \tag{3.30}$$

with the **same** metric factors C_1, C_2 and a universal finite-size scaling function $S(x_1, x_2)$. The new feature of hyperuniversality is the absence of a non-universal prefactor in the finite-size scaling for both f^{sin} and ξ. For later use, we define the finite-size scaling variables

$$z := tL^{y_t} \quad , \quad \mu := hL^{y_h} \tag{3.31}$$

where $y_t = 1/\nu$ and $y_h = (\beta+\gamma)/\nu$ are the renormalization group eigenvalues. In particular, we have the following universal amplitudes

$$Y(0,0), \;\; S(0,0), \;\; \lim_{z\to\pm\infty} Y(z,0)S^{-d}(z,0), \;\; \lim_{\mu\to\pm\infty} Y(0,\mu)S^{-d}(0,\mu) \tag{3.32}$$

which is the main result of this section. Note that scaling considerations alone do not imply a relationship between the amplitudes evaluated in the limits $z(\mu) \to \pm\infty$. From the derivation of these results, it is clear that a confirmation of universality of the amplitudes in eq. (3.32) is equivalent to a confirmation of hyperscaling. It will be one of the objectives of conformal invariance to relate these amplitudes to critical exponents.

Generally, one can say that a universality class is characterized by its set of critical exponents along with its set of universal amplitudes or amplitude ratios.

If one restricts attention to **bulk** critical phenomena, one can show that there are only two independent amplitudes (or amplitude ratios). Consideration of surface critical phenomena, for example, will introduce further independent exponents as well as amplitudes.

We summarize the assumptions implicitly contained in the discussion (see also [282]):

1. There are no dangerous irrelevant variables. This means that there are no singularities in neither f^{sin} nor ξ^{-1} as the irrelevant couplings go to zero. This requires $d < d^*$. For many models, the upper critical dimension $d^* = 4$. Then hyperscaling is implied.
2. There is only a single length scale ξ_∞. Strong anisotropies are excluded.
3. There are no "accidental" zeroes in Y or S. (If they would occur, they could affect the apparent exponents. For a $2D$ example, see [176].)
4. The exponent $\alpha \neq 0, -1, -2, \ldots$. Otherwise, logarithmic corrections may appear, see [83, 281].
5. If further relevant fields different from t, h are present, additional metric factors will enter. This does happen, for example, for tricritical points, surface anisotropies, etc. as we shall see later.

The finite-size scaling form can be used to locate the critical point T_c. Its position is given by the maximum of the specific heat, i.e.

$$\left.\frac{\partial^3 Y(z,\mu)}{\partial z^3}\right|_{z=z_0,\mu=0} = 0 \tag{3.33}$$

where z_0 is universal. The shift in the critical temperature is then $T_c(L) - T_c(\infty) = C_1^{-1} z_0 L^{-1/\nu}$ and one recovers the value for the shift exponent $\lambda = 1/\nu$ of eq. (3.9).

Although the finite-size scaling functions $Y(z,\mu)$ and $S(z,\mu)$ are universal, they do depend on the shape of the finite domain the system is confined to and also on the boundary conditions. This will become explicit from the study of examples below. Only when the bulk limit $z \to \infty$ or $\mu \to \infty$ is taken, these dependencies can disappear.

So far only the leading finite-size term was considered. On any lattice of finite extent L, there are corrections which may come from two sources.

1. **Non-linear scaling fields**. The scaling form written in eq. (3.19) only includes the linear part of the scaling fields corresponding to t and h. While this is sufficient close enough to the critical point, it is sometimes necessary to include the full nonlinear scaling fields

$$u_t = C_1 t L^{1/\nu} \; + \; \mathcal{O}(t,h)$$
$$u_h = C_2 h L^{(\beta+\gamma)/\nu} \; + \; \mathcal{O}(t,h) \tag{3.34}$$

   This will lead to analytic correction terms. Conformal invariance will show that these are related to the secondary operators in the conformal tower.
2. **Irrelevant scaling fields**. These have a **negative** RG eigenvalue $y_{ir} = -\theta_{ir}/\nu < 0$ and do not change the critical point, but they may affect how the critical point is approached. For the correlation length, for example,

$$\begin{aligned}\xi^{-1} &\sim L^{-1}\bar{S}(C_1 t L^{1/\nu}, C_2 h L^{(\beta+\gamma)/\nu}, C_{ir} g L^{-\theta_{ir}/\nu}) \\ &\simeq L^{-1} S(C_1 t L^{1/\nu}, C_2 h L^{(\beta+\gamma)/\nu}) \; + \; C_{ir} g L^{-1-\theta_{ir}/\nu} S_{ir}(C_1 t L^{1/\nu}, C_2 h L^{(\beta+\gamma)/\nu})\end{aligned} \tag{3.35}$$

   where the second term contains the non-universal amplitude $C_{ir}g$ and $S_{ir}(z,\mu) = \frac{\partial}{\partial g}\bar{S}(z,\mu,0)$ is universal. This type of corrections gives rise to non-analytic corrections, of whom the leading one may be described by Wegner's correction exponent [323] $\omega = \theta_{ir}/\nu$. From the point of view of conformal invariance, these are related to the presence of irrelevant primary fields.

3.4 Phenomenological renormalization

We now describe a practical way to find the critical point and the critical exponents from given finite-lattice data. Consider the correlation length and take $h = 0$ for simplicity. According to finite-size scaling it has the form, to leading order in $1/L$

$$\xi^{-1}(T;L) \sim L^{-1} S(z,0) \tag{3.36}$$

Consider two lattices of sizes L, L' and with temperatures T, T' such that the following condition is satisfied

$$\xi(T;L)/L = \xi(T';L')/L' \tag{3.37}$$

We can take this as describing a relationship between T and T' via

$$T' - T_c = (T - T_c)\left(\frac{L}{L'}\right)^{\theta} \tag{3.38}$$

where $\theta = 1/\nu$ is the rounding exponent. Nightingale [261, 263] proposed to reinterpret eq. (3.37) as giving a renormalization group mapping from a temperature T to T'

$$T \to T' = \mathcal{R}_{L,b}(T) \tag{3.39}$$

where $b = L/L'$ is the rescaling factor. This can be done exactly only in the limit $L, L' \to \infty$, but the definition of $\mathcal{R}$ is considered to make sense for L finite at least to a good approximation. Explicit model calculations show this to be normally the case.

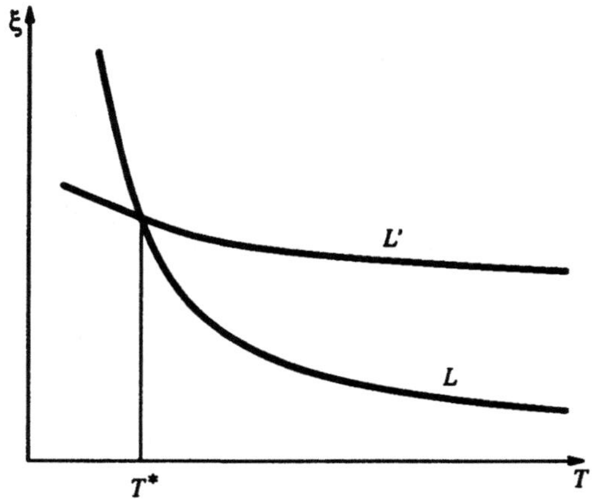

Fig. 16. Solving the phenomenological renormalization condition to find T^*.

One now looks for a fixed point $T^*(L, L')$ of the RG-mapping $\mathcal{R}$ which is given by

$$\xi(T^*; L)/L = \xi(T^*; L')/L' \tag{3.40}$$

and one expects that $T^*(L, L') \to T_c(\infty)$ as $L, L' \to \infty$. The solution of this equation is sketched in Figure 16. This yields a sequence of estimates $T^*(L, L')$ expected to converge to some limit T_c. Indeed, this is the case in exactly solvable models like the $2D$ Ising model [101]. The explicit calculation of the finite-size corrections also suggests that the convergence of the sequence of the T^*'s is fastest if $|L/L' - 1|$ is made as small as possible, e.g. $L' = L + a$, where a is the lattice constant.

Alternatively, one can use the results of the last section. Since $\xi^{-1}(T; L)L = S(0,0)$ is universal, one can define an estimate T^* by using a single lattice size. This may be advantageous if the value of $S(0,0)$ is known. Then, one can show the convergence to be of order $\sim L^{-(1/\nu + |y_{irr}|)}$, while one expects a convergence of order $\sim L^{-1/\nu}$ otherwise. Since only information from one lattice is used, we obtain for a maximal size L_{max}/a a sequence with $L_{max}/a - 2$ entries (the lattices with one or two sites in one direction do not contain much information about the $L \to \infty$ behaviour) while the conventional method yields a sequence with $L_{max}/a - 3$ entries. This might become important if only very few distinct lattices are available.

From the scaling form of the correlation length the exponent ν can be estimated

$$\left(\frac{L'}{L}\right)^{1+1/\nu} = \frac{\partial\xi(T^*;L)}{\partial T} \Big/ \frac{\partial\xi(T^*;L')}{\partial T} \tag{3.41}$$

and one obtains a sequence of estimates $\nu(T^*;L,L')$ converging to the exponent ν. It may seem that the final result for ν depends on the value chosen for T^*. However, since all what is needed is the finite-size scaling form which is the same inside the entire finite-size scaling region, one can replace T^* by any sequence of estimates converging towards T_c. This can be done in such a way to improve the apparent convergence of the sequence. In particular, the final estimate for ν does not depend on the final estimate for T_c (unless, of course, the estimates $\nu(T_c(\infty);L,L')$ are used).

The calculation of other critical exponents like α or γ follows the same lines, using the results given in Table 4. In chapter 9, we shall discuss numerical algorithms for the extrapolation towards the limit $L \to \infty$ of the finite-size sequences.

3.5 Consequences of conformal invariance

The finite-size scaling results presented are valid for all dimensions below the upper critical dimension, $d < d^*$. We now turn exclusively to the case $d = 2$ and impose the stronger constraints of conformal invariance.

Consider a scalar primary operator ϕ with scaling dimension x. It transforms covariantly under a analytic transformation $z \to w(z)$, $\bar{z} \to \bar{w}(\bar{z})$ as shown in (2.21).

$$< \phi(z_1,\bar{z}_1)\phi(z_2,\bar{z}_2) >_z = \left|\frac{dw}{dz}(z_1)\right|^x \left|\frac{dw}{dz}(z_2)\right|^x < \phi(w_1,\bar{w}_1)\phi(w_2,\bar{w}_2) >_w \tag{3.42}$$

The indices of the expectation values remind that the conformal transformation $w(z)$ changes in general the geometry of the space.

Here, we consider the effect of the *logarithmic transformation*

$$w = \frac{L}{2\pi} \ln z \tag{3.43}$$

Under this transformation, the infinite complex plane parametrized by z is transformed into an infinitely long strip of finite width L (see Figure 17). Writing $w = u + iv$ and $z = \rho e^{i\phi}$, one sees that v corresponds to the polar angle $v = \frac{L}{2\pi}\phi$, while u measures the radius via $u = \ln \rho$. In particular, since the operator ϕ is continuous in the infinite z-plane, it should also be so in the w-strip. This implies that one has **periodic boundary conditions** in the strip geometry.

Scale invariance alone is enough to fix the two-point correlation function of ϕ in the infinite z-plane (see eq. (2.28))

$$< \phi(z_1,\bar{z}_1)\phi(z_2,\bar{z}_2) >_z = \frac{1}{(z_1-z_2)^x(\bar{z}_1-\bar{z}_2)^x} \tag{3.44}$$

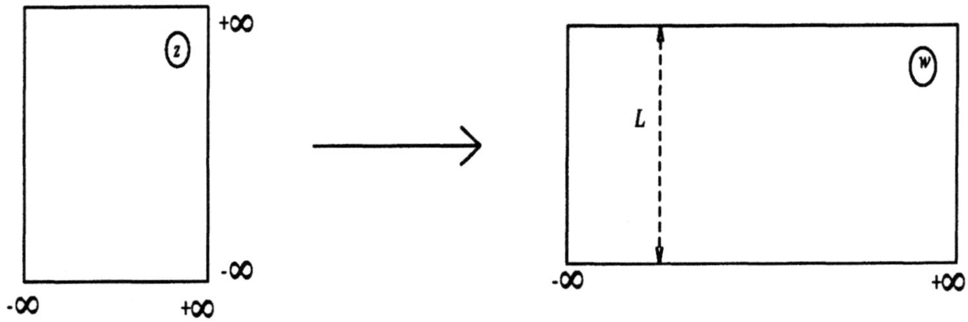

Fig. 17. Logarithmic transformation.

Let us calculate the corresponding correlation function in the strip geometry. Taking the correspondence $z = \exp(2\pi L^{-1} w) = \exp(2\pi L^{-1}(u + iv))$ into account, we have

$$< \phi(w_1, \bar{w}_1)\phi(w_2, \bar{w}_2) >_w = \left(\frac{2\pi}{L}\right)^{2x} \left(\frac{|z_1 z_2|}{|z_1 - z_2|}\right)^x$$

$$= \left(\frac{2\pi}{L}\right)^{2x} \exp\left(\frac{2\pi}{L}(u_1 + u_2)x\right) \times$$

$$\times \left(e^{\frac{2\pi}{L}(2u_1)} + e^{\frac{2\pi}{L}(2u_2)} - e^{\frac{2\pi}{L}(u_1+u_2+iv_1-iv_2)} - e^{\frac{2\pi}{L}(u_1+u_2-iv_1+iv_2)}\right)^{-x}$$

$$= \left(\frac{2\pi}{L}\right)^{2x} \left(2\cosh\frac{2\pi}{L}(u_1 - u_2) \; - 2\cos\frac{2\pi}{L}(v_1 - v_2)\right)^{-x} \qquad (3.45)$$

This gives the exact two-point function of a scalar primary operator in the strip geometry with periodic boundary conditions. To understand the result, consider two limit cases

1. $u_1 - u_2 \ll L$. By a rotation of the coordinate system in the plane, one can always arrange for $v_1 = v_2$. Then one recovers the form valid for the infinite plane.
2. $u_1 - u_2 \gg L$. Then one finds an exponential decay

$$< \phi(u_1, v_1), \phi(u_2, v_2) >_{\text{strip}} \simeq \left(\frac{2\pi}{L}\right)^{2x} \exp\left[-\frac{2\pi}{L}(u_1 - u_2)x\right] \qquad (3.46)$$

which is the usual way to define a correlation length ξ via $< \phi(u,v)\phi(0,0) > \sim \exp -u/\xi$. Comparison yields the result, due to Cardy [74]

$$\boxed{\xi = \frac{L}{2\pi x}} \qquad (3.47)$$

It is not exaggerated to call eq. (3.47) one of the most important results for the application of conformal invariance to critical phenomena.

Let us make the connection with the finite-size scaling functions. Consider a set of scaling operators ϕ_i (like magnetization density, energy density, ...) which have RG eigenvalues y_i. Take the y_i to be ordered, e.g. $y_1 \geq y_2 \geq y_3 \geq y_4 \geq \ldots$. The corresponding scaling dimensions x_i are $x_i + y_i = d$ ($d = 2$ here). To each scaling operator one has a correlation length ξ_i. Then finite-size scaling theory had shown that

$$\xi_i^{-1} = L^{-1} S_i(C_1 t L^{y_t}, C_2 h L^{y_h}) \tag{3.48}$$

with $y_t = y_2$ and $y_h = y_1$ for a conventional ferromagnet. Comparison with eq. (3.47) shows that, for an infinitely long strip of finite width L and with periodic boundary conditions

$$\boxed{S_i(0,0) = 2\pi x_i} \tag{3.49}$$

A few comments are in order:

1. Conformal invariance relates universal scaling amplitudes explicitly to universal critical exponents. This is an a posteriori confirmation of the finite-size scaling hypothesis. One could also say that the renormalization group tends to emphasize **qualitative** aspects in identifying universal quantities, while conformal invariance emphasizes the **quantitative** aspect by giving relations between amplitudes and critical exponents.
2. Equation (3.49) gives a simple means to calculate the exponents of many non-leading operators, since correlation lengths are easily obtained from the eigenvalues of the transfer matrix. From the point of view of conformal invariance there is a clear interest in calculating more than one eigenvalue of the transfer matrix, in distinction to the usual approach, which is satisfied by knowing the largest eigenvalue of the transfer matrix, related to the free energy.
3. As it stands, eq. (3.47) only applies to fully isotropic lattice systems. This restriction can be removed, if only the RG fixed point corresponds to an isotropic system. It is an easy exercise to generalize the derivation to non-scalar operators. We remark that this result can be rederived from conformally mapping strips of different widths onto each other [279].
4. Historically, it is interesting that eq. (3.47) was conjectured from model calculations quite some time before the proof could be given [270, 101, 241, 262, 272, 137].
5. The derivation assumed the isotropy in space as described by the complex coordinates z and $\bar{z}$. It may occur, however, that the (in a terminology borrowed from string theory) left-moving (corresponding to z) and right-moving sectors of the theory are normalized by $\zeta_L \neq \zeta_R$, where $\zeta_{L,R}$ are the respective speeds of sound in the lattice model (their determination will be described in chapters 8 and 9). The above result can be generalized to this situation, see [9] for details.

A similar result can be obtained for the singular free energy density. We know from the last chapter how the partition function $\mathcal{Z}$ transforms under conformal transformations, in particular from the complex plane to the strip geometry (eqs. (2.62) and (2.63)). Consider now a strip of width L and a new coordinate transformation, which changes the width of the strip $L \to L + \delta\epsilon$. Then the derivatives in (2.63) reduce to $\delta\epsilon$. Since for an infinite strip $\ln \mathcal{Z}$ is infinite, we consider the well-defined free energy per unit longitudinal length

$$\mathcal{F}(L) := -\lim_{M\to\infty} \frac{1}{M} \ln \mathcal{Z}(L, M) \tag{3.50}$$

Combining with eqs. (2.62) and (2.63), one finds

$$\delta\mathcal{F}(L) = \delta\epsilon L \frac{d\mathcal{F}(L)}{dL} = \frac{c}{12}\frac{2\pi}{L}\delta\epsilon \tag{3.51}$$

yielding [4, 50]

$$\mathcal{F}(L) = -\frac{\pi c}{6}\frac{1}{L} \tag{3.52}$$

This result for $\mathcal{F}$ does not contain a bulk contribution. Bulk terms were already subtracted in (2.62) since for its derivation the infinite plane expectation value

$$< T_{\text{plane}}(z) >_{\text{plane}} = 0 \tag{3.53}$$

We can rephrase the result in terms of the finite-size scaling function $Y(x_1, x_2)$ of the singular free energy density $f^{sin} = L^{-1}\mathcal{F}$, for an infinitely long strip of width L and periodic boundary conditions

$$\boxed{Y(0,0) = -\frac{\pi c}{6}} \tag{3.54}$$

This result defines an algorithm to compute the central charge c from a lattice Hamiltonian. The comments made in connection with the correlation length also apply here.

To conclude, we have seen how conformal invariance can be used to make quantitative statements on universal scaling amplitudes. The results obtained so far give rise to two questions. First, one would like to use eqs. (3.49) and (3.54) as practical tools for the calculation of both c and the x_i in a given lattice system. Second, what can be said about the finite-size scaling functions $Y(z, \mu)$ and $S_i(z, \mu)$? We will return frequently to these questions in the following chapters.

4 Representation Theory of the Virasoro Algebra

In this and the following chapters, we shall develop the necessary techniques for the explicit calculation of four-point correlation functions. The tools needed are the representation theory of the Virasoro algebra and the operator product algebra and we now give a detailed presentation of the technical points required. The first application of this will be to the Ising model in chapter 6.

The major difference between a critical two-dimensional system and any other physical system lies in the presence of the enlarged symmetry algebra given by the two infinite dimensional Virasoro algebras defined in (2.56). In this chapter, we want to discuss the consequences of such a large algebra. For the sake of simplicity, we will restrict most of our discussion to the representation theory of the single Virasoro algebra generated by $\{L_n\}$

$$[L_n, L_m] = (n-m)\, L_{n+m} + \frac{1}{12} c\, n(n^2-1)\delta_{n+m,0} \tag{4.1}$$

The full algebra is of course generated by the tensor product of the Virasoro algebras $\{L_n\} \otimes \{\bar{L}_m\}$ but because holomorphic and anti-holomorphic algebras are similar and commute, we can discard the anti-holomorphic part while discussing algebraic properties. We have to consider the full algebra when we want to implement physical analyticity conditions in the correlation functions or in the partition function. We will return in detail to these physical conditions in subsequent chapters. Unless we state it explicitly, everything we show in this chapter for $\{L_n\}$ remains true for $\{\bar{L}_m\}$.

Fundamental references for this chapter are the seminal work of Belavin, Polyakov and A. Zamolodchikov [37] and for the last two sections those of Kac [207, 208] and of Feigin and Fuks [119].

4.1 Verma module

Numerous properties of the representations of the Virasoro algebra (4.1) do not depend on a specific physical model. Let us first reverse the defining equation (2.49) for the Virasoro generators

$$T(z)A(z_1, \bar{z}_1) = \sum_{n=-\infty}^{\infty} (z-z_1)^{-n-2} L_n A(z_1, \bar{z}_1) \tag{4.2}$$

The operator $A(z_1, \bar{z}_1)$ can be either a primary or a secondary operator. Suppose now that $A(z_1, \bar{z}_1)$ is the primary operator $\phi(z_1, \bar{z}_1)$. We know that the holomorphic part of the stress-energy tensor T generates infinitesimal transformations which are encoded in the Ward identities eq. (2.41). The singular part of (2.41) is given by

$$T(z)\phi(z_1, \bar{z}_1) = \frac{\Delta}{(z-z_1)^2}\phi(z_1, \bar{z}_1) + \frac{1}{z-z_1}\partial_{z_1}\phi(z_1, \bar{z}_1) + \text{regular terms} \tag{4.3}$$

The "regular terms" include the parts of the correlation function where T couples with the other operators. Comparing the leading terms in (4.3) and in (4.2), we conclude that an algebraic definition of a primary operator is given by

$$\begin{aligned} L_n\phi &= 0 \qquad \text{if } n \geq 1 \\ L_0\phi &= \Delta\,\phi \end{aligned} \tag{4.4}$$

In fact, it is sufficient for a primary operator to satisfy

$$L_1\phi = 0 \qquad \text{and} \qquad L_2\phi = 0 \tag{4.5}$$

The other constraints $L_n\phi = 0$ can be deduced by recursion from eq. (4.5), since $nL_n\phi = [L_{n-1}, L_1]\phi$. L_0 measures the holomorphic part of the scaling dimension of the operator. This holds for primary as well as for secondary operators. Notice also that L_{-1} generates the translations. Therefore we have the identification

$$L_{-1} = \partial_z \tag{4.6}$$

A primary operator defines the ***highest weight*** of a representation of the conformal symmetry. An infinite set of secondary descendant operators are defined by the negatively indexed operators L_{-n}. They build up the *Verma module* of ϕ whose elements are given by

$$\phi^{(-n_k,\ldots,-n_1)} = L_{-n_k}\cdots L_{-n_1}\phi \tag{4.7}$$

The commutation relations (4.1) introduce equivalences between descendant operators. A simple way to reduce the equivalences is to impose some ordering on the indices of $\phi^{(-n_k,\ldots,-n_1)}$. For example we can impose $n_1 \leq n_2 \leq \ldots \leq n_k$.

The secondary operators are all eigenvectors of L_0 with eigenvalue $\Delta + \sum n_i$, since

$$\begin{aligned}
&L_0\phi^{(-n_k,\ldots,-n_1)} = \\
&= L_0 L_{-n_k}\cdots L_{-n_1}\phi \\
&= n_k\, L_{-n_k}\cdots L_{-n_1}\phi + L_{-n_k}L_0\cdots L_{-n_1}\phi \\
&= (n_k + \ldots + n_1 + \Delta)\,\phi^{(-n_k,\ldots,-n_1)}
\end{aligned} \tag{4.8}$$

The primary operator ϕ is the only element of the Verma module with conformal weight Δ and $L_{-1}\phi$ the only one with conformal weight $\Delta + 1$. At conformal weight $\Delta + 2$, we have two elements, $L_{-2}\phi$ and $(L_{-1})^2\phi$.

We introduce now a particular primary operator into eq. (2.49): the identity operator $\mathbf{1}$. Due to the regularity of the integral for $n \geq -1$, the identity operator is primary with $L_0\mathbf{1} = 0$ and $L_{-1}\mathbf{1} = 0$. However, $L_{-2}\mathbf{1}$ is different from zero and we obtain

$$L_{-2}\mathbf{1} = T(z) \tag{4.9}$$

The stress-energy tensor T is therefore a secondary operator which belongs to the conformal tower of the identity. The fact that T is quasi-primary is reflected in the fact that $L_1T = L_1L_{-2}\mathbf{1} = 3L_{-1}\mathbf{1} + L_{-2}L_1\mathbf{1} = 0$. We recall that a quasi-primary operator Φ transforms like a primary one under the projective group. As a consequence, with respect to L_1 it satisfies

$$L_1\Phi = 0 \tag{4.10}$$

This is the defining algebraic condition for the quasi-primary operators. Notice that $L_n\Phi = 0$ must be true for some n positive. In the case of T, we have $L_2T = L_2L_{-2}\mathbf{1} = 4L_0\mathbf{1} + (c/2)\mathbf{1} = (c/2)\mathbf{1}$ and $L_nT = 0$ for $n \geq 3$. We will now show that any quasi-primary operator is the secondary of a primary one, thus making the primary operators the most fundamental quantities of a conformal theory. This follows because the operator L_0 should be bounded from below. If L_0 were unbounded, this would imply the existence of operators with conformal weights $\Delta = -\infty$ with a everywhere infinite two-point function. Such an operator would be difficult to interpret physically and therefore we can suppose L_0 bounded from below by $\Delta_{\min}$. Any quasi-primary operator Φ with conformal weight Δ_Φ at least verifies $L_{n_1} \ldots L_{n_k} \Phi = 0$ for any k and any positive $n_1, \ldots, n_k$ such that $\Delta_\Phi - n_1 - \cdots - n_k < \Delta_{\min}$. There is therefore by construction at least one operator[4] ϕ with conformal weight $\Delta_\Phi \geq \Delta_\phi \geq \Delta_{\min}$ which is primary and contains Φ in its Verma module.

To summarize, an infinite-dimensional representation of the Virasoro algebra, called Verma module, has a highest weight which is a primary operator. Descendant operators of higher conformal weights are either quasi-primary operators or derivatives of the primary or quasiprimary operators. We will show later in this chapter that some descendant quasi-primary operators may be primary as well.

Let us first show some examples of quasi-primary operators. The highest weight ϕ is of course quasi-primary. At level 2, it is easy to verify that

$$\Phi^{(2)} = \left(L_{-2} - \frac{3}{2(2\Delta+1)} L_{-1} L_{-1} \right) \phi \tag{4.11}$$

satisfies $L_1 \Phi^{(2)} = 0$. At level 3, there are two derivative operators $L_{-1}L_{-1}L_{-1}\phi$, $L_{-1}\Phi^{(2)}$, and a new quasi-primary given is by

$$\Phi^{(3)} = \left(L_{-3} - \frac{2}{\Delta+2} L_{-1} L_{-2} + \frac{1}{(\Delta+2)(\Delta+1)} L_{-1} L_{-1} L_{-1} \right) \phi \tag{4.12}$$

Two quasi-primary operators appear at level 4, one containing $L_{-4}\phi$, the other with $L_{-2}L_{-2}\phi$. The number of quasi-operators obviously increases with the level n.

Up to now, we have not specified the possible values for Δ and c. A conformal field theory contains Verma modules of primary operators with conformal weights Δ. In a first step, we assume the the infinite dimensional representation made out of the descendant operators $L_{-n_k} \cdots L_{-n_1}\phi$ with $n_k \geq \cdots \geq n_1 > 0$ is irreducible. This means that all descendant operators at a given *level* $n = n_k + \cdots + n_1$ are linearly independent. We will see in the next subsections that the most interesting representations are not irreducible but it is simpler to show the structure of the Verma modules starting from the generic case. In this case and at a given level n, the number of independent operators is the number of partitions $p(n)$ of n. We define the *generic character* (for $\Delta \neq 0$) of a Verma module by

$$\chi(\Delta, c) = q^{\Delta - \frac{c}{24}} \sum_{n=0}^{\infty} p(n) q^n = q^{\Delta - \frac{c}{24}} \prod_{i=0}^{\infty} \frac{1}{1-q^i} \tag{4.13}$$

The character encodes the information on the structure of the Verma module. The importance of the factor $q^{\Delta - c/24}$ lies in the modular properties of the character, which will be discussed later. We merely remark that the product in eq. (4.13) is up to a factor $q^{1/24}$ the inverse Dedekind's function $\eta(\tau)$ with $q = \exp(2\pi i \tau)$. The formula (4.13) does not hold for $\Delta = 0$ because $p(1) = 1$. In this case $L_{-1}\mathbf{1} = 0$ and the number of operator

[4] There could be many operators if degeneracies occur. Φ can be a linear combination of quasi-primary operators which are secondary to primary operators with conformal weights equal or differing by integers.

at level 1 is zero. This is also true for any operator $L_{-n_k}\dots L_{-n_1}L_{-1}\mathbf{1}=0$. Therefore, the generic character of the identity operator is

$$\chi(0,c)=(1-q)\,q^{-\frac{c}{24}}\prod_{i=0}^{\infty}\frac{1}{1-q^i}=q^{-\frac{c}{24}}\prod_{i=1}^{\infty}\frac{1}{1-q^i} \tag{4.14}$$

The factor $(1-q)$ suppresses the counting of all vanishing operators. We say that $L_{-1}\mathbf{1}$ is a *null operator* and the Verma module is called *degenerate.*

Let us finally notice that the quasi-primary operators often deserve to be counted separately. Suppose that we have p operators $A_1,\dots,A_p$ at level n, then at level $n+1$ we will have p derivative operators $L_{-1}A_1,\dots,L_{-1}A_p$. The remaining ones at level $n+1$ are the new quasi-primary operators. The difference in the number of operators between two successive levels counts the number of quasi-primary operators. In the character formula (4.13), two successive levels differ by one power in q. Therefore a function which counts the number of quasi-primaries at each level is given by

$$\chi_{QP}(\Delta,c)=(1-q)\chi(\Delta,c)+q\delta_{\Delta,0} \tag{4.15}$$

An explicit verification for the first terms of $\chi(0,c)$ is necessary to understand the last term of eq. (4.15).

4.2 Hilbert space structure

It is often more convenient to study algebras through their action on a space of states. Most relations of the previous subsection do not depend on a particular coordinate z_1. We can therefore define the *vacuum state* as the highest weight of the identity operator evaluated at $z=0$.

$$|0\rangle:=\mathbf{1}(z=0) \tag{4.16}$$

This state is defined by $L_n|0\rangle=0$ $(n\geq -1)$. The true physical vacuum, which recombines the z and $\bar{z}$ dependence, is the tensor product $|0\rangle\otimes\overline{|0\rangle}$. The word operator, employed up to now arbitrarily, begins to make sense as a map from the vacuum state into a highest weight state defined by

$$|\Delta\rangle:=\phi(0)|0\rangle \tag{4.17}$$

with the properties $L_n|\Delta\rangle=0$ $(n>0)$ and $L_0|\Delta\rangle=\Delta|\Delta\rangle$. We can connect this state at $z_1=0$ to any value of z_1 with the Taylor series

$$\phi(z_1)|0\rangle=e^{z_1L_{-1}}|\Delta\rangle \tag{4.18}$$

We need to define a dual state to the vacuum state to obtain an Hilbert space structure. The hermitian conjugate operator L_n^+ of L_n must verify

$$\langle 0|L_n^+=0 \qquad \text{if } n\geq 0 \tag{4.19}$$

In our case, a natural dual transformation is given by the projective map $z'=-1/z$ for which we can verify

$$T(z)\,z^2=T(z')\,z'^2 \tag{4.20}$$

Though interchanging the points at zero and at infinity, such a map transforms the complex plane into itself and does not modify the theory. Equation (4.2) taken for example with the identity operator at $z_1 = 0$ can be written

$$z^2 T(z)\mathbf{1} = \sum_{n=-\infty}^{\infty} z^{-n} L_n \mathbf{1} = z'^2 T(z')\mathbf{1} = \sum_{n=-\infty}^{\infty} z'^{+n} L_{-n}^{+} \mathbf{1} \tag{4.21}$$

The consequence for the generators of the Virasoro algebra is expressed by

$$\boxed{L_n^{+} = L_{-n}} \tag{4.22}$$

This is natural because the commutator $[L_n, L_{-n}]$ belongs to the Cartan subalgebra spanned by $(L_0, \{c\})$. This means that $L_0^{+} = L_0$ and the spectrum of conformal weights is real as expected in a physical theory. Notice that for an infinite-dimensional algebra, the hermiticity of the algebra does not imply the unitarity of the representations (i.e. the positivity of the states). Unitarity will be discussed in the next section.

We now show how hermitian conjugation acts on primary operators. The map $z' = -1/z$ modifies the primary operators as follows

$$\phi(z') = z^{2\Delta} \phi(z) \tag{4.23}$$

By duality, the state $|\Delta\rangle$ is mapped to a dual state located at infinity

$$\langle\Delta| = \lim_{z_1 \to \infty} \langle 0|\phi(z_1)\,(z_1)^{2\Delta} \tag{4.24}$$

which verifies the orthogonality property $\langle\Delta|\Delta'\rangle = \delta_{\Delta,\Delta'}$. It is interesting to see that everything is consistent with the two-point function. The Taylor expansion at infinity gives the dual state at finite value of z_1

$$\langle 0|\phi(z_1) = \langle\Delta| z_1^{-2L_0} e^{(1/z_1)L_1} \tag{4.25}$$

Combining this dual state with eq. (4.18), and computing the formula

$$\langle\Delta|(L_1)^n (L_{-1})^m|\Delta\rangle = \delta_{n,m}\, n! \frac{(2\Delta - 1 + n)!}{(2\Delta - 1)!} \tag{4.26}$$

we verify the two point function

$$\langle 0|\phi(z_1)\phi(z_2)|0\rangle = \frac{1}{(z_1 - z_2)^{2\Delta}} = < \phi(z_1)\phi(z_2) > \tag{4.27}$$

The algebraic language encodes a complete information on the operators even if the z dependence is not explicit. For example in the two-point function, the explicit z dependence does not really matter. What really matters is on one side the conformal weight Δ given as eigenvalue of L_0 and on the other side the condition that the two-point function does not vanish. This last condition is encoded in the hermitian product $\langle\Delta|\Delta'\rangle$.

Though not explicitly mentioned in the beginning of this section, there is more in the definition of eq. (4.16): while introducing the hermitian product, we included an information on the two-point function. Later, we will investigate tensor product representations and show how n point functions can be obtained. Notice that the brackets $\langle\rangle$ have now taken another meaning as the vacuum state and its dual in a quantum field theory.

4.3 Null-vectors

It is well-known that for an infinite-dimensional algebra, hermiticity eq. (4.22) does not imply anymore unitarity. This is not a problem in statistical mechanics because unitarity is not a basic axiom and is violated in well-known examples. However, many physical systems satisfy unitarity.

The first models we are going to discuss are the *minimal models* which, unitary or not, are characterized by the fact that the number of distinct Verma modules is finite. *Unitarity* is characterized by the absence of negative norm states in the theory. Without restriction of generality, we can normalize $\langle\Delta|\Delta\rangle = 1$. At level one we have the state $L_{-1}|\Delta\rangle$ with norm

$$\langle\Delta|L_1L_{-1}|\Delta\rangle = 2\Delta \tag{4.28}$$

Therefore, a necessary condition for unitarity is $\Delta \geq 0$. Another simple condition is given by the two-point function of the stress-energy tensor. The norm is given by

$$\langle 0|L_2L_{-2}|0\rangle = \frac{c}{2} \tag{4.29}$$

Therefore $c \geq 0$ is necessary to ensure unitarity. To gain new information, we have to investigate deeper levels in the the Verma module. At level 2, we have two independent vectors $L_{-2}|\Delta\rangle$ and $(L_{-1})^2|\Delta\rangle$. The unitarity condition requires the positivity of the zeroes of the determinant $\det_2(c,\Delta)$

$$\begin{aligned}\det_2(c,\Delta) &:= \begin{vmatrix} \langle\Delta|L_2L_{-2}|\Delta\rangle & \langle\Delta|L_2(L_{-1})^2|\Delta\rangle \\ \langle\Delta|(L_1)^2L_{-2}|\Delta\rangle & \langle\Delta|(L_1)^2(L_{-1})^2|\Delta\rangle \end{vmatrix} \\ &= \begin{vmatrix} 4\Delta + \frac{1}{2}c & 6\Delta \\ 6\Delta & 4\Delta(2\Delta+1) \end{vmatrix} \\ &= 2\Delta\left(16\Delta^2 + 2(c-5)\Delta + c\right)\end{aligned} \tag{4.30}$$

An arbitrary choice of the conformal weight Δ and the central charge c may result in negative eigenvalues at level 2, therefore breaking unitarity. An interesting case occurs if Δ takes one of the values

$$\Delta = \frac{5 - c \pm \sqrt{(1-c)(25-c)}}{16} \tag{4.31}$$

because one of the eigenvalues will be zero and the other positive. It follows that there is a linear combination of $L_{-2}|\Delta\rangle$ and $L_{-1}^2|\Delta\rangle$ with zero norm. This combination must be a quasi-primary operator due to the presence of L_{-2}. Notice that the norm of the derivative of a quasi-primary operator is always proportional to the norm of the quasi-primary operator itself, which is located at a lower level. The vanishing norm of a derivative operator always implies that the norm of its quasi-primary ascendant is zero. At level 2 the only quasi-primary state is $|\Phi^{(2)}\rangle$, defined in eq. (4.11), which norm is now zero. It is easy to verify that, using (4.31)

$$L_2|\Phi^{(2)}\rangle = 0 \tag{4.32}$$

The state $|\Phi^{(2)}\rangle$ is primary. Besides having zero norm, it is orthogonal to any other operator in the Verma module (even its own derivatives).

$$\langle \Delta | L_{n_1} \dots L_{n_k} | \Phi^{(2)} \rangle = 0 \tag{4.33}$$

with $n_1, \dots, n_k$ positive. The spectrum for Δ and c is restricted if we want to have semi-positivity, i.e. at most null-vectors in the Verma module. Such a Verma module is called *degenerate*. Null-vectors can appear at any level. They are always primary and their submodule is orthogonal with themselves and with any other operator. Before discussing the general case, it is interesting to discuss further the special case where Δ and c satisfy eq. (4.31) with $|\Phi^{(2)}\rangle$ as a null-vector. This will show why null-vectors are the ingredients which makes it possible to solve a model.

4.4 Null-vectors and correlation functions

Due to the particular orthogonality property, the Verma module is reduced by the null submodule which is orthogonal to all other vectors. This has important consequences for the correlation functions. Though we did not discuss the tensor product of primary operators, it is clear that a correlation function $< (\phi_1 \dots \phi_k)\phi >$ will be computed from the projection of $|\Delta_1\rangle \otimes \dots \otimes |\Delta_n\rangle$ onto the vector $|\Delta\rangle$ related to ϕ. If we introduce a zero primary vector, we have

$$< \phi_1(z_1) \dots \phi_k(z_k) \Phi^{(2)}(z) >= 0 \tag{4.34}$$

This introduces a constraint which can be expressed as a differential operator acting on $< \phi_1 \dots \phi_n \phi >$ which can be solved. It is first necessary to show how a correlation function containing a secondary operator like $L_{-p}\phi(z)$ is related to the one with the primary operator only. By definition we have

$$< \phi_1(z_1) \dots \phi_k(z_k) L_{-p}\phi(z) >= \frac{1}{2\pi i} \oint_{C_z} dz'(z'-z)^{-p+1} < \phi_1(z_1) \dots \phi_k(z_k) T(z') \phi(z) > \tag{4.35}$$

We analytically deform the contour C_z into the combination of a large contour C surrounding all the points and k smaller ones C_i surrounding each point z_i $(i = 1 \dots k)$. The orientations of the contour is shown in Figure 18. Let us discuss first the integration over the large contour C. This integral is very similar to the one in eq. (2.40) from which we obtained the Ward identities. The Ward identities hold for any infinitesimal deformation $z \to z + \epsilon$. However, we did not discuss further the actual value of the integral. Suppose $\epsilon(z')$ behaves at large distance as $(z')^{-p+1}$ like in eq. (4.35). We know that $T(z')$ behaves like $(z')^{-4}$ asymptotically. Therefore the integrand in (4.35) will behave as $(z')^{-p-3}$ at large distances. The integrand is therefore regular at infinity if $p \geq -1$ for which the integral over C vanishes. This is the case we are interested in because null-vectors are built up with L_{-p} with $p > 0$. Therefore the integral along C is zero and the residue of the integral along C_z is minus the sum of the integrals along each C_i. We compute now these other residues by introducing the Ward identities (2.41) in eq. (4.35). The

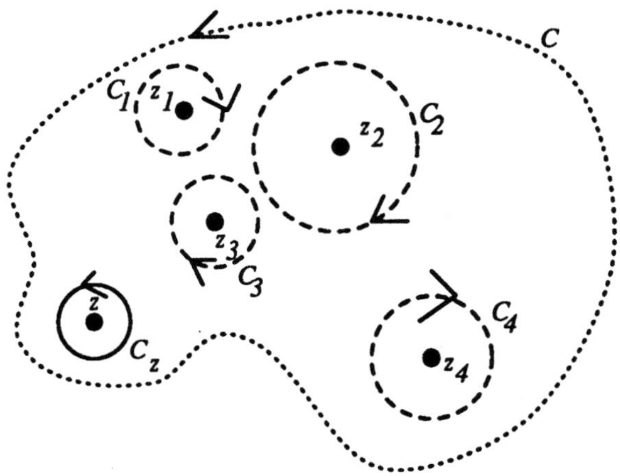

Fig. 18. Deformation of a oriented contour: $\oint_{C_z} = \oint_C - \sum_i \oint_{C_i}$.

computation of the residues simplifies to

$$< \phi_1(z_1)\dots\phi_k(z_k)L_{-p}\phi(z) >= \tag{4.36}$$
$$-\frac{1}{2\pi i}\sum_{i=1}^{k}\oint_{C_i} dz'(z'-z)^{-p+1}\left(\frac{\Delta_i}{(z'-z_i)^2}+\frac{1}{z'-z_i}\partial_{z_i}\right) < \phi_1(z_1)\dots\phi_k(z_k)\phi(z) >$$

This can be integrated out and we define the differential operator $\mathcal{L}_{-p}$ by

$$< \phi_1(z_1)\dots\phi_k(z_k)L_{-p}\phi(z) >= \mathcal{L}_{-p} < \phi_1(z_1)\dots\phi_k(z_k)\phi(z) > \tag{4.37}$$

with

$$\mathcal{L}_{-p} = \sum_{i=1}^{k}\left(\frac{(p-1)\Delta_i}{(z_i-z)^p} - \frac{1}{(z_i-z)^{p-1}}\partial_{z_i}\right) \tag{4.38}$$

The same procedure can be applied by recursion to $< \phi_1\dots\phi_k L_{-p_r}\dots L_{-p_1}\phi(z) >$. We first deform the contour corresponding to L_{-p_r}, i.e. the one surrounding all the others. The procedure is obviously the same. The poles between the different stress-energy tensors do not contribute. We successively deform the contours and compute the residues until we have done this for the most interior one related to L_{-p_1}. The result is

$$< \phi_1\dots\phi_k L_{-p_r}\dots L_{-p_1}\phi(z) >= \mathcal{L}_{-p_1}\cdots\mathcal{L}_{-p_r} < \phi_1\dots\phi_k\phi(z) > \tag{4.39}$$

Each $\mathcal{L}$ acts on everything placed to its right. The central charge does not appear because it comes from poles which are not considered with the new set of contours $\{C_i\}$. The c-dependence is hidden in the correlation function of the primary operators. Notice that it is sometimes interesting to further work out $\mathcal{L}$. For example

$$\mathcal{L}_{-1} = -(\partial_{z_1} + \dots + \partial_{z_k}) = \partial_z \tag{4.40}$$

as a consequence of the projective Ward identities (eq. (2.29)).

We are now ready to return to the null-vector problem (4.34). We have proven that the correlation function of a descendant operator with several primary ones can be expressed as a differential operator. Consequently, a correlation function with a null-vector like (4.34) can be expressed as a differential equation acting on the correlation function of the primary operators. In this case, we have

$$\left(\mathcal{L}_{-2} - \frac{3}{2(2\Delta+1)}\partial_z^2\right) < \phi_1(z_1)\dots\phi_k(z_k)\phi(z) >= \tag{4.41}$$
$$\left[-\frac{3}{2(2\Delta+1)}\partial_z^2 + \sum_{i=1}^{k}\left(\frac{1}{z-z_i}\partial_{z_i} + \frac{\Delta_i}{(z-z_i)^2}\right)\right] < \phi_1(z_1)\dots\phi_k(z_k)\phi(z) >= 0$$

It is easy to verify that this differential equation is compatible with the two and three-point functions. More important is the fact that this equation provides solutions to the four-point function. We know that we can introduce the projective invariant $\eta = (z-z_1)(z_2-z_3)/(z-z_3)(z_2-z_1)$ and write

$$< \phi_1(z_1)\phi_2(z_2)\phi_3(z_3)\phi(z) >= \prod_{i=1}^{3}(z-z_i)^{-\gamma_{0i}}\prod_{i<j}(z_i-z_j)^{-\gamma_{ij}}\,F(\eta) \tag{4.42}$$

We gave in chapter 2, eq. (2.33) the general constraint on the exponents γ. Suppose that we take the particular choice

$$\begin{aligned} &\gamma_{01} = \gamma_{02} = 0, \qquad \gamma_{03} = 2\Delta, \qquad &&\gamma_{12} = +\Delta + \Delta_1 + \Delta_2 - \Delta_3, \\ &\gamma_{13} = -\Delta + \Delta_1 - \Delta_2 + \Delta_3, &&\gamma_{23} = -\Delta - \Delta_1 + \Delta_2 + \Delta_3 \end{aligned} \tag{4.43}$$

Notice that $F(\eta)$ can be defined as the limit

$$F(\eta) = \lim_{\xi\to\infty} \xi^{2\Delta_3} < \phi_1(0)\phi_2(1)\phi_3(\xi)\phi(\eta) > \tag{4.44}$$

It is then straightforward to show that $F(\eta)$ must be a solution to the differential equation

$$\left[-\frac{3}{2(2\Delta+1)}\partial_\eta^2 - \left(\frac{1}{\eta} - \frac{1}{1-\eta}\right)\partial_\eta + \frac{\Delta_1}{\eta^2} + \frac{\Delta+\Delta_1+\Delta_2-\Delta_3}{\eta(1-\eta)} + \frac{\Delta_2}{(1-\eta)^2}\right]F(\eta) = 0 \tag{4.45}$$

We have shown how the null-vectors of degenerate Verma module imply solvable differential equations for the correlation functions. Notice that a differential equation like (4.41) is bound to have several distinct solutions, for instance two, F_1 and F_2 for (4.45). However, we expect the primary operator $\phi(z,\bar{z})$ to have a null vector which will constrain the $\bar{z}$-dependence. The most general solution for the four-point function is therefore given by

$$< \phi_1(z_1,\bar{z}_1)\phi_2(z_2,\bar{z}_2)\phi_3(z_3,\bar{z}_3)\phi(z,\bar{z}) >= \tag{4.46}$$
$$\prod_{i=1}^{3}(z-z_i)^{-\gamma_{0i}}(\bar{z}-\bar{z}_i)^{-\bar{\gamma}_{0i}}\prod_{i<j}(z_i-z_j)^{-\gamma_{ij}}(\bar{z}_i-\bar{z}_j)^{-\bar{\gamma}_{ij}}\sum_{k,l}\mathcal{N}_{kl}F_k(\eta)\bar{F}_l(\bar{\eta})$$

The physical four-point function is of course evaluated at Euclidean coordinates where $\bar{z}$ is complex conjugate to z. Invariance under physical constraints, such as rotation invariance, induces strong constraints on the coefficients $\mathcal{N}_{kl}$. This fundamental problem called *monodromy* problem will be discussed in the next chapter.

4.5 Kac formula and unitarity

Let us finish with more remarks on the degenerate Verma modules, which contain null-vectors. The general discussion of the null-vectors at any level was done by Kac [207]. Since the mathematical proofs [119] are quite complicated, we merely summarize some important results. The determinant computed in eq. (4.30) at level 2 can be generalized to any level. At level n, Kac found the following formula

$$\det_n(c, \Delta) \sim \prod_{r,s=1;\, 1\leq rs\leq n}^{n} (\Delta - \Delta_{r,s})^{p(n-rs)} \tag{4.47}$$

A particular parametrization was introduced to write this formula. The central charge is defined as

$$c := 1 - \frac{6}{m(m+1)} \tag{4.48}$$

where m is real or complex. A model is called *minimal* if m is a rational number. The zeroes of the determinant take the values

$$\Delta_{r,s} = \Delta_{m-r,m+1-s} = \frac{[r(m+1) - sm]^2 - 1}{4m(m+1)} \tag{4.49}$$

where r and s are positive integers. $p(a)$ is the number of partitions of the integer number a. For r, s given, the first null-vector appears at level rs and propagates with increasing multiplicity at all levels larger than rs. For example $\det_2(c, \Delta)$ (eq. (4.30)) is proportional to Δ, and $\det_1(c, \Delta) = \Delta$ because $\Delta_{1,1} = 0$. This reflects the fact that $\det_n(c, \Delta)$ always divides $\det_{n+1}(c, \Delta)$. The proof of the Kac formula was given by Feigin and Fuchs. In the particular framework of unitary **irreducible** representations, where all vectors have semi-positive norms, Friedan, Qiu, and Shenker [133] found a very interesting result. There are two cases to be distinguished.

$c \geq 1$: unitarity only implies $\Delta \geq 0$. However, the number of primary operators is infinite.

$c < 1$: unitarity can be enforced for $c = 1 - 6/m(m+1)$ with $m \in \mathbb{N}$ starting from $m = 2$. In this case, we have a finite number of primary operators with conformal weights $\Delta_{r,s}$ with $1 \leq r \leq m-1$ and $1 \leq s \leq m$. This is called a *unitary minimal model.* Notice that each operator is counted twice here because of the symmetry $\Delta_{r,s} = \Delta_{m-r,m+1-s}$.

The table of possible conformal weights $\Delta_{r,s}$ is called the *Kac table*. Let us discuss the first examples. With $m = 2$, we have $c = 0$ and the only unitary representation is the vacuum one with conformal weight $\Delta_{1,1} = \Delta_{1,2} = 0$. However we can see from our computation of $\det_2(0,0)$ that we have there a double zero. This is equivalent to say that in this case only the trivial one-dimensional representation $|0\rangle$ exists. Or to say that $T = 0$. More interesting is the case $m = 3$ where $c = 1/2$. In this case, we have the vacuum representation and two non-trivial representations. There are $|\Delta_{1,2}\rangle$ with $\Delta_{1,2} = \Delta_{2,2} = 1/16$ and $|\Delta_{1,3}\rangle$ with $\Delta_{1,3} = \Delta_{2,1} = 1/2$ (Table 5). We recognize the spectrum of the Ising model where $|\Delta_{1,2}\rangle$ corresponds to the holomorphic part of the spin-density operator and $|\Delta_{1,3}\rangle$ represents the holomorphic part of the energy-density operator. In chapter 3, we have already shown how to calculate the central charge from the finite-size scaling of the

Table 5. Kac table for the Ising model ($m = 3$ and $c = 1/2$) on the left and for the RSOS model with $m = 5$ ($c = 4/5$) on the right. In this latter case, extra boxes indicate the sub-table corresponding to the three-state Potts model.

$r \backslash s$	1	2	3
2	$\frac{1}{2}$	$\frac{1}{16}$	0
1	0	$\frac{1}{16}$	$\frac{1}{2}$

$r \backslash s$	1	2	3	4	5
4	3	$\frac{13}{8}$	$\frac{2}{3}$	$\frac{1}{8}$	0
3	$\frac{7}{5}$	$\frac{21}{40}$	$\frac{1}{15}$	$\frac{1}{40}$	$\frac{2}{5}$
2	$\frac{2}{5}$	$\frac{1}{40}$	$\frac{1}{15}$	$\frac{21}{40}$	$\frac{7}{5}$
1	0	$\frac{1}{8}$	$\frac{2}{3}$	$\frac{13}{8}$	3

free energy density of a lattice model and we shall present the complete calculation of c for the Ising model in chapter 10.

The three-state Potts model is obtained with $m = 5$ ($c = 4/5$) together with the A_5-RSOS model. The respective Kac tables are given in Table 5. Notice that the four-state Potts model is reached in the limit $m \to \infty$. In fact, all critical A_m-RSOS models yield examples of unitary conformal theories for all values of m permitted by the Kac formula.

Notice that the Kac formula applies for any values of m. The interesting case when m is rational defines the *minimal models*. They are always non-unitary, unless m is integer, but are characterized by a finite number of primary operators. Suppose $m = p/(p' - p)$, p and p' coprime, then the irreducible highest weight representations are given in the Kac table where $\Delta_{r,s} = \Delta_{p-r,p'-s}$ and $1 \leq r \leq p-1$, $1 \leq s \leq p'-1$. The non-unitary Yang-Lee model corresponds to $p = 2$, $p' = 5$. Therefore $c = -22/5$ and the two distinct primary operators have highest weights $\Delta_{1,1} = 0$ and $\Delta_{1,2} = -1/5$. We shall denote the minimal models by $\mathcal{M}_{p,p'}$, the unitary series corresponding to $\mathcal{M}_{m,m+1}$.

Though we do not discuss the second Verma module generated by $\bar{L}$, we know that $\bar{c} = c$ because the two components of the stress-energy tensor must behave like the complex conjugates of each other in a physical theory. However, physical operators can be a mixture of non-equivalent holomorphic and anti-holomorphic representations. For example, this is the case for $m = 5$ where we have two different physical models: the RSOS model and the three-state Potts model. The determination of physical operators is related to the monodromy problem introduced in the last section.

In conclusion, several different physical models may be possible for a given value of the central charge. The conformal weights of all minimal models must belong to the set defined by the Kac formula (4.49). Conversely, unitary conformal field theories can be constructed for each integer value $m \geq 3$ via the *coset construction* [150]. Lattice realizations of these are provided by the A_m and D_m RSOS models introduced in chapter 1.

4.6 Minimal characters

Let us finally discuss the characters of the degenerate representations because they tell us the content in secondary operators, level by level. The Kac determinant eq. (4.47) determines more than the values of Δ for the highest weight representations. It tells us the structure of null-vectors. Suppose we start with a primary operator of conformal weight $\Delta_{r,s}$. We know that there is a null-vector at level rs with conformal weight $\Delta_{r,s} + rs = \Delta_{r+m,m+1-s}$. Because of the symmetry $\Delta_{r,s} = \Delta_{m-r,m+1-s}$, there is a second null-vector at level $(m-r)(m+1-s)$ with conformal weight $\Delta_{r,s} + (m-r)(m+1-s) = \Delta_{2m-r,s}$. Both null-vectors are primary and their Kac determinants can be computed with the Kac formula. Up to now, we only discussed the consequences of one null-vector. The existence of an intricated pattern of null-vectors increases the complexity of the Verma module. However, the number of non-trivial vectors is reduced and we have to analyse the Kac formula to determine what is left from the generic Verma module.

A careful study shows that the corresponding null-modules of the two null-vectors intersect and therefore they have some common descendants. The intersection can be decomposed into two new Verma null-submodules the primary highest weights of which are of course new primary null-vectors. These latter two null-vectors have once again an intersection given by new null-submodules. The same pattern takes place step by step. The sequence of inclusions is depicted in Figure 19. At the top, we indicate by r, s

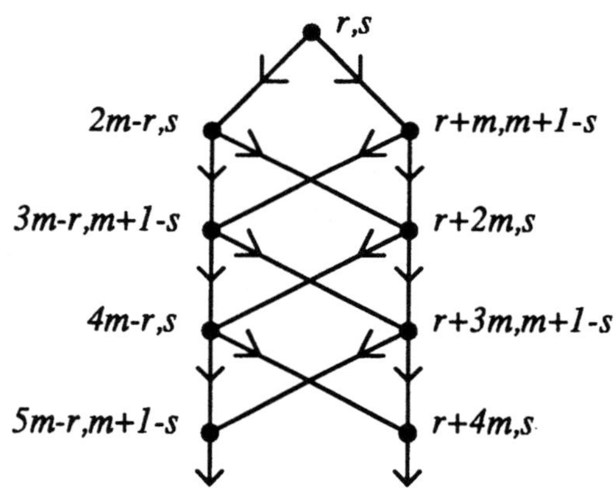

Fig. 19. Sequence of Verma submodules with conformal weights $\Delta_{a,b}$ where a, b are listed to the sides of the graph.

the highest weight primary vector with conformal weight $\Delta_{r,s}$. It has two null-vectors indexed by $2m-r, s$ and $r+m, m+1-s$. Both null-vectors intersect at the null-vectors $3m-r, m+1-s$ and $r+2m, s$. Once again these new null-vectors intersect at new null-vectors. The never-ending structure of null submodules of $|\Delta_{r,s}\rangle$ can be computed recursively with the Kac formula.

The character of the degenerate Verma module tells us how many non-trivial vectors we have at a given level. The generic formula (4.13) gives the content of a non-degenerate Verma module. Each null-module has basically the structure of a non-degenerate module. The fact that they intersect in a complicated way has to be taken into account.

From the generic character $\chi(\Delta_{r,s}, c)$ computed in (4.13), we have to subtract $\chi(\Delta_{2m-r,s}, c)$ and $\chi(\Delta_{r+m,m+1-s}, c)$ to discard in the counting of independent vectors the contributions from the null-vectors. Unfortunately, while doing this we subtracted twice the content of the intersection of the submodules with highest weights $\Delta_{3m-r,m+1-s}$ and $\Delta_{r+2m,s}$. To compensate, we have to add once $\chi(\Delta_{3m-r,m+1-s}, c)$ and $\chi(\Delta_{r+2m,s}, c)$. Up to now, we have done

$$\chi(\Delta_{r,s}, c) - \chi(\Delta_{2m-r,s}, c) - \chi(\Delta_{r+m,m+1-s}, c) + \chi(\Delta_{3m-r,m+1-s}, c) + \chi(\Delta_{r+2m,s}, c) \tag{4.50}$$

From Figure 19, we know where the two last submodules considered will intersect and this means that this intersection has now been counted twice. The submodules describing this intersection have to be subtracted. The computation goes on infinitely many times. The character of the degenerate Verma module becomes an infinite sum and subtraction of generic characters. We recognize in Figure 19 four kinds of series of indices and we can write the character, found by Rocha-Caridi [293]

$$\begin{aligned}\chi_{r,s}^{(m)} &= \chi(\Delta_{r,s}, c) - \sum_{k=1}^{\infty} \left[\chi(\Delta_{2km-r,s}, c) + \chi(\Delta_{r+m(2k-1),m+1-s}, c)\right. \\ &\qquad \left. - \chi(\Delta_{(2k+1)m-r,m+1-s}, c) - \chi(\Delta_{r+2km,s}, c)\right] \\ &= \sum_{k=-\infty}^{+\infty} \left[\chi(\Delta_{r+2km,s}, c) - \chi(\Delta_{r+m(2k-1),m+1-s}, c)\right]\end{aligned} \tag{4.51}$$

Notice how positive and negative contributions to the character were written in the last line of (4.51). It is only a question of practical purpose whether to choose the expression with a summation over positive values of k or the more compact expression with k running from $-\infty$ to $+\infty$. All generic characters essentially differ by their leading terms $q^{\Delta - c/24}$, which can be factorized out. We can write (4.51) explicitly as the infinite sum

$$\begin{aligned}\chi_{r,s}^{(m)} &= q^{\Delta_{r,s} - \frac{c(m)}{24}} \prod_{i=0}^{\infty} \frac{1}{1-q^i} \\ &\quad \times \left(1 - \sum_{k=1}^{\infty} \left[q^{(km-r)(k(m+1)-s)} + q^{((k-1)m+r)((k-1)(m+1)+s)}\right.\right. \\ &\qquad \left.\left. - q^{k(km(m+1)-(m+1)r+ms)} - q^{k(km(m+1)+(m+1)r-ms)}\right]\right) \\ &= q^{\Delta_{r,s} - \frac{c(m)}{24}} \prod_{i=0}^{\infty} \frac{1}{1-q^i}\end{aligned}$$

$$\times \sum_{k=-\infty}^{+\infty} \left[q^{k(km(m+1)+(m+1)r-ms)} - q^{((k-1)m+r)((k-1)(m+1)+s)} \right] \tag{4.52}$$

The summation integer k is the same as in (4.51). The dependence of the character in the central charge is introduced because of the particular modular properties of the characters. In chapter 11, we will show how the different characters of a minimal model build up a representation of the modular group. They are therefore strongly related to each other.

Notice finally that the character formulas for the general minimal models $\mathcal{M}_{p,p'}$ are very similar to the unitary characters. It is sufficient to systematically replace the values m and $m+1$ by p and p' respectively.

5 Operator Algebra and Correlation Functions

Physical operators have both a holomorphic part and an antiholomorphic part. Though each part is submitted to independent constraints, such as the ones given by differential equations due to null-vectors, they connect to each other by the fact that the coordinates z and $\bar{z}$ are complex conjugate in the physical geometrical space. Physical correlation functions of scalar operators ($\Delta = \overline{\Delta}$) must be well-defined single-valued functions. Correlation functions of operators with non-trivial spins $s = \Delta - \overline{\Delta}$ can gain a phase term under the exchange of two operators. However such a change must be consistent all over the possible physical n-point functions. The consistency conditions for correlation functions are the subject of this chapter. We take here advantage of a particular property of two-dimensional conformal theories: the *operator product algebra* [327]. It is an extension of the short-distance expansion of the product of two operators to conformal theories. The particular features of the latter theories results in the existence of finite algebras of operators whose correlation functions can be exactly computed.

5.1 Operator algebra and associativity

We have seen how the two Virasoro algebras, which reflect the z and $\bar{z}$ behaviour of the theory, can be independently studied and why physical primary operators formally decompose as

$$\phi(z,\bar{z}) \approx \phi(z) \otimes \overline{\phi}(\bar{z}) \tag{5.1}$$

However, the tensor product of highest weight states of each Virasoro algebra is exactly the heighest weight state of a physical primary operators and we write

$$|\Delta, \overline{\Delta}\rangle = |\Delta\rangle \otimes |\overline{\Delta}\rangle \tag{5.2}$$

The decomposition of physical operators (5.1) is more rigorous for the (secondary) stress-energy tensor where we can actually write $T(z) = T(z) \otimes \overline{1}$ and $\overline{T}(\bar{z}) = \mathbf{1} \otimes \overline{T}(\bar{z})$. However, the stress-energy tensor is not the only operator which generates an algebra. A new hypothesis of conformal field theories is the *Operator Algebra Hypothesis* [37].

Operator Algebra Hypothesis: *The set of operators of the theory is complete in the sense that the product of any pair of operators can be expressed as the convergent infinite sum*

$$A_i(z,\bar{z}) A_j(0,0) = \sum_k C_{i,j}^{k}(z,\bar{z}) A_k(0,0) \tag{5.3}$$

The operator product coefficients $C_{i,j}^{k}(z,\bar{z})$ must be single-valued functions to ensure well-defined local properties of correlation functions

$$< A_i(z,\bar{z}) A_j(0,0) X >= \sum_k C_{i,j}^{k}(z,\bar{z}) < A_k(0,0) X > \tag{5.4}$$

where X stands for any functional of local operators $\{A_s(z_s,\bar{z}_s)\}$ of the theory.

The operator algebra hypothesis extend Wilson's [327] *Operator Product Expansion* (OPE) by adding stronger properties. It is assumed that such a sum is convergent for a certain range of the local coordinates. Using translation invariance, we can write the

OPE where two arbitrary points appear in (5.3). We already made the assumption that all operators of the theory are primary operators or descendants of them. We can express the OPE of primary operators as

$$\phi_i(z,\bar{z})\phi_j(0,0) = \sum_k \sum_{(n_1,\ldots,n_p)} \sum_{(\bar{n}_1,\ldots,\bar{n}_p)} \mathbf{C}_{i,j}^{k;(n_1,\ldots,n_p;\bar{n}_1,\ldots,\bar{n}_p)}$$
$$\times z^{\Delta_k-\Delta_i-\Delta_j+\sum_s n_s} \bar{z}^{\bar{\Delta}_k-\bar{\Delta}_i-\bar{\Delta}_j+\sum_s \bar{n}_s} \phi_k^{(-n_1,\ldots,-n_p;-\bar{n}_1,\ldots,-\bar{n}_p)}(0,0) \quad (5.5)$$

The operators on the r.h.s. of eq. (5.5) are descendant operators of the primary operator ϕ_k.

$$\phi_k^{(-n_1,\ldots,-n_p;-\bar{n}_1,\ldots,-\bar{n}_p)}(0,0) = L_{-n_1}\cdots L_{-n_p}\bar{L}_{-\bar{n}_1}\cdots\bar{L}_{-\bar{n}_p}|\Delta_k,\overline{\Delta}_k\rangle \quad (5.6)$$

The powers in z or $\bar{z}$ result from dimensional considerations. The operator product coefficients $\mathbf{C}_{i,j}^{k;(n_1,\ldots,n_p;\bar{n}_1,\ldots,\bar{n}_p)}$ are now real or complex numbers. It may seem untractable to compute this infinite number of coefficients. This is fortunately not the case. Actually, without any restriction of the generality, we can write

$$\mathbf{C}_{i,j}^{k;(n_1,\ldots,n_p;\bar{n}_1,\ldots,\bar{n}_p)} = \mathbf{C}_{i,j}^k\, \omega_{i,j}^{k;(n_1,\ldots,n_p)}\, \bar{\omega}_{i,j}^{k;(\bar{n}_1,\ldots,\bar{n}_p)} \quad (5.7)$$

where $\mathbf{C}_{i,j}^k$ is the coupling to the primary operator ϕ_k. The corrections to this value due to descendant operators factorizes in the two ω-terms. This is the consequence of the full decomposition of the stress-energy tensor into holomorphic and anti-holomorphic parts. The coefficients ω and $\bar{\omega}$, though tricky to compute, are fixed by conformal invariance. To show this, notice that the coefficient $\mathbf{C}_{i,j}^k$ appears is the three-point function in the limit

$$\lim_{\xi,\bar{\xi}\to\infty} \xi^{2\Delta_k}\bar{\xi}^{2\overline{\Delta}_k} < \phi_k(\xi,\bar{\xi})\phi_i(1,1)\phi_j(0,0) >= \langle\Delta_k,\overline{\Delta}_k|\phi_i(1,1)|\Delta_j,\overline{\Delta}_j\rangle = \mathbf{C}_{i,j}^k \quad (5.8)$$

Suppose that we take a similar limit with a descendant operator of ϕ_k. We recall the differential equation (4.39) to relate such a three-point function to (5.8). Obviously the result will be proportional to $\mathbf{C}_{i,j}^k$. The factorization of the z and $\bar{z}$-dependent differential operators $\mathcal{L}_{-n}$ and $\overline{\mathcal{L}}_{-\bar{n}}$ implies the factorization of the other terms, proportional to ω, $\bar{\omega}$, and to the norm of the secondary operator. This argument is sufficient to prove that $\{\omega,\bar{\omega}\}$ are fixed by conformal invariance. A practical way to calculate them explicitly can be found in Appendix B of [37]. However, it is in most cases not necessary to compute them.

The OPE of a primary operator with itself couples to the identity. This is the same as writing

$$\langle\Delta_i,\overline{\Delta}_i|\Delta_j,\overline{\Delta}_j\rangle = \mathbf{C}_{i,j}^0 = \delta_{i,j} \quad (5.9)$$

We assume here that we deal with real operators. A generalization to operators which appear in complex conjugate pairs is obvious. The same care must be applied to $\mathbf{C}_{i,j}^k$ which is completely symmetric only for real operators. Besides the central charge and the critical exponents, the expansion coefficient $\mathbf{C}_{i,j}^k$ are the last fundamental parameters. They are not determined by conformal invariance only. A more popular way to write the OPE is the formal expression

$$\phi_i\cdot\phi_j = N_{i,j}^k\,\phi_k \quad (5.10)$$

where $N_{i,j}^k$ is zero if $\mathbf{C}_{i,j}^k$ is zero and an integer if $\mathbf{C}_{i,j}^k \neq 0$. This integer is in general one for minimal models. However, there are models where an operator is degenerate and $N_{i,j}^k$ is related to this degeneracy. In particular in models with a continuous symmetry, where each operator carries a representation of the symmetry algebra, the OPE is related to the irreducible decomposition of the tensor product of representations and $N_{i,j}^k$ counts the multiplicity of a representation in the tensor product.

We have seen in chapter 2 some examples of OPE. The Ward identities state what appears in the OPE of the secondary operator $T(z)$ with a primary operator or with itself (eqs. (2.41) and (2.52)). These examples can be summarized by

$$\mathbf{1} \cdot \phi = \phi, \qquad \mathbf{1} \cdot \mathbf{1} = \mathbf{1} \tag{5.11}$$

The determination of the OPE coefficients is fundamental for solving all correlation functions. Two and three-point functions are unsufficient to determine them. The required information is given by the solutions to the four-point functions. The condition that four-point functions must be analytically well-defined means that we have to discuss their properties under analytic continuations in their variables z and $\bar{z}$. For the sake of simplicity and to avoid redundant variables, it is better to study the analytical properties of four-point functions expressed with the projectively invariant quantities $\eta = (z_j - z_i)(z_k - z_l)/(z_k - z_i)(z_j - z_l)$ and $\bar{\eta} = (\bar{z}_j - \bar{z}_i)(\bar{z}_k - \bar{z}_l)/(\bar{z}_k - \bar{z}_i)(\bar{z}_j - \bar{z}_l)$. We make use of eq. (4.44) to define

$$\begin{aligned} G_{i,j}^{l,k}(\eta,\bar{\eta}) &:= \lim_{\xi,\bar{\xi}\to\infty} \xi^{2\Delta_l}\bar{\xi}^{2\overline{\Delta}_l} < \phi_l(\xi,\bar{\xi})\phi_k(1,1)\phi_j(\eta,\bar{\eta})\phi_i(0,0) > \\ &= \langle \Delta_l, \overline{\Delta}_l | \phi_k(1,1)\phi_j(\eta,\bar{\eta}) | \Delta_i, \overline{\Delta}_i \rangle \end{aligned} \tag{5.12}$$

This function will be very useful to discuss the neighbourhood of $\eta = \bar{\eta} = 0$. The choice made in defining $G_{i,j}^{l,k}$ is of course arbitrary. It is useful to discuss other choices and to relate them to each other. For example we can introduce $\eta' = (z_j - z_k)(z_i - z_l)/(z_i - z_k)(z_j - z_l) = 1 - \eta$ and its conjugate $\bar{\eta}' = 1 - \bar{\eta}$. This is the same as evaluating ϕ_k at 0 and ϕ_i at 1 or permuting both. The limit taken in (5.12) defines in a similar manner a new function

$$G_{k,j}^{l,i}(1-\eta, 1-\bar{\eta}) = \langle \Delta_l, \overline{\Delta}_l | \phi_i(1,1)\phi_j(1-\eta, 1-\bar{\eta}) | \Delta_k, \overline{\Delta}_k \rangle \tag{5.13}$$

related to the previous one by

$$G_{k,j}^{l,i}(1-\eta, 1-\bar{\eta}) = G_{i,j}^{l,k}(\eta,\bar{\eta}) \tag{5.14}$$

This second choice modifies the choice of the asymptotic states $|\Delta, \overline{\Delta}\rangle$. We obviously have a third choice with the projective transformation which permutes 0 and ∞. It is used to define a third function of the projective invariant parameters $(1/\eta, 1/\bar{\eta})$

$$G_{l,j}^{i,k}\left(\frac{1}{\eta}, \frac{1}{\bar{\eta}}\right) = \langle \Delta_i, \overline{\Delta}_i | \phi_k(1,1)\phi_j(1/\eta, 1/\bar{\eta}) | \Delta_l, \overline{\Delta}_l \rangle \tag{5.15}$$

This last function G is related to the first one by

$$\eta^{-2\Delta_j}\bar{\eta}^{-2\overline{\Delta}_j}\, G_{l,j}^{i,k}\left(\frac{1}{\eta}, \frac{1}{\bar{\eta}}\right) = G_{i,j}^{l,k}(\eta,\bar{\eta}) \tag{5.16}$$

The two formulas (5.14) and (5.16) are called *crossing conditions*. The prefactor on the l.h.s. of (5.16) is due to the particular definition of the asymptotic state at infinity (eq. (4.25)). Each of the three functions G is completely equivalent to the same four-point function. However, each of them will be most useful to discuss the specific singular behaviours as two of the four points get closer. This occurs in the formalism with the variables $(\eta, \bar{\eta})$ as $\eta = \bar{\eta} \to 0, 1$ and ∞.

We can now replace $\phi_j(\eta,\bar{\eta})|\Delta_i, \overline{\Delta}_i\rangle$ in $G^{l,k}_{i,j}(\eta,\bar{\eta})$ by the OPE $\phi_j \cdot \phi_i = N^s_{i,j}\phi_s$. We obtain the infinite sum over three-point functions which can be written

$$G^{l,k}_{i,j}(\eta,\bar{\eta}) = \sum_s \mathbf{C}^l_{k,s}\mathbf{C}^s_{j,i}\,\mathcal{A}^{l,k}_{i,j}(s;\eta,\bar{\eta}) \tag{5.17}$$

We define here the functions $\mathcal{A}^{l,k}_{i,j}(s;\eta,\bar{\eta})$ which collect the contributions from the descendant operators of ϕ_s in the OPE $\phi_j \cdot \phi_i$. Due to the decoupling between T and $\overline{T}$, we can decompose $\mathcal{A}^{l,k}_{i,j}$ into holomorphic and anti-homomorphic parts $\mathcal{F}$ and $\overline{\mathcal{F}}$ called *conformal blocks*.

$$\mathcal{A}^{l,k}_{i,j}(s;\eta,\bar{\eta}) = \mathcal{F}^{l,k}_{i,j}(s;\eta)\,\overline{\mathcal{F}}^{l,k}_{i,j}(s;\bar{\eta}) \tag{5.18}$$

The same dimensional reasons as in (5.5) permit to write

$$\mathcal{F}^{l,k}_{i,j}(s;\eta) = \eta^{\Delta_s - \Delta_i - \Delta_j}\,F^{l,k}_{i,j}(s;\eta) \tag{5.19}$$

where $F^{l,k}_{i,j}(s;\eta)$ is an infinite power series in η which takes into account the contributions of the holomorphic part of the descendant operators. The anti-holomorphic behaviour is similarly.

This reduction of a four-point function into a sum of three-point functions can be done with $G^{l,i}_{k,j}(1-\eta, 1-\bar{\eta})$ and the OPE $\phi_j \cdot \phi_k = N^t_{i,k}\phi_t$ for $\phi_j(1-\eta,1-\bar{\eta})|\Delta_k, \overline{\Delta}_k\rangle$. We obtain a new decomposition

$$G^{l,i}_{k,j}(1-\eta,1-\bar{\eta}) = \sum_t \mathbf{C}^l_{i,t}\mathbf{C}^t_{j,k}\,\mathcal{A}^{l,i}_{k,j}(t;1-\eta,1-\bar{\eta}) \tag{5.20}$$

where

$$\mathcal{A}^{l,i}_{k,j}(t;1-\eta,1-\bar{\eta}) = \mathcal{F}^{l,i}_{k,j}(t;1-\eta)\,\overline{\mathcal{F}}^{l,i}_{k,j}(t;1-\bar{\eta}) \tag{5.21}$$

and we can extract the asymptotic behaviour from the conformal blocks with

$$\mathcal{F}^{l,i}_{k,j}(t;1-\eta) = (1-\eta)^{\Delta_t - \Delta_k - \Delta_j}\,F^{l,i}_{k,j}(t;1-\eta) \tag{5.22}$$

The crossing condition (5.14) implies the equality

$$\sum_s \mathbf{C}^l_{k,s}\mathbf{C}^s_{j,i}\,\mathcal{A}^{l,k}_{i,j}(s;\eta,\bar{\eta}) = \sum_t \mathbf{C}^l_{i,t}\mathbf{C}^t_{j,k}\,\mathcal{A}^{l,i}_{k,j}(t;1-\eta,1-\bar{\eta}) \tag{5.23}$$

It is on purpose that we chose the indices s and t for the functions $\mathcal{A}$. This is just the s and t-channel decomposition of scattering theory. Equation (5.23) is the *crossing symmetry condition* graphically depicted in Figure 20. Three-point functions are described by the vertices and the two-point functions by propagators.

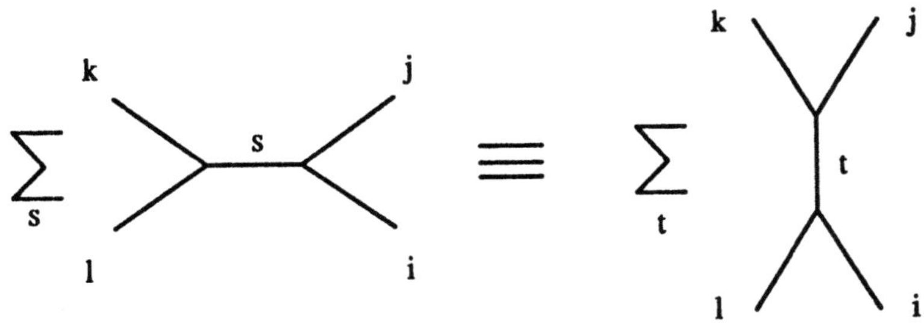

Fig. 20. Crossing symmetry condition for the four-point functions.

An equivalent statement to the crossing symmetry condition is the requirement of *associativity* for the formal OPE

$$\sum_s N^l_{k,s} N^s_{j,i} = \sum_t N^l_{i,t} N^t_{j,k} \tag{5.24}$$

We can introduce the OPE in any n-point function and replace it by a infinite sum over $(n-1)$-point functions. Introducing step by step OPEs into correlation functions results in multiple infinite sums which are graphically represented in Figure 21. The choice of OPEs is arbitrary and therefore crossing symmetry is analogous to the duality condition in string theory. Figure 21 only shows one possible computation of a n-point function. The implementation of the crossing condition for the four-point functions implies the

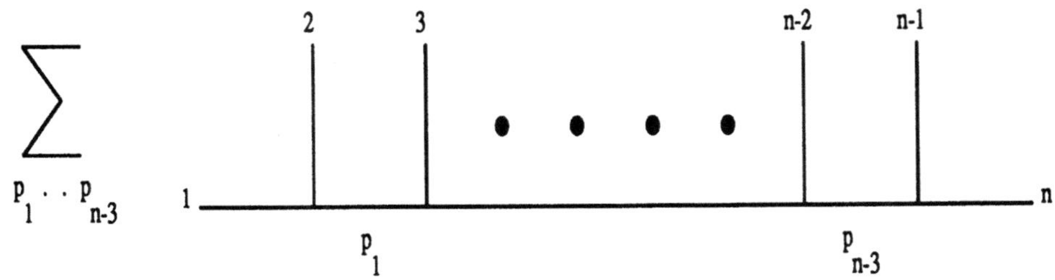

Fig. 21. One possible decomposition of a n-point function.

duality of any n-point function. It is therefore fundamental but sufficient to consistently solve all four-point functions of all primary operators.

Let us finally notice that we can discuss the u-channel and introduce the OPE $\phi_j \cdot \phi_l = N^u_{j,l}\phi_u$ into $G^{i,k}_{l,j}(1/\eta, 1/\bar{\eta})$ to introduce the last decomposition

$$G^{i,k}_{l,j}\left(\frac{1}{\eta}, \frac{1}{\bar{\eta}}\right) = \sum_u \mathbf{C}^i_{k,u} \mathbf{C}^u_{j,l} \mathcal{A}^{i,k}_{l,j}\left(u; \frac{1}{\eta}, \frac{1}{\bar{\eta}}\right) \tag{5.25}$$

As before, we can decompose $\mathcal{A}_{l,j}^{i,k}$ into its conformal blocks

$$\mathcal{A}_{l,j}^{i,k}\left(u;\frac{1}{\eta},\frac{1}{\bar{\eta}}\right)=\mathcal{F}_{l,j}^{i,k}\left(u;\frac{1}{\eta}\right)\overline{\mathcal{F}}_{l,j}^{i,k}\left(u;\frac{1}{\bar{\eta}}\right) \tag{5.26}$$

However, from the crossing condition (5.16), we have the particular asymptotic behaviour

$$\mathcal{F}_{l,j}^{i,k}\left(u;\frac{1}{\eta}\right)=\left(\frac{1}{\eta}\right)^{\Delta_u-\Delta_l+\Delta_j}F_{l,j}^{i,k}\left(u;\frac{1}{\eta}\right) \tag{5.27}$$

The possible crossing conditions can be straightforwardly derived.

5.2 Analyticity and the monodromy problem

The decomposition of a four-point function into its conformal blocks gives the necessary insight into the analytic structure of the correlation function. The conformal blocks have always a particularly simple behaviour under specific analytic continuations in their parameter η. Recall for example the holomorphic conformal blocks in the s-channel which have the structure

$$\mathcal{F}_{i,j}^{l,k}(s;\eta)=\eta^{\Delta_s-\Delta_i-\Delta_j}F_{i,j}^{l,k}(s;\eta) \tag{5.28}$$

The power series $F_{i,j}^{l,k}$ provides an explicit expansion of the correlation function in the neighbourhood of zero. The analytic continuation of $\mathcal{F}_{i,j}^{l,k}(s;\eta)$ along a curve surrounding the point zero (Figure 22) modifies the conformal block by a phase $\Omega_{s,s}^{(0)}=e^{2\pi i(\Delta_s-\Delta_i-\Delta_j)}$. This is called a *monodromy transformation* [188, 189]. All conformal blocks $\{\mathcal{F}_{i,j}^{l,k}(s;\eta)\}_s$ in the s-channel are eigenvectors of the monodromy transformation $\Omega^{(0)}$.

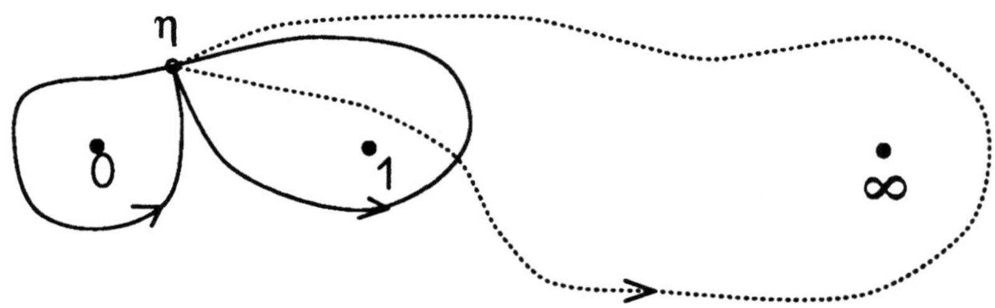

Fig. 22. Monodromy transformations: η moves around 0 (solid line), 1 (dashed line), and ∞ (dotted line).

As a result, such an analytic continuation modifies $\mathcal{A}_{i,j}^{l,k}(s;\eta,\bar{\eta})$ by a phase term.

$$\mathcal{A}_{i,j}^{l,k}(s;\eta,\bar{\eta})\xrightarrow{\Omega^{(0)}}e^{2\pi i(\Delta_s-\Delta_i-\Delta_j-\overline{\Delta}_s+\overline{\Delta}_i+\overline{\Delta}_j)}\mathcal{A}_{i,j}^{l,k}(s;\eta,\bar{\eta}) \tag{5.29}$$

The contributions from $\overline{\mathcal{F}}$ have a negative sign with respect to the ones from $\mathcal{F}$ because the corresponding complex conjugate monodromy transformations winds up around zero in the other direction. Different phase terms cause the correlation function to be ill-defined. A well-defined correlation function requires the s-dependent phases of $\mathcal{A}_{i,j}^{l,k}$ to be all identical (up to multiples of $2\pi i$).

Single-valued two-, three-, and four-point correlation functions need a trivial phase. This is the case if the spin of all primary operators are integer.

$$|\Delta_a - \overline{\Delta}_a| \in \mathbb{N} \tag{5.30}$$

Physical theories where all primary operators are scalars or have integer spins are called *local*, as all their correlation functions are locally well-defined.

However, it is possible to relax somehow this statement because we know that statistical systems with particular boundary conditions have correlators which are multi-valued and gain a phase term under analytic continuation. This means that it is physically possible that a global s-independent phase term appears. However, such phase terms must be consistent throughout all possible correlation functions. This situation occurs in parafermionic sectors of the theory where fractional spins appear. It is however reasonable to investigate first the strict monodromy invariance because any physical models possesses a local sector which already contains the essence of the model.

Moreover, if the analytic continuation around zero are simple, the analytic continuation $\Omega^{(1)}$ of $\mathcal{F}_{i,j}^{l,k}(s;\eta)$ along a curve surrounding the point one (Figure 22) is more complicated. This transformation is not diagonal but becomes the linear combination

$$\mathcal{F}_{i,j}^{l,k}(s;\eta) \stackrel{\Omega^{(1)}}{\longrightarrow} \sum_{s'} \Omega_{s,s'}^{(1)} \mathcal{F}_{i,j}^{l,k}(s';\eta) \tag{5.31}$$

The s-channel basis $\{\mathcal{F}_{i,j}^{l,k}(s;\eta)\}_s$ is not the appropriate basis to discuss this monodromy transformation. The t-channel basis of conformal block is more appropriate because it provides a natural expansion in the parameter $(1-\eta)$ as shown in eq. (5.22). This basis diagonalizes the monodromy transformation given by the matrix $\tilde{\Omega}^{(1)}$ and we have

$$\mathcal{F}_{k,j}^{l,i}(t;1-\eta) \stackrel{\tilde{\Omega}^{(1)}}{\longrightarrow} \sum_{t'} \tilde{\Omega}_{t,t'}^{(1)} \mathcal{F}_{k,j}^{l,i}(t';1-\eta) = \delta_{t,t'}\, e^{2\pi i(\Delta_t - \Delta_k - \Delta_j)}\, \mathcal{F}_{k,j}^{l,i}(t';1-\eta) \tag{5.32}$$

The eigenvalues of $\tilde{\Omega}^{(1)}$ and $\Omega^{(1)}$ are the same because the two sets of functions $\{\mathcal{F}_{i,j}^{l,k}(s;\eta)\}_s$ and $\{\mathcal{F}_{k,j}^{l,i}(t;1-\eta)\}_t$ are linearly dependent and describe the different solutions to the same differential operator (due to a specific null vector).

The space of conformal blocks in a given channel are the eigenfunctions of a given monodromy transformation. The monodromy group contains $\Omega^{(0)}$, $\Omega^{(1)}$ and of course $\Omega^{(\infty)}$ which is diagonal in the u-channel (eq. (5.27)). However the latter transformation is not independent of the others because the points zero, one and infinity are the only singular points of the correlation functions. Performing successively the three monodromy transformations is equivalent to analytically continue the conformal blocks along a curve surrounding a portion of the space empty of singular points. The conformal blocks are invariant under such a continuation and it follows that

$$\left(\Omega^{(\infty)}\Omega^{(1)}\Omega^{(0)}\right)_{s,s'} = \delta_{s,s'} \tag{5.33}$$

We know that both sides of eq. (5.23) describe the same correlation functions and we know that a necessary condition to implement locality is the integer-spin condition (5.30). However such a condition is not sufficient to ensure single-valued correlation functions because the other monodromy transformations will mix up the different conformal blocks and modify the pairings between holomorphic and anti-holomorphic parts.

To ensure locality and to satisfy the condition of crossing, we need the four-point functions to be invariant under **any** *analytic continuation in η along curves surrounding the points zero, one, and the point at infinity.*

This is the precise statement of the monodromy problem mentioned in chapter 4. The solution lies in the determination of the expansion coefficients $\mathbf{C}_{i,j}^{k}$. This is still a complicated task which gives rise to connections with *quantum groups.* We will determine the expansion coefficients for minimal models in the next two chapters.

Notice that if we want to relax the condition of strict monodromy invariance to let correlation functions change by a phase factor under the exchange of two operators, we can impose the invariance under a specific subgroup of the monodromy group. Such a discussion can be done once the strict monodromy invariance problem has been fully solved. We will illustrate this for the critical Ising model in the next chapter.

5.3 The Riemann point of view

Let us finish this chapter by a beautiful discussion due to Riemann, which enlightens the conditions under which the monodromy problem has a chance to be solvable, see [7]. The following method was extensively used in [91, 92].

A four-point correlation function is given by OPE coefficients $\mathbf{C}_{i,j}^{k}$ and a set of intricated functions $\{\mathcal{F}_{i,j}^{l,k}(s;\eta)\}$ for which we know the leading behaviour as η goes to zero. Due to the crossing conditions we can use equivalent other families $\{\mathcal{F}_{k,j}^{l,i}(t;1-\eta)\}$ and $\{\mathcal{F}_{l,j}^{i,k}(u;1/\eta)\}$ where we know their asymptotic behaviour as η goes respectively to one and infinity. The elements of these families are the different solutions of the differential equation $D_N\mathcal{F}=0$ of order N which constrains the correlation function. N is the level of the corresponding null vector. Therefore the number of elements in the three possible families must the same and equal to N. The details of these functions or of the differential equation are very complicated. However, *Riemann held the view that such a complicated family is described by its singular behaviour complemented by some general properties.* In our four-point function, we know the singular behaviour at zero, one and infinity given respectively by η^{α_s}, $(1-\eta)^{\beta_t}$, $(1/\eta)^{\gamma_u}$ where the *singular exponents* are related to the conformal weights by

$$\alpha_s=\Delta_s-\Delta_i-\Delta_j\,,\qquad \beta_t=\Delta_t-\Delta_k-\Delta_j\,,\qquad \gamma_u=\Delta_u-\Delta_l+\Delta_j \tag{5.34}$$

for $1\leq s,t,u\leq N$. We make the assumption that the singular exponents in each channel are distinct and do not differ by integers. This is important to avoid logarithmic behaviour like $\eta^{\alpha}\ln\eta$. We want to show the following:

Proposition: *The vector space generated by any basis of functions (we know one in each channel) must transform into itself under any analytic continuation along paths which do not cross one of the singular points* 0, 1, *and* ∞. *Therefore the choice of the* *singular exponents α, β, and γ is restricted by the condition of closure under analytic continuations and must satisfy the sum rule*

$$\boxed{\sum_{r=1}^{N}(\alpha_r+\beta_r+\gamma_r)=\frac{N(N-1)}{2}-\sum_a R_a} \tag{5.35}$$

where $\{R_a\}$ forms a finite set of positive integers. Moreover, none of the numbers $\alpha_r-\alpha_{r'}$, $\beta_r-\beta_{r'}$, $\gamma_r-\gamma_{r'}$ may be an integer.[5]
This proposition is obviously necessary to build correlation functions which preserve their values but also their interpretations under monodromy transformations. A generalization of the proposition to a n-point function, related to differential operators having more singular points $(n-1)$ is straightforward. To simplify the notation, we rename

$$\begin{aligned} f_s^{(0)}(\eta) &= \mathcal{F}_{i,j}^{l,k}(s;\eta) \\ f_t^{(1)}(\eta) &= \mathcal{F}_{k,j}^{l,i}(t;1-\eta) \\ f_u^{(\infty)}(\eta) &= \mathcal{F}_{l,j}^{i,k}(u;1/\eta) \end{aligned} \tag{5.36}$$

The three bases of functions $f^{(\ell)}$, $\ell=0,1,\infty$, describe the same family. A major information comes from the study of the Wronski determinant defined in any basis by

$$W(\eta)=\begin{vmatrix} f_1^{(\ell)} & \cdot\cdot & f_N^{(\ell)} \\ \partial_\eta f_1^{(\ell)} & \cdot\cdot & \partial_\eta f_N^{(\ell)} \\ \cdot & \cdot\cdot & \cdot \\ \partial_\eta^{N-1} f_1^{(\ell)} & \cdot\cdot & \partial_\eta^{N-1} f_N^{(\ell)} \end{vmatrix} \tag{5.37}$$

Notice that $W(\eta)$ is independent of ℓ because all three bases of functions are linearly related. The logarithmic derivative $\partial_\eta \ln W(\eta)$ is a meromorphic function with single poles located at zero, one, and infinity because of the corresponding singular behaviour of the family. Moreover, the Wronski determinant may vanish for some regular values $\eta=\eta_a$ which contribute to the pole structure. There is no way to specify more closely the values of the parameters η_a which are arbitrary and have no physical interpretation. The residues are given by the leading behaviour of the elements of the family. At finite values of η we can compute

$$\partial_\eta \ln W(\eta)=\frac{(\sum_r \alpha_r)-\frac{N(N-1)}{2}}{\eta}+\frac{(\sum_r \beta_r)-\frac{N(N-1)}{2}}{\eta-1}+\sum_a \frac{R_a}{\eta-\eta_a} \tag{5.38}$$

The coefficients R_a may be quite complicated. They are positive integers because the zeroes are regular. We can compare the limit $\eta\to\infty$ of eq. (5.38) with the asymptotic behaviour given by $\{f_s^{(\infty)}\}$ which gives

$$\partial_\eta \ln W(\eta)_{\eta \to \infty}^{\sim} -\frac{(\sum_r \gamma_r)+\frac{N(N-1)}{2}}{\eta}+\mathcal{O}\left(\frac{1}{\eta^2}\right) \tag{5.39}$$

---

[5] If $\alpha_r-\alpha_{r'}$ etc. happens to be an integer, one should consider logarithmic behaviour of the type $\eta^\alpha \ln\eta$. The singularities must be regular singular (see [188, 189]).

The result is the equality given by the proposition (5.35)

$$\sum_{r=1}^{N}(\alpha_r+\beta_r+\gamma_r)=\frac{N(N-1)}{2}-\sum_a R_a \tag{5.40}$$

We can introduce the definition (5.34) to express the last equation as a consistency *sum rule* for the conformal weights

$$\sum_{s=1}^{N}\Delta_s+\sum_{t=1}^{N}\Delta_t+\sum_{u=1}^{N}\Delta_u=N(\Delta_i+\Delta_j+\Delta_k+\Delta_l)+\frac{N(N-1)}{2}-\sum_a R_a \tag{5.41}$$

This equation is very powerful because it restricts the conformal weights of the possible operators which contribute to a non-trivial correlation function. The restriction actually constrains the possible operator content of the OPE of two primary operators.

Notice that the differential operator D_N is partially determined by the singular behaviour. Any solution $f(\eta)$ is a linear combination of the basis elements $f^{(\ell)}$, $\ell=1,\ldots,N$. Therefore the following determinant vanishes

$$\begin{vmatrix} f & f^{(1)} & .. & f^{(N)} \\ \partial_\eta f & \partial_\eta f^{(1)} & .. & \partial_\eta f^{(N)} \\ . & . & .. & \\ \partial_\eta^N f & \partial_\eta^N f^{(1)} & .. & \partial_\eta^N f^{(N)} \end{vmatrix}=0 \tag{5.42}$$

This can be expressed as the differential equation, where we have normalized the coefficient in $\partial_\eta^N f(\eta)$ to one

$$\partial_\eta^N f(\eta)+\partial_\eta \ln W(\eta)\partial_\eta^{N-1}f(\eta)+\cdots+\widehat{W}_0(\eta)f(\eta)=0 \tag{5.43}$$

The coefficient in front of $\partial_\eta^{N-1}f(\eta)$ is given in (5.38). The other coefficients $\widehat{W}_j$ are not completely determined by the singular behaviour. The points $\eta=0$, 1 and ∞ are called ***singular regular points*** if the singular exponents in each channel do not differ by integers. The other points with residues R_a are *regular* points which do not affect the monodromy properties. Notice that by construction of the null-vector related differential operators which follow from eqs. (4.37) and (4.38), we know that they are always free of such regular singularities. The differential operator D_N is called *Fuchs differential operator*. In the particular situation where $N=2$ and there are no regular points, the differential equation is completely determined by the singular exponents, which satisfy $\alpha_1+\alpha_2+\beta_1+\beta_2+\gamma_1+\gamma_2=1$. We have

$$D_2 f=\left[\partial_\eta^2-\left(\frac{\alpha_1+\alpha_2-1}{\eta}-\frac{\beta_1+\beta_2-1}{1-\eta}\right)\partial_\eta\right.$$
$$\left.+\frac{\alpha_1\alpha_2}{\eta^2}-\frac{\gamma_1\gamma_2-\alpha_1\alpha_2-\beta_1\beta_2}{\eta(1-\eta)}+\frac{\beta_1\beta_2}{(1-\eta)^2}\right]f(\eta)=0 \tag{5.44}$$

After the transformation

$$f(\eta)=\eta^{\alpha_1}(1-\eta)^{\beta_1}g(\eta) \tag{5.45}$$

we recognize the hypergeometric differential equation for $g(\eta)$.

$$\left[\partial_\eta^2 - \left(\frac{\alpha_2-\alpha_1-1}{\eta} - \frac{\beta_2-\beta_1-1}{1-\eta}\right)\partial_\eta - \frac{(\gamma_1+\alpha_1+\beta_1)(\gamma_2+\alpha_1+\beta_1)}{\eta(1-\eta)}\right] g(\eta) = 0 \tag{5.46}$$

We write the two solutions which diagonalize the monodromy transformation $\Omega^{(0)}$ in the usual notation

$$g(\eta) = \begin{cases} {}_2F_1(\gamma_1+\alpha_1+\beta_1, \gamma_2+\alpha_1+\beta_1, 1+\alpha_1-\alpha_2; \eta) \\ \eta^{\alpha_2-\alpha_1}{}_2F_1(\gamma_1+\alpha_2+\beta_1, \gamma_2+\alpha_2+\beta_1, 1+\alpha_2-\alpha_1; \eta) \end{cases} \tag{5.47}$$

This shows that any four-point function containing the primary operator $\phi_{(1,2)}$ or $\phi_{(2,1)}$ with conformal weights $\Delta_{1,2}$ and $\Delta_{2,1}$ is up to a prefactor made out of hypergeometric functions ${}_2F_1$.

It is not possible to tell much more about D_N for $N > 2$. However, the sum rule (5.41) remains a very powerful tool. Equation (5.41) can detect possible analytically closed subfamilies. We know from chapter 4 that a unitary primary operators ϕ has many null-vectors. Each of them implies a differential equation of the same order as its Verma module level. Obviously the null-vector at the lowest possible level, say N, and its differential operator, say D_N, must completely determine the correlation functions containing ϕ, and therefore the OPEs of ϕ with any other primary operator. To be consistent with each other, another differential operator $D_{N'}$ of order $N' > N$ (related to a higher level null-vector in the same Verma module) must admit a N-dimensional subspace of solutions which overlap with the N-dimensional space of solutions of D_N. The following decomposition is therefore necessary

$$D_{N'} = D_{N'-N} \cdot D_N \tag{5.48}$$

Such a decomposition is detectable in the N'-dimensional family of $D_{N'}$ with the sum rule (5.41) which will show if a N-dimensional subfamily may satisfy the sum rule and therefore be analytically closed. In this case, both $D_{N'-N}$ and D_N will be free of regular singularities.

In the next chapter, we will discuss the Ising model and we will show that there are decompositions of differential operators into a product of differential operators with this time regular singularities which have nothing to do with the null-vector structure. Such decomposition are related to the fact that a subset of the conformal blocks may form a analytically closed subfamily which can be detected by the sum rule (5.41). The corresponding decomposition of differential operators cannot be foreseen in the study of the Verma module and the Riemann point of view provides a very simple tool to gain more information on the possible operator content and OPEs of a physical system.

6 The Ising Model Correlation Functions

We have now all the necessary tools to completely discuss a specific example: the celebrated critical Ising model. We recall that the content in primary operators, read from the Kac table (Table 5), reduces to the identity operator $\mathbf{1} = \phi_{(1,1)}$ (conformal weights $\Delta_{1,1} = \overline{\Delta}_{1,1} = 0$) which contains the stress-energy tensor, the spin-density $\sigma = \phi_{(1,2)} = \phi_{(2,2)}$ ($\Delta_{1,2} = \overline{\Delta}_{1,2} = 1/16$), and the energy-density $\varepsilon = \phi_{(2,1)} = \phi_{(1,3)}$ ($\Delta_{2,1} = \overline{\Delta}_{2,1} = 1/2$). The central charge is $c = 1/2$ and the model is labeled by $m = 3$ in the Kac notation (eqs. (4.48) and (4.49)). We will calculate the most important four-point functions and determine exactly the monodromy group and the operator product coefficients. The fermionic operators will appear when we reduce the constraint of locality to a subgroup of the monodromy group. We stress however, that the particular virtue of these techniques is that the exact solution of a lattice system is not required. This allows to discuss more complicated systems along exactly the same lines.

6.1 Four-point function of the spin-density operator

We first discuss the correlation functions of the spin-density operator σ. We can now collect the results of the previous chapters. Holomorphic and anti-holomorphic parts are both constrained by the null-vectors at level 2, explicitly given by (4.11). Along the same line as in the previous chapter, we discuss the four-point function reduced to the projectively invariant form.

$$G_\sigma(\eta,\bar\eta) = \langle 1/16, 1/16|\sigma(1,1)\sigma(\eta,\bar\eta)|1/16,1/16\rangle \tag{6.1}$$

Let us first discuss the holomorphic behaviour. We know that eq. (4.45) expresses the consequence of the level 2 null vector of $\sigma(\eta,\bar\eta)$ and we have to solve for η

$$\left[-\frac{4}{3}\partial_\eta^2 - \left(\frac{1}{\eta} - \frac{1}{1-\eta}\right)\partial_\eta + \frac{(1/16)}{\eta^2} + \frac{(1/8)}{\eta(1-\eta)} + \frac{(1/16)}{(1-\eta)^2}\right] G_\sigma(\eta) = 0 \tag{6.2}$$

We can discuss the asymptotic behaviour of the solutions as η becomes close to 0, 1, or ∞. It is straightforward to determine the leading powers of the conformal blocks $\eta^\alpha F(\eta)$, where $\alpha = \Delta - 2\Delta_{12}$ (according to eq. (5.19)) and Δ stands for the conformal weights of the operators which enter into the OPE $\sigma\cdot\sigma$. The leading order $\eta^{\alpha-2}$ of the solution $\eta^\alpha F(\eta)$ to eq. (6.2) is proportional to

$$-\frac{4}{3}\alpha(\alpha-1) - \alpha + \frac{1}{16} = 0 \tag{6.3}$$

Simple calculation gives the two solutions

$$\Delta = 0 = \Delta_{1,1}, \qquad \text{and} \qquad \Delta = \frac{1}{2} = \Delta_{2,1} \tag{6.4}$$

and we obtain the OPE

$$\sigma\cdot\sigma = \mathbf{1} + \varepsilon \tag{6.5}$$

This result is obviously confirmed by the study of the asymptotic behaviour as $\eta \to 1$, which is completely similar. The conformal blocks for $\eta \to \infty$ have slightly different leading powers given in eq. (5.27), but as expected in agreement with the OPE (6.5). Notice that the sum rule (5.41) is verified with no additive term R_a.

Notice now that (6.2) becomes the well-known hypergeometric differential equation for the function $g_\sigma(\eta)$ defined by

$$G_\sigma(\eta) = \left(\frac{1}{\eta(1-\eta)}\right)^{\frac{1}{8}} g_\sigma(\eta) \tag{6.6}$$

The hypergeometric differential equation for $g_\sigma(\eta)$ reads

$$\left[\partial_\eta^2 + \left(\frac{1/2}{\eta} - \frac{1/2}{1-\eta}\right)\partial_\eta + \frac{1/16}{\eta(1-\eta)}\right] g_\sigma(\eta) = 0 \tag{6.7}$$

This confirms the result obtained with the Riemann point of view of chapter 5. Notice that the second-order differential equations related to the level 2 null-vectors of $\phi_{(1,2)}$ and $\phi_{(2,1)}$ of **any** minimal model will always be transformable into hypergeometric differential operators, see eqs. (5.44) to (5.47).

The two solutions to this equation and the monodromy properties can be found in any table of integrals, see [1, 153]. They are normally expressed as power series or as Euler contour integrals. Fortunately the Ising model has the exceptional property that all correlation functions are algebraic and it is simple to verify the following pair of solutions.

$$g_\sigma^\pm(\eta) = \sqrt{1 \pm \sqrt{1-\eta}} \tag{6.8}$$

Notice that the invariance of the differential operator (6.7) under the exchange $\eta \leftrightarrow 1-\eta$ implies the following second basis of solutions

$$g_\sigma^+(\eta) \pm g_\sigma^-(\eta) = \sqrt{2}\, g_\sigma^\pm(1-\eta) = \sqrt{2}\sqrt{1 \pm \sqrt{\eta}} \tag{6.9}$$

Both solutions $g_\sigma^\pm(\eta)$ are invariant under the analytic continuations along a curve surrounding the origin, i.e. $\Omega^{(0)}$, and their leading behaviour in this limit is

$$\begin{aligned} g_\sigma^+(\eta) &= \sqrt{2}\,(1+\mathcal{O}(\eta)) \\ g_\sigma^-(\eta) &= \frac{1}{\sqrt{2}}\eta^{1/2}\,(1+\mathcal{O}(\eta)) \end{aligned} \tag{6.10}$$

Up to the term factorized in (6.6) and a normalization, the two solutions are the two conformal blocks of the s-channel. The monodromy matrix $\Omega^{(0)}$ is therefore

$$\Omega^{(0)} = \begin{pmatrix} 1 & 0 \\ 0 & -1 \end{pmatrix} \tag{6.11}$$

The transformations of the two solutions is more complicated under a monodromy transformation surrounding the point 1. We expand the solutions in $(1-\eta)$ to derive

$$g_\sigma^\pm(\eta) = \left(1 \pm \frac{1}{2}(1-\eta)^{1/2}\right)\left(1+\mathcal{O}(1-\eta)\right) \tag{6.12}$$

The analytic continuation along a curve surrounding the point one exchanges the two functions

$$\Omega^{(1)} = \begin{pmatrix} 0 & 1 \\ 1 & 0 \end{pmatrix} \tag{6.13}$$

The second basis of solutions (6.9) has therefore a particular interpretation. The linear combination $(g_\sigma^+ + g_\sigma^-)$ is regular at $\eta = 1$ while the other, $(g_\sigma^+ - g_\sigma^-)$, behaves like $(1-\eta)^{1/2}$ as $\eta \to 1$. These two combinations diagonalize $\Omega^{(1)}$ and are proportional to the two conformal blocks defined in (5.22) for the t-channel. It is finally interesting to derive the two leading behaviours as $\eta \to \infty$

$$\begin{aligned} g_\sigma^\pm(\eta) &= \sqrt{1 \pm i\eta^{1/2}\sqrt{1-\frac{1}{\eta}}} = e^{\pm i\frac{\pi}{4}}\eta^{1/4}\sqrt{1-\frac{1}{2\eta} \mp \frac{i}{\eta^{1/2}} + \mathcal{O}\left(\frac{1}{\eta^2}\right)} \\ &= e^{\pm i\frac{\pi}{4}}\eta^{1/4}\left(1 \mp \frac{i}{2\eta^{1/2}}\right)\left(1+\mathcal{O}\left(\frac{1}{\eta}\right)\right) \\ &= e^{\pm i\frac{\pi}{4}}\left(\frac{1}{\eta}\right)^{-1/4} + \frac{1}{2}e^{\mp i\frac{\pi}{4}}\left(\frac{1}{\eta}\right)^{1/4} + \dots \end{aligned} \tag{6.14}$$

The monodromy transformation $\Omega^{(\infty)}$ is therefore given by

$$\Omega^{(\infty)} = \begin{pmatrix} 0 & 1 \\ -1 & 0 \end{pmatrix} \tag{6.15}$$

The linear combinations $g_\sigma^+ \pm i g_\sigma^-$ diagonalize $\Omega^{(\infty)}$ and are proportional to the u-channel conformal blocks given in eq. (5.27). As expected, the three monodromy matrices satisfy eq. (5.33).

We are now in position to restore the anti-holomorphic part. The physical monodromy invariant four-point function is logically the linear combination

$$G_\sigma(\eta,\bar\eta) = \frac{1}{2}\left(\frac{1}{\eta\bar\eta(1-\eta)(1-\bar\eta)}\right)^{\frac{1}{8}} \left(g_\sigma^+(\eta)g_\sigma^+(\bar\eta) + g_\sigma^-(\eta)g_\sigma^-(\bar\eta)\right) \tag{6.16}$$

It is easy to verify that this solution satisfies the crossing condition and is therefore monodromy invariant. The amplitude 1/2 in front of (6.16) is fixed by the normalization of the two point functions given by $\mathbf{C}^{\mathbf{1}}_{\sigma,\sigma} = 1$. The square of this coefficient appears in eq. (5.17) for the contribution of the identity operator to the OPE $\sigma\cdot\sigma$. This contribution is given in (6.16) by the leading behaviour in $1/(\eta\bar\eta)^{1/8}$ coming from the term $g_\sigma^+(\eta)g_\sigma^+(\bar\eta)$. We deduce from the expansion (6.10) the amplitude 1/2. The expansion of (6.16) for small values of η and $\bar\eta$ gives the leading terms of the full decomposition into conformal blocks given in (5.17).

$$G_\sigma(\eta,\bar\eta) = \left(\frac{1}{\eta\bar\eta(1-\eta)(1-\bar\eta)}\right)^{\frac{1}{8}} \left(1 + \frac{1}{4}(\eta\bar\eta)^{\frac{1}{2}}\right)\left(1 + \mathcal{O}(\eta,\bar\eta)\right) \tag{6.17}$$

We can read off the expansion coefficients. The contribution from ϵ to the OPE $\sigma\cdot\sigma$ appears in the $g_\sigma^-(\eta)g_\sigma^-(\bar\eta)$ term. The coefficient in front of the corresponding leading term $\eta^{1/2-1/8}\bar\eta^{1/2-1/8}$ is equal to $(\mathbf{C}^{\epsilon}_{\sigma,\sigma})^2$. We obtain therefore

$$\mathbf{C}^{\mathbf{1}}_{\sigma,\sigma} = 1 \qquad \mathbf{C}^{\epsilon}_{\sigma,\sigma} = \frac{1}{2} \tag{6.18}$$

Our solution is now complete. Notice that $\mathbf{C}^{\sigma}_{\sigma,\sigma} = 0$ because the contribution of σ to $\sigma \cdot \sigma$ would imply a singular behaviour $\eta^{-1/16}$ in the correlation function as $\eta \to 0$. Such a term is obviously absent.

We can restore the z_1, z_2, z_3, z_4 and $\bar{z}_1, \bar{z}_2, \bar{z}_3, \bar{z}_4$-dependence to write the correlation function $< \sigma(z_4, \bar{z}_4)\sigma(z_3, \bar{z}_3)\sigma(z_1, \bar{z}_1)\sigma(z_2, \bar{z}_2) >$. The relation was given by eqs. (4.42) and (4.43) and we can write

$$< \sigma(z_4, \bar{z}_4)\sigma(z_3, \bar{z}_3)\sigma(z_1, \bar{z}_1)\sigma(z_2, \bar{z}_2) >=$$

$$\frac{1}{|z_1 - z_4|^{1/4}} \frac{1}{|z_3 - z_2|^{1/4}} G_\sigma \left(\frac{(z_1 - z_2)(z_3 - z_4)}{(z_3 - z_2)(z_1 - z_4)}, \frac{(\bar{z}_1 - \bar{z}_2)(\bar{z}_3 - \bar{z}_4)}{(\bar{z}_3 - \bar{z}_2)(\bar{z}_1 - \bar{z}_4)} \right) \quad (6.19)$$

This is in agreement with the exact calculation in the scaling limit of the Ising model [331, G10].

6.2 Four-point function of the energy-density operator

We proceed along the same lines as in the last section to discuss the four-point function

$$G_\varepsilon(\eta, \bar{\eta}) = \langle 1/2, 1/2 | \varepsilon(1,1) \varepsilon(\eta, \bar{\eta}) | 1/2, 1/2 \rangle \quad (6.20)$$

Due to the level 2 null-vector of $\phi_{(2,1)}$, we have again to solve second-order differential equations (given by eq. (4.45)) which holomorphic part is

$$\left[-\frac{3}{4}\partial_\eta^2 - \left(\frac{1}{\eta} - \frac{1}{1-\eta} \right) \partial_\eta + \frac{(1/2)}{\eta^2} + \frac{1}{\eta(1-\eta)} + \frac{(1/2)}{(1-\eta)^2} \right] G_\varepsilon(\eta) = 0 \quad (6.21)$$

We can again determine the asymptotic behaviour of the s-channel conformal blocks $\eta^\alpha F(\eta)$ to compute the OPE $\varepsilon \cdot \varepsilon$. With $\alpha = \Delta - 2\Delta_{21}$, we obtain

$$\Delta = 0 = \Delta_{1,1}, \qquad \text{and} \qquad \Delta = \frac{5}{3} = ? \quad (6.22)$$

The second solution does not correspond to an operator of the Kac Table (Table 5). In chapter 13, we shall describe a failed attempt [51] to establish its presence in an antiferromagnetic Ising model. However we can examine the one-dimensional subfamily from the Riemann point of view. Taking only the first solution amounts to have the OPE

$$\varepsilon \cdot \varepsilon = \mathbf{1} \quad (6.23)$$

which is of course expected, due to the well-known free fermionic interpretation of the critical Ising model [G10, G11]. The sum rule (5.41) for the $N = 1$ subfamily can be written

$$0 = 2 - \sum_a R_a \quad (6.24)$$

which is perfectly solvable, R_a being positive integers. Therefore, we expect the differential operator (6.21) to decompose into a product $D'_1 D_1$ of two first order differential operators. The two zeroes[6)] of the Wronski determinant (5.37) which give rise to the

[6)] Or the single zero of order two.

additive terms R_a appear in the first order differential operators a simple poles. After some computations, we obtain

$$D'_1 = -\frac{1}{4}\left[3\,\partial_\eta + \left(\frac{1}{\eta} - \frac{1}{1-\eta}\right) + 3\left(\frac{1}{\eta-\eta_1} + \frac{1}{\eta-\eta_1^*}\right)\right]$$

$$D_1 = \partial_\eta + \left(\frac{1}{\eta} - \frac{1}{1-\eta}\right) - \left(\frac{1}{\eta-\eta_1} + \frac{1}{\eta-\eta_1^*}\right) \tag{6.25}$$

where (η_1, η_1^*) are the complex conjugate solutions of

$$\eta^2 - \eta + 1 = 0 \tag{6.26}$$

This is the first example of the regular points discussed in chapter 5. They appear as expected in a decomposition of a differential operator which cannot be explained from the knowledge of the Verma module of ε only. The solution, common to (6.21) and to $D_1 G_\varepsilon(\eta) = 0$ is given by

$$G_\varepsilon(\eta) = \left(\frac{1}{\eta} + \frac{1}{1-\eta} - 1\right) = \frac{\eta^2 - \eta + 1}{\eta(1-\eta)} \tag{6.27}$$

Due to the one-dimensionality of the solution, $G_\varepsilon(\eta)$ is the Wronski determinant (5.37) which numerator has regular zeroes. The physical local four-point function is finally

$$G_\varepsilon(\eta, \bar\eta) = \left(\frac{1}{\eta} + \frac{1}{1-\eta} - 1\right)\left(\frac{1}{\bar\eta} + \frac{1}{1-\bar\eta} - 1\right) \tag{6.28}$$

The expression is easier to recognize if we restore general coordinates. Taking eqs. (4.42) and (4.43) into account results in the well-known expression

$$< \varepsilon(z_4, \bar z_4)\varepsilon(z_3, \bar z_3)\varepsilon(z_1, \bar z_1)\varepsilon(z_2, \bar z_2) >=$$

$$\left(\frac{1}{(z_4-z_3)(z_1-z_2)} - \frac{1}{(z_4-z_1)(z_3-z_2)} + \frac{1}{(z_4-z_2)(z_3-z_1)}\right) \times$$
$$\left(\frac{1}{(\bar z_4-\bar z_3)(\bar z_1-\bar z_2)} - \frac{1}{(\bar z_4-\bar z_1)(\bar z_3-\bar z_2)} + \frac{1}{(\bar z_4-\bar z_2)(\bar z_3-\bar z_1)}\right) \tag{6.29}$$

We recognize the Wick Theorem for the free massless fermions with which we can decompose the energy-density operator

$$\varepsilon(z, \bar z) = i\psi(z)\psi(\bar z) \tag{6.30}$$

The factor i takes into account that $\psi(z)$ and $\psi(\bar z)$ anticommute and that ε is a real scalar operator.

This simple example illustrates a decomposition of a differential operator which cannot be predicted from an elementary overview of the structure of the Verma module of ε.

6.3 Mixed four-point function

We deduce from the OPEs (6.5) and (6.23) that the last possible non-zero four-point function involves two operators σ and two ε. We define

$$G_{\sigma,\varepsilon}(\eta,\bar{\eta}) = \langle 1/16, 1/16|\varepsilon(1,1)\varepsilon(\eta,\bar{\eta})|1/16, 1/16\rangle \tag{6.31}$$

The differential equation for the holomorphic part is again given by the level 2 null-vector of $\varepsilon(\eta)$.

$$\left[-\frac{3}{4}\partial_\eta^2 - \left(\frac{1}{\eta} - \frac{1}{1-\eta}\right)\partial_\eta + \frac{(1/16)}{\eta^2} + \frac{1}{\eta(1-\eta)} + \frac{(1/2)}{(1-\eta)^2}\right] G_{\sigma,\varepsilon}(\eta) = 0 \tag{6.32}$$

We can proceed this time with another method. We know from the OPEs (6.5) and (6.23) the operator content of $\varepsilon \cdot \sigma$ which is given by

$$\varepsilon \cdot \sigma = \sigma \tag{6.33}$$

This means that only one conformal block enters in the description of each channel of $G_{\sigma,\varepsilon}$. The asymptotic behaviours as $\eta \to 0$, 1 and ∞ are read from eqs. (5.19), (5.22), and (5.27).

$$\begin{aligned} \eta \to 0: &\quad \eta^{\Delta_\sigma - \Delta_\varepsilon - \Delta_\sigma} = \eta^{-\frac{1}{2}} \\ \eta \to 1: &\quad (1-\eta)^{\Delta_1 - 2\Delta_\varepsilon} = (1-\eta)^{-1} \\ \eta \to \infty: &\quad \left(\frac{1}{\eta}\right)^{\Delta_\sigma + \Delta_\varepsilon - \Delta_\sigma} = \left(\frac{1}{\eta}\right)^{\frac{1}{2}} \end{aligned} \tag{6.34}$$

The one dimensional family having such asymptotic singularities satisfies the sum rule eq. (5.41) with one zero with residue $R_1 = 1$. It is easy to verify that this asymptotic behaviour is compatible with the differential equation (6.32), whose second possible solution does not correspond to a contribution in the OPE from any operator in the Kac table (Table 5). The differential operator (6.32) must once again decompose into the product of two first-order differential operators $D'_1 D_1$. Some simple calculations show that

$$\begin{aligned} D'_1 &= -\frac{1}{4}\left[3\,\partial_\eta + \left(\frac{(5/2)}{\eta} - \frac{1}{1-\eta} + \frac{3}{\eta+1}\right)\right] \\ D_1 &= \partial_\eta + \left(\frac{(1/2)}{\eta} - \frac{1}{1-\eta} - \frac{1}{\eta+1}\right) \end{aligned} \tag{6.35}$$

and the holomorphic solution to $D_1 G_{\sigma,\varepsilon}$ is

$$G_{\sigma,\varepsilon}(\eta) = \frac{\eta+1}{\eta^{1/2}(1-\eta)} \tag{6.36}$$

The properly normalized crossing symmetric four-point function is finally

$$G_{\sigma,\varepsilon}(\eta,\bar{\eta}) = \frac{1}{4}\left|\frac{\eta+1}{\eta^{1/2}(1-\eta)}\right|^2 \tag{6.37}$$

The factor 1/4 corresponds to $(\mathbf{C}^{\sigma}_{\varepsilon,\sigma})^2$ in the asymptotic limits $\eta \to 0$ or ∞. As $\eta \to 1$, we have $1+\eta \to 2$ and the asymptotic behaviour is normalized by $\mathbf{C}^{\mathbf{1}}_{\varepsilon,\varepsilon}\mathbf{C}^{\mathbf{1}}_{\sigma,\sigma} = 1$, which is the correct result. The four-point function expressed in terms of general coordinates is

$$< \sigma(z_4,\bar{z}_4)\varepsilon(z_3,\bar{z}_3)\varepsilon(z_1,\bar{z}_1)\sigma(z_2,\bar{z}_2) >= \tag{6.38}$$

$$\left(\frac{1}{4}\right)\frac{1}{|z_1-z_3|^2|z_2-z_4|^{1/4}}\left|\left(\frac{(z_1-z_2)(z_3-z_4)}{(z_3-z_2)(z_1-z_4)}\right)^{\frac{1}{2}}+\left(\frac{(z_3-z_2)(z_1-z_4)}{(z_1-z_2)(z_3-z_4)}\right)^{\frac{1}{2}}\right|^2$$

This local solution obviously agrees with the crossing conditions (5.14) and (5.16).

To summarize, we have computed all possible four-point functions of primary operators and all expansion coefficients. The OPEs are given by

$$\boxed{\begin{aligned}&\sigma\cdot\sigma = \mathbf{1}+\varepsilon\\ &\varepsilon\cdot\varepsilon = \mathbf{1}\\ &\sigma\cdot\varepsilon = \sigma\end{aligned}} \tag{6.39}$$

The remaining n-point functions, which are all algebraic, can be computed along the same lines.

6.4 Semi-local four-point functions

The critical algebra $\{\mathbf{1},\sigma,\varepsilon\}$ is obtained from the high-temperature phase where the spin-density operator σ plays the role of order operator. In the low-temperature phase, the disorder operator μ, dual to the order operator, becomes a distinct operator at the critical temperature with conformal weights $\Delta_\mu = \overline{\Delta}_\mu = 1/16$. The critical theory contains therefore another critical algebra $\{\mathbf{1},\mu,\varepsilon\}$ which is completely isomorphic to the former one. OPEs and correlation functions within this new algebra are totally identical to the ones computed in the previous sections. However, it is well-known that order and disorder operators are non-local with respect to each other. This means that the exchange of such operators results in a phase factor in their correlation functions. Consider the following four-point function:

$$G_{\mu,\sigma}(\eta,\bar{\eta}) = \langle\mu|\sigma(1,1)\sigma(\eta,\bar{\eta})|\mu\rangle \tag{6.40}$$

The existence of null-vector implies a differential equations, which is the same as in eq. (6.2). The conformal blocks discussed in that section are the same. However, the correlation function must gain a $(-)$ sign when analytically continued in η along curves surrounding the points 0 and ∞ where disorder operators μ are located. Correlation functions are therefore invariant under the subgroup of the monodromy group generated by $(\Omega^{(0)})^2$, $\Omega^{(1)}$, and $(\Omega^{(\infty)})^2$. The only possible correlation function is the combination

$$G_{\mu,\sigma}(\eta,\bar{\eta}) = \frac{1}{2}\left(\frac{1}{\eta\bar{\eta}(1-\eta)(1-\bar{\eta})}\right)^{\frac{1}{8}}\left(g^+_\sigma(\eta)g^-_\sigma(\bar{\eta})+g^-_\sigma(\eta)g^+_\sigma(\bar{\eta})\right) \tag{6.41}$$

which has eigenvalue -1 under $\Omega^{(0)}$ and $\Omega^{(\infty)}$. The normalization factor $1/2$ is determined by the asymptotic limit $\eta \to 1$ which should be normalized to $\mathbf{C}^{\mathbf{1}}_{\sigma,\sigma}\mathbf{C}^{\mathbf{1}}_{\mu,\mu} = 1$. The expansion (6.12) gives the necessary information to expand (6.41).

$$G_{\mu,\sigma}(\eta,\bar\eta) = \left(\frac{1}{\eta\bar\eta(1-\eta)(1-\bar\eta)}\right)^{\frac{1}{8}} \left(1 - \frac{1}{4}\left((1-\eta)(1-\bar\eta)\right)^{\frac{1}{2}}\right) \left(1 + \mathcal{O}(1-\eta, 1-\bar\eta)\right) \tag{6.42}$$

The factor $-1/4$ is equal to $\mathbf{C}^{\epsilon}_{\sigma,\sigma}\mathbf{C}^{\epsilon}_{\mu,\mu}$, giving therefore $\mathbf{C}^{\epsilon}_{\mu,\mu} = -\frac{1}{2}$. In the previous sections of this chapter, we computed the square of the expansion coefficients and we arbitrarily chose positive operator product coefficients, leaving a phase freedom undiscussed. The reason is that, as in quantum mechanics, exact phases are in general not important, while relative phases obey very strong constraints. This last calculation shows the importance of relative phases.

We can now discuss the new operators which appears in the limit as $\eta \to 0$ where we have

$$G_{\mu,\sigma}(\eta,\bar\eta) = \frac{1}{2}\left(\frac{1}{\eta\bar\eta}\right)^{\frac{1}{8}} \left(\bar\eta^{\frac{1}{2}} + \eta^{\frac{1}{2}} + \ldots\right) \tag{6.43}$$

The non-local poles are due to the presence of fermionic operators in the OPE $\sigma \cdot \mu$ which can be written

$$\sigma \cdot \mu = \bar\psi + \psi \tag{6.44}$$

where $\psi(z)$ and $\psi(\bar z)$ are the free fermions which build up the energy-density operator. They imply the new expansion coefficients $\mathbf{C}^{\psi}_{\mu,\sigma} = \mathbf{C}^{\bar\psi}_{\mu,\sigma} = 1/\sqrt{2}$. The fermionic operators couple locally with the identity and with themselves. There is therefore a last consistent local operator algebra $\{\mathbf{1}, \psi, \bar\psi, \epsilon\}$ which is nothing but the free fermion theory. The OPEs are:

$$\boxed{\begin{array}{l} \psi \cdot \psi = \mathbf{1} \\ \bar\psi \cdot \bar\psi = \mathbf{1} \\ \psi \cdot \bar\psi = \epsilon \\ \psi \cdot \epsilon = \bar\psi \\ \bar\psi \cdot \epsilon = \psi \end{array}} \tag{6.45}$$

By construction, $\mathbf{C}^{\epsilon}_{\psi,\bar\psi} = -\mathbf{C}^{\epsilon}_{\bar\psi,\psi} = i$. Such an operator spectrum can be found in the Ising model on the torus with mixed periodic and anti-periodic boundary conditions along the two homological cycles, as will be shown in chapter 10. Notice that the largest possible semi-local algebra is naturally given by the operators $\{\mathbf{1}, \psi, \bar\psi, \epsilon, \sigma, \mu\}$. We do not discuss any further the remaining correlation functions which can be deduced from the OPEs and from the correlation functions discussed so far.

7 Coulomb Gas Realization

The Ising model correlation functions are too peculiar and too simple to provide a general and efficient method to compute correlation functions of minimal models. It is however known that many models have strong connections with the Coulomb gas in two dimensions. We will explain here a general method which permits to realize any minimal model as a Coulomb gas constrained by a charge placed at infinity. This chapter intends to be an introduction rather than an exhaustive discussion and the amount of necessary field theory knowledge will be kept at the lowest possible level. The purpose of this chapter is to show how to compute any correlation function of minimal models and to give a general formula for the operator product expansion coefficients. The technique was developed in the remarkable papers of Vl. Dotsenko and Fateev [109, 110].

7.1 Free massless boson gas

The theory of a free massless boson is a somehow ill-defined theory and it is important to specify what is meant in the actual framework. The action of such a theory is defined as

$$S[\varphi] \sim \int dz\, d\bar{z}\ \partial_z \varphi \partial_{\bar{z}} \varphi \tag{7.1}$$

Infrared and ultraviolet problems can be formally resolved by the introduction of a infrared cut-off a and a ultraviolet cut-off R, which of course introduce a (mass) scale in the model. The equation of motion deduced from the action means that the propagator satisfies the equation

$$\partial_z \partial_{\bar{z}} < \varphi(z,\bar{z})\, \varphi(0,0) > \sim \delta^{(2)}(z,\bar{z}) \tag{7.2}$$

The propagator is not defined in the limit where we remove the cut-offs, and breaks the positivity of the two-point function. With the particular normalization chosen in this chapter, we have

$$< \varphi(z,\bar{z})\, \varphi(0,0) > = -4 \ln(|z|/R) = -2 \ln(z/R) - 2 \ln(\bar{z}/R) \tag{7.3}$$

which is nothing else than Coulomb's law in two dimensions, hence the name *Coulomb gas* for this model. A massless bosonic theory is often claimed to be inconsistent because of the fundamental problem of the propagator which gives rise to inconsistencies if we try to compute the correlation function for any functional of the field φ. However, there is a subsector of the model which is well-defined in the asymptotic limit. The simplest operators which are free from the renormalization problems are

$$J = \left(\frac{i}{2}\right) \partial_z \varphi, \qquad \text{and} \qquad \bar{J} = -\left(\frac{i}{2}\right) \partial_{\bar{z}} \varphi. \tag{7.4}$$

Their two-point functions are meromorphic

$$\begin{aligned} < J(z_1,\bar{z}_1)\, J(z_2,\bar{z}_2) > &= \frac{1/2}{(z_1 - z_2)^2} \\ < \bar{J}(z_1,\bar{z}_1)\, \bar{J}(z_2,\bar{z}_2) > &= \frac{1/2}{(\bar{z}_1 - \bar{z}_2)^2} \end{aligned} \tag{7.5}$$

The free boson theory obeys the Wick theorem for the n-point functions, whose calculation can thus be reduced to the calculation of the two-point functions. The fact that the theory is massless and the particular structure of its propagator permits to separate φ into its holomorphic and anti-holomorphic components

$$\varphi(z,\bar{z}) = \varphi(z) + \overline{\varphi}(\bar{z}) \tag{7.6}$$

The two operators J and $\bar{J}$ are respectively holomorphic or antiholomorphic operators with scaling dimensions one and spin ± 1. Both are conserved currents ($\partial_{\bar{z}} J = \partial_z \bar{J} = 0$) which generate $U(1)$ current algebras. Consider first the holomorphic operator $J(z)$. We can repeat the operations made in chapter 2 to derive the Virasoro algebra from the stress-energy tensor T. The moments of the current J are defined as in eq. (2.49).

$$J_n A(z_1,\bar{z}_1) = \frac{1}{2\pi i}\oint_{C_{z_1}} dz\,(z-z_1)^n J(z) A(z_1,\bar{z}_1) \tag{7.7}$$

where C_{z_1} is a closed curve surrounding the point z_1. The OPE of two currents is simpler than in eq. (2.52) for the stress-energy tensor. Due to the Wick theorem, we have

$$J(z')\,J(z) = \frac{1/2}{(z'-z)^2} + \text{regular terms} \tag{7.8}$$

Along the same lines as in chapter 2, we can derive the commutation relations of the so-called $\hat{U}(1)$ *affine Kac-Moody algebra.*

$$[J_n, J_m] = \frac{n}{2}\,\delta_{n+m,0} \tag{7.9}$$

Generally, one can obtain from a Lie algebra $\mathbf{g}$ with commutation relations $[J^a, J^b] = if^{ab}{}_c J^c$ its *affine extension* $\hat{\mathbf{g}}$ of level k or Kac-Moody algebra [149]

$$[J^a_n, J^b_m] = if^{ab}{}_c J^c_{n+m} + kn\delta_{n+m,0} \tag{7.10}$$

where n, m are integers. Kac-Moody algebras are infinite-dimensional extensions of the familiar finite-dimensional simple Lie algebras. The Cartan matrix of a Kac-Moody algebra has to be positive semidefinite, rather than strictly positive definite as required for the finite-dimensional simple Lie algebras. For detailed introductions, see [149, 208]. The currents can be used to express the stress-energy tensor which has the so-called *Sugawara-Sommerfield* form [310, 307]

$$T(z) =: JJ : (z) = -\frac{1}{4} : \partial_z\varphi\partial_z\varphi : (z) = \lim_{\epsilon\to 0}\left(J(z+\epsilon)J(z) - \frac{1/2}{\epsilon^2}\right) \tag{7.11}$$

The *Normal Ordering* :: subtracts the self-interaction of the field φ with itself. This amounts to subtract the singular part in the OPE of J with itself. The OPE of T with J shows that J is primary

$$T(z)J(0) = \frac{1}{z^2}\,J(0) + \frac{1}{z}\,\partial_z J(0) + \text{regular terms} \tag{7.12}$$

The commutation relations between the Virasoro generators L_n with the $\hat{U}(1)$ generators J_n can be computed from OPE (7.8) and definition (7.11).

$$[L_n, J_m] = -m\, J_{n+m} \tag{7.13}$$

The $c = 1$ Virasoro central charge can be read from the OPE of T with itself

$$T(z)T(0) = \frac{1/2}{z^4} + \frac{2}{z^2}\, T(0) + \frac{1}{z}\, \partial_z T(0) + \text{regular terms} \tag{7.14}$$

The current part of the free massless boson theory is built up with the generator of a $\hat{U}(1)$ current which is a Virasoro primary operator. However, we expect at $c = 1$ infinitely many Virasoro primary operators. The Kac Moody algebra will be used to refine the notion of a primary operator. Some Virasoro primary operators will be moreover $\hat{U}(1)$ primary

$$J_n \phi = 0 \qquad \text{if} \qquad n > 0 \tag{7.15}$$

and one requires a primary operator of the theory to be primary with respect to both the Virasoro and the Kac-Moody algebra. From the point of view of representation theory, several irreducible representations of the Virasoro algebra are combined into a single irreducible representation of the two combined algebras. This concludes the discussion of the algebraic part of the model.

$\hat{U}(1)$ primary operators, which are always Virasoro primary, are given by the second kind of operator which can be renormalized to a cut-off independent quantity: the *vertex operator*

$$V_\alpha(z) \sim : e^{i\alpha\varphi(z)} : \tag{7.16}$$

We discuss here only the holomorphic part of the theory because we are not primarily interested in this section by the problem of locality. Conserved quantities generate proper Ward identities and for the current algebra, we have the following relation

$$< J(z) V_{\alpha_1}(z_1) \cdots V_{\alpha_n}(z_n) > = \left(\sum_{i=1}^{n} \frac{\alpha_i}{z - z_i} \right) < V_{\alpha_1}(z_1) \cdots V_{\alpha_n}(z_n) > \tag{7.17}$$

The $U(1)$ *charge* of V_α is α. Conserved quantities must preserve their canonical dimensions at any scale and a consistency condition to have $J(z) \sim 1/z^2$ at large distance is obtained with the condition that the total charge $\alpha_1 + \cdots + \alpha_n$ must be zero for physical correlation functions

$$< V_{\alpha_1}(z_1) \cdots V_{\alpha_n}(z_n) > = 0 \qquad \text{if} \qquad \alpha_1 + \cdots + \alpha_n \neq 0 \tag{7.18}$$

This *neutrality condition* is more traditionally obtained with a careful study of the ultraviolet R dependence of exponentials of the free field φ. Using the definition (7.16) and the Wick theorem, the neutral two-point function is given by

$$< V_\alpha(z)\, V_{-\alpha}(0) > = \frac{1}{z^{2\alpha^2}} \tag{7.19}$$

The conformal weight is $\Delta_\alpha^{free} = \alpha^2$. A general n-point function of Vertex operators is given by the infinite product

$$< V_{\alpha_1}(z_1) \cdots V_{\alpha_n}(z_n) > = \prod_{i<j} (z_i - z_j)^{2\alpha_i \alpha_j}\, \delta_{\alpha_1 + \cdots + \alpha_n, 0} \tag{7.20}$$

This concludes the general discussion of the free massless bosonic theory. Models with $c = 1$ contribute to statistical mechanics systems where a $U(1)$ charge is preserved. Due to the existence of the marginal operator $J\bar{J}$, it is possible to move continuously from a critical model to another critical model with different critical exponents. Such a discussion lies beyond the scope of this chapter, but we shall come back to this in chapters 11 and 12 when discussing several spin systems with central charge $c = 1$ like the XY model or the Ashkin-Teller model.

7.2 Screened Coulomb gas

The free Coulomb gas is defined at the value $c = 1$ of the central charge. The method used to reduce the degrees of freedom consists in a modification of the boundary conditions on the free field φ. Suppose that we introduce a vertex operator with charge $-2\alpha_0$ and conformal weight $\Delta^{free}_{-2\alpha_0} = 4\alpha_0^2$ placed at infinity in each correlator and we take the well-defined limit

$$\ll V_{\alpha_1}(z_1)\cdots V_{\alpha_n}(z_n) \gg = \lim_{R\to\infty} R^{8\alpha_0^2} < V_{\alpha_1}(z_1)\cdots V_{\alpha_n}(z_n)\, V_{-2\alpha_0}(R) > \qquad (7.21)$$

The limit is well-defined because the correlation function on the r.h.s. of (7.21) diverge like $R^{-2\Delta^{free}_{-2\alpha_0}}$ for large R. These new correlation functions have the neutrality condition replaced by

$$\sum_{i=1}^{n} \alpha_i = 2\alpha_0 \qquad (7.22)$$

In particular, the two-point functions of the modified theory have to satisfy this new charge condition

$$\ll V_\alpha(z)\, V_{2\alpha_0-\alpha}(0) \gg = \frac{1}{z^{2\alpha(\alpha-2\alpha_0)}} \qquad (7.23)$$

The conformal weight of the vertex operator V_α or $V_{2\alpha_0-\alpha}$ becomes

$$\Delta_\alpha = \Delta_{2\alpha_0-\alpha} = \alpha(\alpha - 2\alpha_0) \qquad (7.24)$$

The difficult part is now to show that a consistent theory comes out of this constraint. Both operators V_α and $V_{2\alpha_0-\alpha}$ must be shown to be equivalent and to have isomorphic Verma modules. A full discussion would lead to the BRST structure of this model but this introductory chapter is only supposed to give the recipe necessary to compute correlation functions.

The charge at infinity modifies some geometric properties of the vertices but preserves the conformal invariance of the model. The stress-energy tensor must be modified to take the modifications into account. We also want that V_α remains a Virasoro primary operator. The operators of the $c = 1$ theory are sufficient to completely take the effects of the charge at infinity into account and the modified holomorphic part of the stress-energy tensor is

$$T(z) =: JJ : (z) + 2\alpha_0 \partial_z J(z) = -\frac{1}{4} : \partial_z\varphi\partial_z\varphi : (z) + i\alpha_0\partial_z^2\varphi(z) \qquad (7.25)$$

It is straightforward to verify that both V_α and $V_{2\alpha_0-\alpha}$ are primary operators. The new term in the definition of T should not surprise because it is the only other possible conserved quantity with scaling dimension 2 of the theory. The factor $2\alpha_0$ is adjusted to keep the vertex operators primary. The modified T gives rise to a Virasoro algebra with central charge

$$c = 1 - 24\alpha_0^2 \tag{7.26}$$

This is the first spectacular result: the charge at infinity results in a lowering of the central charge. The second spectacular result is the existence of non-trivial *screening current* with conformal weight unity

$$S_\pm(z) = V_{\alpha_\pm}(z) \qquad \text{with} \qquad \alpha_\pm = \alpha_0 \pm \sqrt{\alpha_0^2 + 1} \qquad \text{i.e.} \qquad \Delta_{\alpha_\pm} = 1 \tag{7.27}$$

These currents are used to create charges which can screen the charge at infinity. They are defined by

$$Q_\pm = \frac{1}{2\pi i} \oint_C dz\, S_\pm(z) \tag{7.28}$$

The choices for the contour of integration will be discussed later. The charges have scaling dimension zero and therefore they will not affect the conformal properties of the correlation functions. However, they cause a shift in the charge by a amount of α_+ or α_-. Suppose now that we restrict the subset of permissible vertex operators to the ones whose charge is quantized and belongs to the subset

$$\alpha_{r,s} = \frac{1}{2}(1-r)\,\alpha_+ + \frac{1}{2}(1-s)\,\alpha_- \qquad \text{with} \qquad r, s \in \mathbb{N} \tag{7.29}$$

We introduce as in chapter 4 the parametrization $c = 1 - 6/m(m+1)$ and we compute the conformal weight of $V_{r,s} := V_{\alpha_{r,s}}$ in terms of r, s, and m. We recognize the celebrated Kac formula

$$\Delta_{\alpha_{r,s}} = \Delta_{r,s} = \frac{(r(m+1) - sm)^2 - 1}{4m(m+1)} \tag{7.30}$$

This is the beginning of the connection to the minimal models. For the sake of simplicity, we restrict the discussion to the integer values of m, related to the unitary minimal models. It is straightforward to extend the discussion to rational values $m = (p-q)/q$ where p, q are coprime integers. This latter case represents the space of minimal models. With $m \in \mathbb{N}$, the values of r and s are restricted to $1 \le r \le (m-1)$ and $1 \le s \le m$[7]. We denote $V_{\alpha_{r,s}} = V_{r,s}$ and $V_{2\alpha_0 - \alpha_{r,s}} = V_{-r,-s}$. With this notation, $S_+ = V_{-1,1}$ and $S_- = V_{1,-1}$.

We want now to associate the holomorphic part $\phi_{(r,s)}$ of a primary operator with the vertex operator $V_{r,s}$. By convention we often denote the scalar operator $\phi(z,\bar{z}) \sim \phi_{(r,s)}(z) \otimes \phi_{(r,s)}(\bar{z})$ simply by $\phi_{(r,s)}$. If, however, the spin is not zero and the primary operator is made out of different representations, $\phi(z,\bar{z}) \sim \phi_{(r,s)}(z) \otimes \phi_{(r',s')}(\bar{z})$, we will write $\phi_{(r,s)(r',s')}$. In the Ising model, the fermionic operators should be written $\psi = \phi_{(2,1)(1,1)}$ and $\bar{\psi} = \phi_{(1,1)(2,1)}$, while the energy-density operator can be simply identified by $\varepsilon = \phi_{(2,1)}$. We recall now that the symmetry of the Kac Table imposes $\phi_{(r,s)} = \phi_{(m-r,m+1-s)}$. To associate rigorously vertex operators to unitary primary operators, we

---

[7] In the general case: $1 \le r \le (q-1)$ and $1 \le s \le (p-1)$.

should understand the sequence of relations between $V_{r,s}$ and $V_{-r,-s}$ given by (since $m\alpha_+ + (m+1)\alpha_- = 0$)

$$V_{r,s} \approx V_{m-r,m+1-s} \approx Q_+^m Q_-^{m+1} V_{-r,-s} \approx V_{-r,-s} \tag{7.31}$$

The exact meaning of the symbol $\approx$ requires a long discussion of the BRST structure lying behind the *screened Coulomb gas* [120, 121, 53], see also the review [135]. Instead, we will show how to build consistently the correlation functions and argue about the consistency of this construction.

7.3 Minimal correlation functions

We begin this section by a the calculation of a particular four-point correlation function. We consider the scalar operator $\phi_{(1,2)}$ of any minimal theory, which is known to have a null-vector at level 2. The four-point function

$$< \phi_{(1,2)}(z_4,\bar{z}_4)\phi_{(1,2)}(z_3,\bar{z}_3)\phi_{(1,2)}(z_1,\bar{z}_1)\phi_{(1,2)}(z_2,\bar{z}_2) > \tag{7.32}$$

has both the holomorphic and anti-holomorphic parts constrained by second-order differential operators. Consider first the holomorphic part and with the equivalence $V_{r,s} \equiv V_{-r,-s}$ we investigate

$$V_{1,2}(z_4)V_{1,2}(z_3)V_{-1,-2}(z_1)V_{1,2}(z_2) \tag{7.33}$$

The total charge of this combination is $2\alpha_0 - \alpha_-$. The extra charge $-\alpha_-$ must be screened by a charge Q_-, which will not affect the conformal properties but only balances the charge deficit. Notice that we chose the indices of the vertices in (7.33) to keep the charge deficit as small as possible. Our aim is to establish the relation

$$< \phi_{(1,2)}(z_4)\phi_{(1,2)}(z_3)\phi_{(1,2)}(z_1)\phi_{(1,2)}(z_2) > \sim \ll Q_- V_{1,2}(z_4)V_{1,2}(z_3)V_{-1,-2}(z_1)V_{1,2}(z_2) \gg \tag{7.34}$$

It is now important to specify the contour of integration due to the screening charge Q_-. We write explicitly the r.h.s. of (7.34) using the free Coulomb gas propagators and the definition (7.21).

$$\begin{aligned}
&\ll Q_- V_{1,2}(z_4)V_{1,2}(z_3)V_{-1,-2}(z_1)V_{1,2}(z_2) \gg = \\
&= \frac{1}{2\pi i}\oint_C dz \ll V_{1,-1}(z)V_{1,2}(z_4)V_{1,2}(z_3)V_{-1,-2}(z_1)V_{1,2}(z_2) \gg \\
&= (z_{23}z_{24}z_{34})^{2\alpha_{1,2}^2}\left(z_{12}z_{13}z_{14})\right)^{2\alpha_{1,2}(\alpha_{1,2}-2\alpha_0)}\left(\frac{1}{2\pi i}\right) \\
&\times \oint_C dz\,(z-z_4)^{2\alpha-\alpha_{1,2}}(z-z_3)^{2\alpha-\alpha_{1,2}}(z-z_2)^{2\alpha-\alpha_{1,2}}(z-z_1)^{2\alpha-(2\alpha_0-\alpha_{1,2})}
\end{aligned} \tag{7.35}$$

As in the previous chapters, we consider instead the function of the projective invariant parameter $\eta = (z_1 - z_2)(z_3 - z_4)/(z_3 - z_2)(z_1 - z_4)$ defined in the usual limit given in eq. (4.44). We obtain a term proportional to

$$[\eta(1-\eta)]^{-2\Delta_{1,2}} \int_C dt\,(t-1)^{2\alpha-\alpha_{1,2}} t^{2\alpha-\alpha_{1,2}} (t-\eta)^{2\alpha-(2\alpha_0-\alpha_{1,2})} \tag{7.36}$$

We know from the last chapters how the differential equation for such two-dimensional family of functions reduces to an hypergeometric differential equation. The integral of eq. (7.36) is the general ***Euler integral representation*** of the hypergeometric functions [308]. In particular, the hypergeometric function ${}_2F_1(a,b,c;\eta)$ can be calculated with the integral

$$ {}_2F_1(a,b,c;\eta) = \frac{\Gamma(c)}{\Gamma(b)\Gamma(c-b)} \int_1^\infty dt\,(t-1)^{c-b-1} t^{a-c} (t-\eta)^{-a} \tag{7.37} $$

The choice of the contour determines which solution we consider. If we choose $C = [1,\infty[$, the integral is regular in η in the vicinity of zero. However, if we take $C = [0,\eta]$, the integral is singular and behaves like $\eta^{1+4\alpha-\alpha_0} = \eta^{\Delta_{1,3}}$. This can be seen after the appropriate change of variable $t = t'\eta$. We obtain

$$ \eta^{\Delta_{1,3}} \int_0^1 dt'\,(\eta t'-1)^{2\alpha-\alpha_{1,2}} t'^{2\alpha-\alpha_{1,2}} (t'-1)^{2\alpha-(2\alpha_0-\alpha_{1,2})} \tag{7.38} $$

The integral is now regular at $\eta = 0$. The two first choices $C = [1,\infty[$ and $C = [0,\eta]$ define the basis of solutions which diagonalize the monodromy transformation $\Omega^{(0)}$. This shows that the OPE is given by

$$ \phi_{(1,2)} \cdot \phi_{(1,2)} = \phi_{(1,1)} + \phi_{(1,3)} = \mathbf{1} + \phi_{(1,3)} \tag{7.39} $$

This OPE and its corresponding singular behaviours could be calculated from the differential equation given by the level 2 null-vector and along Riemann's method. We made several calculations of this type in the last chapter, so we will not repeat them here. The uniqueness of the possible second-order differential equation with a given singular behaviour shows that the Coulomb gas realization discussed here makes sense.

The solutions giving the conformal blocks in the t-channel are obtained with the linearly related choice of contours $C =]-\infty, 0]$ and $C = [\eta, 1]$. The two last possible choices $C = [0,1]$ and $C = [\eta, \infty[$ diagonalize $\Omega^{(\infty)}$ and give the u-channel.

A discussion of the singular behaviours related to specific choices of contours can be found in any good table of integrals. Therefore we do not discuss any further this example, but we rather make a general discussion on the construction of the general four-point function of scalar operators

$$ < \phi_{(r_4,s_4)}(z_4,\bar{z}_4)\phi_{(r_3,s_3)}(z_3,\bar{z}_3)\phi_{(r_1,s_1)}(z_1,\bar{z}_1)\phi_{(r_2,s_2)}(z_2,\bar{z}_2) > \tag{7.40} $$

A vertex realization requires to replace each operator $\phi_{(r,s)}$ either by $V_{r,s}$ or $V_{-r,-s}$. The charge has to be balanced by a suitable number of screening charges, say r charges Q_+ and s charges Q_-, and for example we can calculate

$$ \ll Q_+^r Q_-^s V_{r_4,s_4}(z_4) V_{r_3,s_3}(z_3) V_{-r_1,-s_1}(z_1) V_{r_2,s_2}(z_2) \gg \tag{7.41} $$

The contours must be consistently ordered to give a well-defined basis of solutions. The number of screening charges depends on the choice of vertices, $V_{r,s}$ or $V_{-r,-s}$. This means that we have several possible contour integral realizations for the same correlation function. This reflects the fact that a primary operators has many null-vectors, which define many compatible differential equations for the same conformal blocks. The discussion of the compatibility of the different contour integral realizations has been discussed by many authors. We do not enter into more detail. Notice that there are infinitely many contour integral realizations due to the fact that

$$m\alpha_+ + (m+1)\alpha_- = 0 \tag{7.42}$$

and therefore the operator $Q_+^m Q_-^{m+1}$ can be added as many times as wanted into any correlation function. It is neutral with respect to both the conformal algebra and the charge balance.

The main interest of the contour integral representation is that it provides a tool to compute the monodromy matrices, and therefore the expansion coefficients. In any expression, which involves one or several contours, we have for each contour two independent choices for the path of integrations. The first two ones discussed in this section gave the singular behaviour at the origin. The other ones can be obtained by deformation of the former ones. This technical question is simple to implement once we know how to deform contours for a single integral.

7.4 Minimal algebras and OPE coefficients

In this last section, we would like to survey some general results. With the Coulomb gas approach, we can study the solution to the monodromy problem and extract fundamental closed operator algebras. The simplest case is when all operators are spinless (scalars). The OPE of the operators $\{\phi_{(r,s)}(z,\bar{z})\}$ is given by

$$\phi_{(r_1,s_1)} \cdot \phi_{(r_2,s_2)} = \sum_{\substack{r_3=|r_1-r_2|+1,\, r_3=(r_1+r_2-1)\,\mathrm{mod}(2) \\ s_3=|s_1-s_2|+1,\, s_3=(s_1+s_2-1)\,\mathrm{mod}(2)}}^{\substack{\min(r_1+r_2-1,\,2m-r_1-r_2-1) \\ \min(s_1+s_2-1,\,2m-s_1-s_2+1)}} \phi_{(r_3,s_3)} \tag{7.43}$$

The indices r_3 and s_3 run from their lower bounds to their upper ones with the condition that they have the same parity as r_1+r_2-1 and s_1+s_2-1 respectively. The largest complete operator algebra is given by the set of all scalar operators $\{\phi_{(r,s)}(z,\bar{z})\}$. This algebra will be called in chapter 11 the A-algebra. However, this algebra contains several closed subalgebras like the one generated by the two operators $\{\phi_{(1,1)}, \phi_{(m-1,1)}\}$. Another interesting closed subalgebra is generated by the odd-indexed operators $\{\phi_{(r,s)} | r, s \text{ both odd}\}$.

The operator product coefficients are rather complicated and there is no simple way to express them. It is however useful to give explicitly the full formula. The **non-zero** operator product coefficients of the OPE (7.43) are [111]

$$[\mathbf{C}^{(r_3,s_3)}_{(r_1,s_1),(r_2,s_2)}]^2 = \frac{D^{(r_3,s_3)}_{(r_1,s_1),(r_2,s_2)} D^{(r_2,s_2)}_{(r_1,s_1),(r_3,s_3)}}{D^{(1,1)}_{(r_1,s_1),(r_1,s_1)}} \tag{7.44}$$

where the coefficients D are defined by

$$D^{(r_3,s_3)}_{(r_1,s_1),(r_2,s_2)} = \rho^{4(s'-1)(r'-1)} \prod_{j=1}^{r'-1} \prod_{i=1}^{s'-1} \frac{1}{(i-\rho j)^2} \prod_{j=1}^{r'-1} \frac{\Gamma(j\rho)}{\Gamma(1-j\rho)} \prod_{j=1}^{s'-1} \frac{\Gamma(j\rho')}{\Gamma(1-j\rho')}$$

$$\times \prod_{j=0}^{r'-2} \prod_{i=0}^{s'-2} \left(\frac{1}{[(s_1-1-i)-\rho(r_1-1-j)]} \right.$$

$$\left. \frac{1}{[(s_2-1-i)-\rho(r_2-1-j)][(s_3+1+i)-\rho(r_3+1+j)]} \right)^2$$

$$\times \prod_{i=0}^{s'-2} \frac{\Gamma(-\rho'(s_1-1-i)+r_1)\Gamma(-\rho'(s_2-1-i)+r_2)\Gamma(\rho'(s_3+1+i)-r_3)}{\Gamma(1+\rho'(s_1-1-i)-r_1)\Gamma(1+\rho'(s_2-1-i)-r_2)\Gamma(1-\rho'(s_3+1+i)+r_3)}$$

$$\times \prod_{j=0}^{r'-2} \frac{\Gamma(-\rho(r_1-1-j)+s_1)\Gamma(-\rho(r_2-1-j)+s_2)\Gamma(\rho(r_3+1+j)-s_3)}{\Gamma(1+\rho(r_1-1-j)-s_1)\Gamma(1+\rho(r_2-1-j)-s_2)\Gamma(1-\rho(r_3+1+j)+s_3)} \tag{7.45}$$

The few new parameters are

$$r' = \frac{1}{2}(r_1+r_2-r_3+1), \qquad s' = \frac{1}{2}(s_1+s_2-s_3+1),$$
$$\rho = \alpha_+^2 = \frac{m+1}{m}, \qquad \rho' = \alpha_-^2 = \frac{m}{m+1} \tag{7.46}$$

and $\Gamma(n)$ is the Euler Γ-function. The convention on the product is that if $r' < 2$ or $s' < 2$, then $\prod_{i=0}^{-1} h(i) = 1$, whatever $h(i)$ is. Notice that we have to decide carefully if we take the indices of $\phi_{(r,s)}$ or those of $\phi_{(m-r,m+1-s)}$. The reason is that r' and s' must be integers. In practice, this does not impede the calculation of the operator product coefficients. The expansion coefficents enjoy several symmetry properties. Although not obvious from the previous formulas, they are fully symmetric in the indices

$$\mathbf{C}_{(r_1,s_1),(r_2,s_2)}^{(r_3,s_3)} = \mathbf{C}_{(r_1,s_1),(r_3,s_3)}^{(r_2,s_2)} = \mathbf{C}_{(r_2,s_2),(r_1,s_1)}^{(r_3,s_3)} \tag{7.47}$$

and for rational values of m, they have the property

$$\mathbf{C}_{(m-r_1,m+1-s_1),(r_2,s_2)}^{(m-r_3,m+1-s_3)} = \mathbf{C}_{(r_1,s_1),(r_2,s_2)}^{(r_3,s_3)} \tag{7.48}$$

Software packages for the calculation of the OPE coefficients are available, see [314].

Let us finish with some comments on non-scalar operators. Algebras with integer spin operators are a consequence of the fact that the following OPE are identical

$$\phi_{(r_1,s_1)(r_1,s_1)} \cdot \phi_{(r_2,s_2)(r_2,s_2)} = \phi_{(r_1,m+1-s_1)(r_1,m+1-s_1)} \cdot \phi_{(r_2,m+1-s_2)(r_2,m+1-s_2)} \tag{7.49}$$

Natural non-scalar operators are given by $\phi_{(r,m+1-s)(r,s)}$ because their spins are always integer or half-integer.

$$\Delta_{r,m+1-s} - \Delta_{r,s} = \frac{(m-2r)(m+1-2s)}{4} \tag{7.50}$$

If for example $m = 1 \operatorname{mod}(4)$, then there is a closed local algebra made out of the operators $\{\phi_{(r,s)} | r, s \text{ odd}\} \cup \{\phi_{(r,m+1-s)(r,s)} | r, s \text{ odd}\}$. This algebra will appear in chapter 11 as *D*-algebra. The OPEs between non-scalar operators are directly comparable with those for their scalar relatives.

$$\phi_{(r_1,m+1-s_1)(r_1,s_1)} \cdot \phi_{(r_2,m+1-s_2)(r_2,s_2)} = \phi_{(r_1,s_1)(r_1,s_1)} \cdot \phi_{(r_2,s_2)(r_2,s_2)} \tag{7.51}$$

The non-scalar operators imply a $\mathbf{Z}_2$ grading of the operator algebra. Such algebras often implies the doubling of some scalar operators. In the particular example of the three-state Potts model ($m = 5$), the Kac Table 5 shows that the D-algebra contains two operators with highest weights $(2/3, 2/3)$ and two operators with highest weights $(1/15, 1/15)$. In this example, we did not consider the half-integer spin operators given by $\{\phi_{(m-r,s)(r,s)} | r, s \,\text{even}\}$. Closed operator algebras can be obtained with such operators if one imposes the invariance of the correlation functions under a subgroup of the monodromy group. This was illustrated in the case of the Ising model in chapter 6.

It is finally important to notice that the equivalence of the OPEs of two non-scalar operators and of their two scalar relatives does not imply the equality of their expansion coefficients. We have seen that in the Ising model where we relaxed the condition of strict locality. In that example, we encountered fermion representations. The energy-density operator $\varepsilon = \phi_{(2,1)(2,1)}$ was in the local sector and the two fermions $\psi = \phi_{(2,1)(1,1)}$ and $\bar{\psi} = \phi_{(1,1)(2,1)}$ in the semi-local sector. The OPE of the fermions were a direct consequence of the OPE of the energy-density operator. However, the numerical values of the OPE coefficients were different.

8 The Hamiltonian Limit and Universality

We have already discussed in chapter 1 the close analogy between statistical mechanics and field theory. Reinterpreting the transfer matrix as an Euclidean time evolution operator, we were led to define a corresponding quantum Hamiltonian. While in general, this will have a rather complicated form, there is a limit procedure, called the *Hamiltonian limit*, to simplify the structure of the transfer matrix considerably without affecting the universal properties of the critical behaviour under study [311, 129]. As compared to the isotropic transfer matrix, the resulting quantum Hamiltonian is much sparser which is useful for numerical investigations. At the same time, we also show, in the context of the Ising model, that this way the universality between different model realizations of systems in the same universality class becomes explicit in yielding the same quantum Hamiltonian. For reviews, see [227, 177].

Yet further aspects of universality become apparent when examining the Temperley-Lieb algebras and the Landau-Ginzburg classification of conformal models.

8.1 Hamiltonian limit in the Ising model

Having formulated the problem of describing a statistical system in terms of its **quantum** Hamiltonian in chapter 1, we now give an example how the Hamiltonian limit can be taken in practice. Consider the $2D$ Ising model with the classical Hamiltonian

$$\mathcal{H} = -\sum_{r,t} \{J_x \cos(\Theta_{r,t} - \Theta_{r+1,t}) + J_\tau \cos(\Theta_{r,t} - \Theta_{r,t+1})\} \tag{8.1}$$

where the lattice variable Θ can take the values 0 or π. This can be easily rewritten with the ordinary Ising spins $\sigma = \pm 1$, but the formulation chosen here is more readily generalizable to more general systems like a Potts model. We explicitly allow for distinct couplings J_x, J_τ between the Ising spins in the "time" and "space" directions. Take a square lattice with $N \times T$ sites, where N is the number of "spatial" and T the number of "temporal" sites. Then the partition function is (periodic boundary conditions in time direction are understood)

$$\mathcal{Z} = \operatorname{tr} \mathcal{T}^T \tag{8.2}$$

where $\mathcal{T} = \exp(-\tau H)$ is the transfer matrix introduced in chapter 1, see eq. (1.70). We now want to take the limit $\tau \longrightarrow 0$ in a peculiar way. We have

$$\mathcal{T} = e^{-\tau H} = 1 - \tau H + \mathcal{O}(\tau^2) \tag{8.3}$$

$\mathcal{T}$ is a $2^N \times 2^N$ matrix and acts between pairs of configurations $\{\Theta_i\}$, $\{\tilde{\Theta}_i\}$, with $i = 1, \ldots, N$.

For the **diagonal elements**, let $\Theta_i = \tilde{\Theta}_i$ for all $i = 1, \ldots, N$. Then

$$\begin{aligned}\mathcal{T}_{\text{diag}} &= \exp \beta J_x \sum_{i=1}^{N} \cos(\Theta_i - \Theta_{i+1}) \\ &= 1 + \tau\lambda^{-1} \sum_{i=1}^{N} \cos(\Theta_i - \Theta_{i+1}) + \mathcal{O}(\tau^2)\end{aligned} \tag{8.4}$$

where

$$\beta J_x = \tau \lambda^{-1} \tag{8.5}$$

Consider now the **first off-diagonal element**. Choose k such that $\Theta_k \neq \tilde{\Theta}_k$ and the rest of the two configurations is equal: $\Theta_i = \tilde{\Theta}_i$ for all $i \neq k$. We must have $\Theta_k - \tilde{\Theta}_k = \pm\pi$. Then the first class of off-diagonal elements is

$$\begin{aligned} \mathcal{T}_{\text{off},1} &= e^{-2\beta J_\tau} \cdot [\text{ diagonal terms }] \\ &= e^{-2\beta J_\tau} (1 + \mathcal{O}(\tau)) \\ &= \tau + \mathcal{O}(\tau^2) \end{aligned} \tag{8.6}$$

where we have put

$$e^{-2\beta J_\tau} = \tau \tag{8.7}$$

Consider the **second off-diagonal element**. Choose two distinct k_1, k_2 such that the corresponding lattice variables are distinct, while all the others are equal:

$$\Theta_{k_a} \neq \tilde{\Theta}_{k_a} \text{ if } a = 1,2 \quad \text{and } \Theta_i = \tilde{\Theta}_i \text{ for all } i \neq k_a \tag{8.8}$$

Then the corresponding off-diagonal elements of $\mathcal{T}$ are

$$\mathcal{T}_{\text{off},2} = e^{-2(2\beta J_\tau)} (1 + \mathcal{O}(\tau)) = \mathcal{O}(\tau^2) \tag{8.9}$$

Similarly, one sees that the nth **off-diagonal element** is

$$\mathcal{T}_{\text{off},n} = e^{-n(2\beta J_\tau)} (1 + \mathcal{O}(\tau)) = \mathcal{O}(\tau^n) \tag{8.10}$$

provided that the direction-dependent couplings J_x and J_τ are chosen according to eqs. (8.5) and (8.7). This gives the relation

$$e^{-2\beta J_\tau} = \lambda \beta J_x \tag{8.11}$$

where λ is a free parameter.

We now let $\tau \longrightarrow 0$. If eqs. (8.5) and (8.7) are satisfied, the resulting *quantum Hamiltonian* only contains, besides diagonal terms, contributions which involve a **single** spin flip. This is a great simplification as compared to the usual transfer matrix $\mathcal{T}$, which in general contains terms involving up to N simultaneous spin flips. So H is a much sparser matrix and should be expected to be easier to treat, in particular in the context of numerical diagonalization. To achieve this, one is forced into an extremely **anisotropic** limit

$$\beta J_x \longrightarrow 0 \quad , \quad \beta J_\tau \longrightarrow \infty \tag{8.12}$$

but such that the parameter λ of eq. (8.11) remains fixed. This limit is called the *Hamiltonian limit.* Recalling the results of chapter 1, the $\tau \longrightarrow 0$ limit can also be understood as a continuum limit in the time direction, while the discrete lattice in the space directions is retained. If the system under study has an isotropic RG fixed point with sufficiently local interactions, taking the Hamiltonian limit the full information about the universal critical behaviour of the system is preserved. On the other hand, due to the freedom in defining λ in eq. (8.11), the normalization of H is arbitrary. For conformal invariance applications, we have to fix this normalization factor in a sensible way.

We proceed to write down H explicitly. Introduce an operator $\hat{\Theta}_n$ with eigenvalues $\Theta_n = 0, \pi$ and eigenstates $|\Theta_n\rangle$ and a conjugate operator $\hat{p}_n$

$$\begin{aligned} \hat{\Theta}_n |\Theta_n\rangle &= \Theta_n |\Theta_n\rangle \\ \exp(im\,\hat{p}_n)\,|\Theta_n\rangle &= |\Theta_n + m\pi\rangle \end{aligned} \tag{8.13}$$

where m is an integer. Then

$$H = -\sum_{n=1}^{N} \left[\lambda^{-1}\cos\left(\hat{\Theta}_n - \hat{\Theta}_{n+1}\right) + 2\cos\hat{p}_n\right] \tag{8.14}$$

This can be alternatively expressed in terms of Pauli matrices

$$\begin{aligned} \sigma^z(n) &= \exp i\hat{\Theta}_n = \begin{pmatrix} 1 & 0 \\ 0 & -1 \end{pmatrix} \\ \sigma^+(n) &= \exp i\hat{p}_n = \begin{pmatrix} 0 & 1 \\ 0 & 0 \end{pmatrix} \end{aligned} \tag{8.15}$$

Writing $\sigma^{\pm} = (\sigma^x \pm i\sigma^y)/2$, we have $2\cos\hat{p}_n = \sigma^x(n)$ and the quantum Hamiltonian of the $2D$ Ising model becomes

$$H = -\sum_{n=1}^{N} \left[t\sigma^x(n) + \sigma^z(n)\sigma^z(n+1)\right] \tag{8.16}$$

This Hamiltonian will be one of the main objects of study. One often refers to H as the quantum Ising chain or the $(1+1)D$ Ising model. It will serve as a starting point for the application of conformal invariance in the $2D$ Ising model. The *transverse field* $t = \lambda^{-1}$ can be interpreted in two different ways:

1. when ground state properties of H are studied, t corresponds to the temperature of the $2D$ classical Ising spin model.
2. when H is used in turn to construct Boltzmann weights for a $1D$ quantum chain, t becomes a transverse magnetic field, which does not break the global symmetry of the model.

The correspondence between the eigenvalues and eigenvectors of H with the quantities of interest in statistical mechanics is clear from Tables 2 and 3. In particular, the E_i are just the eigenvalues of H and the inverse correlation lengths are simply given by the energy gaps. This is very convenient, because of eq. (3.49) the correlation lengths are those quantities which are the easiest to describe from the point of view of conformal invariance. The generalization to more than one space dimension or to more general models is straightforward. Before analysing the spectrum of H in detail, we shall give a second method to derive H.

8.2 Hubbard-Stratonović transformation

There are many possible realizations of a given universality class. As a further illustration of the correspondence between statistical mechanics and field theory we give a reformulation of the Ising model in terms of a scalar ϕ^4 field theory, following [G1].

Consider a d-dimensional hypercubic lattice with $\mathcal{N}$ sites. The Ising model partition function can be written as

$$\mathcal{Z} = \frac{1}{2^{\mathcal{N}}} \sum_{\{\sigma\}} \exp \sum_{(a,b)} K_{ab}\, \sigma_a \sigma_b \ , \quad \sigma_a^2 = 1 \tag{8.17}$$

where K is a matrix of coupling constants. Define another matrix

$$P = P_0 \cdot \mathbf{1} + K \tag{8.18}$$

where P_0 is some constant. If one chooses P_0 such that $P_0 \geq \max_a \sum_b K_{ab}$, then P is positive definite. Define a vector of spin variables $\vec{\sigma} = (\sigma_1, \ldots, \sigma_{\mathcal{N}})$. Then the partition function is

$$\mathcal{Z} = e^{-\mathcal{N}P_0/2}\, 2^{-\mathcal{N}} \sum_{\{\sigma\}} \exp \frac{1}{2} \vec{\sigma} P \vec{\sigma} \tag{8.19}$$

Now, consider a quadratic form of continuous variables y_a

$$Q(y) := \sum_{a,b=1}^{\mathcal{N}} Q_{ab}\, y_a y_b \tag{8.20}$$

and recall the identity

$$\int_{-\infty}^{\infty} d^{\mathcal{N}}y \, \exp\left(-\frac{1}{2}\vec{y}Q\vec{y}\right) = (2\pi)^{\mathcal{N}/2} (\det Q)^{-1/2} \tag{8.21}$$

Under a change of variables $\vec{y} = \vec{s} + Q^{-1}\vec{\sigma}$, one has

$$Q(y) = \vec{s}Q\vec{s} + 2\vec{s}\vec{\sigma} + \vec{\sigma}Q^{-1}\vec{\sigma} \tag{8.22}$$

If Q is chosen such that

$$Q = P^{-1} = (P_0 \cdot 1 + K)^{-1} \tag{8.23}$$

one sees from eq. (8.21)

$$(2\pi)^{\mathcal{N}/2} (\det P)^{1/2} = \exp\left(-\frac{1}{2}\vec{\sigma}P\vec{\sigma}\right) \int_{-\infty}^{\infty} d^{\mathcal{N}}s \, \exp\left(-\frac{1}{2}\vec{s}Q\vec{s} - \sum_{a=1}^{\mathcal{N}} s_a \sigma_a\right) \tag{8.24}$$

and consequently

$$\mathcal{Z} = e^{-f_0} 2^{-\mathcal{N}} \sum_{\{\sigma\}} \int_{-\infty}^{\infty} d^{\mathcal{N}}s \, \exp\left[-\sum_{(a,b)} Q_{ab} s_a s_b - \frac{1}{2}\sum_a Q_{aa} s_a^2 - \sum_a s_a \sigma_a\right] \tag{8.25}$$

where $e^{-f_0} = e^{-\mathcal{N}P_0/2}(2\pi)^{-\mathcal{N}/2}(\det P)^{-1/2}$ is some smooth background term. The sum can now be carried out and one finally obtains

$$\mathcal{Z} = e^{-f_0} \int_{-\infty}^{\infty} d^{\mathcal{N}}s \, \exp\left[-\sum_{(a,b)} Q_{ab}s_a s_b - \sum_a \left(\frac{1}{2} Q_{aa} s_a^2 - \ln\cosh s_a \right) \right] \tag{8.26}$$

The exponential in eq. (8.26) contains two terms. The first one describes the interactions between the continuous spin variables $\vec{s}$. These are related to the original interaction matrix K

$$-P_0^2 Q_{ab} = -P_0 \cdot 1 + K - P_0^{-1}K^2 + P_0^{-2}K^3 + \cdots \tag{8.27}$$

The second term in the exponential is a weight factor $w(s_a)$ for the continuous spins

$$w(s_a) = -\frac{1}{2}(Q_{aa} - 1)s_a^2 - \frac{1}{12}s_a^4 + \cdots \tag{8.28}$$

We now make the connection with a scalar continuum ϕ^4 theory. Let K describe nearest neighbor interactions. Then, from eq. (8.27) the higher powers of K, which do contribute to Q and describe further interactions terms of longer range, become increasingly suppressed with powers of P_0. Therefore the presence of these terms will not change the predominantly ferromagnetic character of the interactions and they will be irrelevant. Going to a continuum scalar field $\phi(x)$, the interactions terms due to K will create a term $\sim (\partial_\mu \phi)^2$, while the other terms, due to the higher powers of K, will create higher derivatives. At least close to the upper critical dimension $d^* = 4$, standard power counting arguments can be used to show that these further contributions are irrelevant. The same is true for the higher powers of $s_a \sim \phi(x)$ neglected in the weight factor. One thus expects that the critical behaviour of the model should be the same as the one of a scalar ϕ^4 theory with a generating function

$$\mathcal{Z} \sim \int \mathcal{D}\phi \, e^{-\int dx \, \mathcal{L}} \quad \text{where}$$

$$\mathcal{L} = \frac{1}{2}(\partial_\mu \phi)^2 + \frac{m^2}{2}\phi^2 + \frac{g}{4!}\phi^4 \tag{8.29}$$

is the Lagrangian. Of course, the last step in the derivation going from eq. (8.26) to (8.29) is at best heuristic. However, in the next section we shall see that even for $d = 2$, we recover from (8.29) the same quantum Hamiltonian as obtained for the Ising spin system.

8.3 Hamiltonian limit of the scalar ϕ^4 theory

We now pass again from the Lagrangian formulation of the ϕ^4 theory as given in (8.29) to the Hamiltonian formulation, following the original paper of Brower, Furman and Subbarao [59]. We discretize the space direction(s), but the time direction is continuous. For notational simplicity, we take $d = 2$. From a Legendre transformation one obtains in the canonical way the Hamiltonian

$$H = : \sum_n \left[\frac{1}{2}\pi^2(n) + V(\phi(n)) + \lambda \left(\phi(n+1) - \phi(n)\right)^2 \right] : \tag{8.30}$$

where $\pi(n)$ is the canonically conjugate momentum to $\phi(n)$

$$\pi^2(n) = -\frac{\partial^2}{\partial\phi^2(n)} \tag{8.31}$$

and the potential $V(\phi) = \frac{1}{2}m^2\phi^2 + \frac{1}{4!}g\phi^4$. For $m^2 < 0$ and $g > 0$, V has the well-known double-well structure (see Figure 23), where the two minima correspond to the Ising states $s = \pm 1$.

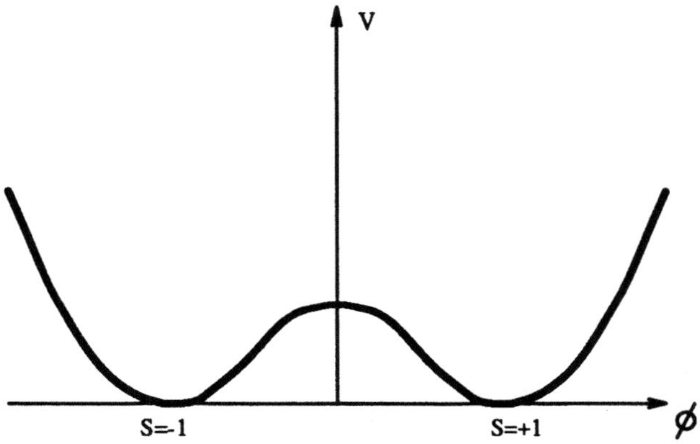

Fig. 23. Effective potential $V(\phi)$.

First, consider the case $\lambda = 0$, restricting attention to a single-site Hamiltonian. Then one has a $1D$ Schrödinger equation for an anharmonic oscillator. In the limit $m^2 \to -\infty$, the two lowest levels become (almost) degenerate, with an exponentially small splitting $\sim \exp m^2$. There is a large gap with all other levels. The main contribution to $\mathcal{Z}$ comes only from two eigenstates, which are concentrated around the Ising values $s = \pm 1$.

Next, reintroduce a small intersite coupling $\lambda \geq 0$. Then the two lowest eigenstates will become degenerate if $-m^2$ is large enough. Since we are only interested in the critical behaviour, which is dominated by the lowest eigenvalues of H, it is sufficient to *truncate the Hilbert space just to these two lowest lying states*! Let ε_0 and ε_1 be the energies of these two states and let

$$|\varepsilon_0\rangle = \begin{pmatrix} 0 \\ 1 \end{pmatrix} \quad \text{and} \quad |\varepsilon_1\rangle = \begin{pmatrix} 1 \\ 0 \end{pmatrix} \tag{8.32}$$

be the corresponding eigenstates. Due to the truncation, H will contain only diagonal terms and those corresponding to a single "flip", i.e. a tunneling between $|\varepsilon_0\rangle$ and $|\varepsilon_1\rangle$,

$$H = H_{diag} + H_{flip} \tag{8.33}$$

Writing $\varepsilon = \varepsilon_1 - \varepsilon_0$, the diagonal term is

$$H_{diag} = \sum_n \left[\frac{\varepsilon}{2}\left(1 + \sigma^z(n)\right) + \varepsilon_0\right] \tag{8.34}$$

where σ^z is a Pauli matrix. For the flip term, observe that the tunneling between the two states is given by an operator with elements $\langle\varepsilon_i|\phi(n)|\varepsilon_j\rangle$, which vanishes if $i = j$. Therefore the tunneling operator must be, in the basis given by eq. (8.32), proportional to σ^x. Using the identity

$$(\phi(n+1) - \phi(n))^2 = \phi^2(n+1) + \phi^2(n) - 2\phi(n+1)\phi(n) \tag{8.35}$$

and from eq. (8.30) we have

$$H = \sum_n \left[\frac{\varepsilon}{2}(1 + \sigma^z(n)) + \Lambda(1 - \sigma^x(n+1)\sigma^x(n))\right] \tag{8.36}$$

with $\Lambda = 2\lambda\,|\langle\varepsilon_0|\phi(n)|\varepsilon_1\rangle|^2$. After a change of basis $\sigma^x \to \sigma^z, \sigma^z \to -\sigma^x$ and setting $t = \varepsilon/(2\Lambda)$, the Hamiltonian is indeed, up to normalisation and an additive constant, identical to the $(1+1)D$ Ising model Hamiltonian eq. (8.16).

We remark that the derivation of the Hamiltonian limit uses features analogous to the Ising spin model derivation.

1. One has to take the simultaneous limit $\lambda \to 0, m^2 \to -\infty$, such that t is kept fixed (compare with eq. (8.11)).
2. Make the truncation to single-flip terms.
3. This is justified because the minima of $V(\phi)$ are very deep and the critical behaviour of the model only depends on the lowest eigenvalues.

We summarize the relationships obtained in Figure 24, which gives a commutative diagram. Therefore, using the quantum Hamiltonian formulation is a convenient technique to describe the long-range correlations of both the classical Ising spin system and the ϕ^4 field theory. The close connection between these two systems is in fact much more general and gives rise to the Landau-Ginzburg classification briefly described below. Figure 24 illustrates again the universality properties of the critical point spectrum of the quantum Hamiltonian H, see also eq. (8.37) below. In particular, it has become apparent that the critical behaviour of the system is quite independent of the detailed form of the interaction potential.

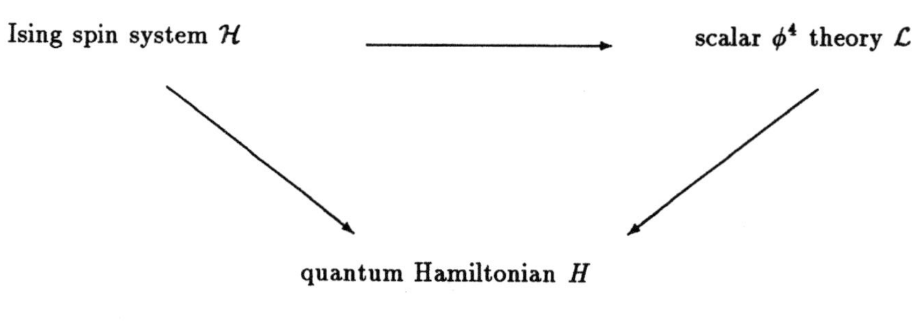

Fig. 24. Relationship between systems in the Ising universality class.

8.4 Hamiltonian spectrum and conformal invariance

We have seen that the conformal invariance results eqs. (3.49) and (3.54) were derived in the strip geometry with periodic boundary conditions. Taking the Hamiltonian limit, however, automatically involves going to the strip geometry. Therefore, with the normalization of H properly chosen, we can directly relate the entire spectrum of H, calculated precisely at the critical point $T = T_c$, to universal exponents x_i and the central charge c. On the other hand, we have already seen that the exponents x_i were obtained as eigenvalues of the states of the Verma module under the Virasoro generators $L_0, \bar{L}_0$. Consequently, the quantum Hamiltonian (with periodic boundary conditions!) can be written as being **a part of the Virasoro algebra** (2.56)

$$H = \frac{2\pi}{N}\left(L_0 + \bar{L}_0\right) - \frac{\pi c}{6}\frac{1}{N} + \text{ const.} \tag{8.37}$$

where the transverse width of the strip is $L = Na$ and a is the spatial lattice constant.

Eq. (8.37) can be used in two ways. First, one can use information on the critical spectrum of H to obtain representations of the Virasoro algebra. Conversely, knowing the Virasoro algebra representations, the **entire** critical point spectrum of H can be computed. Conformal invariance acts therefore as a *dynamical symmetry* of a critical statistical system. In particular, eq. (8.37) gives a sufficient criterion to establish the conformal invariance of the critical point of a given two-dimensional lattice Hamiltonian.

8.5 Temperley-Lieb algebra

In the recent years, several methods were developed to provide exact algebraic approaches to integrable critical or off-critical models. We will show here the particular *Temperley-Lieb algebra* which characterizes many important models. We take the example of the p-state Potts model as an introduction to more general examples. Further details can be found in [313, 267]

It is useful to start with the anisotropic p-state Potts model with different coupling constants along the horizontal "space" direction and the vertical "time" direction. Periodic boundary conditions are assumed. The partition function can be written

$$\mathcal{Z} = \sum_{\{\sigma\}} \exp\left(\sum_{<i,j>_\tau} J_\tau \delta_{\sigma_i,\sigma_j} + \sum_{<i,j>_x} J_x \delta_{\sigma_i,\sigma_j}\right) \tag{8.38}$$

where $< \ >_{\tau,x}$ stands for nearest neighbours in the time and space directions and we have absorbed the temperature into the coupling constants J_x, J_τ. The lattice is chosen to have T temporal rows and N spatial sites. The site variable σ_i takes the values $0, \ldots, p-1$.

A standard way to write the partition function is to introduce the transfer matrix. We decompose the sum in the exponent of (8.38) into parts which take the spatial and temporal dependences into account. Let $\sigma = \{\sigma_1, \ldots, \sigma_N\}$ stand for a configuration of spin in the particular row t. The set of all such configurations spans a p^N- dimensional space. We write $\sigma' = \{\sigma'_1, \ldots, \sigma'_N\}$ for a configuration of the upper adjacent row $t+1$. We define

$$V_{\sigma,\sigma'} = \exp\left(J_x \sum_{i=1}^{N} \delta_{\sigma_i,\sigma_{i+1}}\right) \prod_{j=1}^{N} \delta_{\sigma_j,\sigma'_j}$$

$$W_{\sigma,\sigma} = \exp\left(J_\tau \sum_{i=1}^{N} \delta_{\sigma_i,\sigma'_i}\right) \tag{8.39}$$

V and W are $p^N \times p^N$ matrices. Assuming periodic boundary conditions, we can now express the partition function as

$$\mathcal{Z} = \mathrm{tr}(VW)^M \tag{8.40}$$

Our purpose is to write $\mathcal{Z}$ such as to make apparent the **duality** property between high and low temperatures. We choose the Ising model $p = 2$ throughout the calculation. The slight modifications required to have the correct results for any value of p will become obvious. First note the decomposition

$$W_{\sigma,\sigma'} = \prod_{i=1}^{N} \exp\left(J_\tau \delta_{\sigma_i,\sigma'_i}\right) = \prod_{i=1}^{N} w(i) \tag{8.41}$$

Each term $w(i)$ in the product acts as the identity at all sites except at the sites i and i'. The term $w(i)$ can be written as a $2^N \times 2^N$ matrix connecting the spaces of states of each rows.

$$w(i) = \mathbf{1} \otimes \cdots \otimes \mathbf{1} \otimes \overbrace{\mathrm{w}}^{i} \otimes \cdots \mathbf{1} \tag{8.42}$$

where $\mathbf{1}$ is the identity matrix in 2 dimensions. The element w acts between the sites i and i'. However, its shape is independent of the particular position of the site i. It can be written as the 2×2 matrix

$$\mathrm{w} = \begin{pmatrix} e^{J_\tau} & 1 \\ 1 & e^{J_\tau} \end{pmatrix} \tag{8.43}$$

We introduce the fundamental matrix

$$e_{2i-1} = \frac{1}{\sqrt{p}} \mathbf{1} \otimes \cdots \otimes \overbrace{\begin{pmatrix} 1 & 1 \\ 1 & 1 \end{pmatrix}}^{i} \otimes \cdots \mathbf{1} \tag{8.44}$$

and the modified coupling

$$\Upsilon_\tau = \frac{1}{\sqrt{p}} (e^{J_\tau} - 1) \tag{8.45}$$

with $p = 2$ to complete our calculation. A simple recombination shows

$$w(i) = \sqrt{2}\,(\Upsilon_\tau \mathbf{1}_N + e_{2i-1}) \tag{8.46}$$

where $\mathbf{1}_N = \mathbf{1} \otimes \cdots \otimes \mathbf{1}$ is the identity acting at all sites.

We now work out V in a similar way. We write

$$V_{\sigma,\sigma'} = \prod_{i=1}^{N} \delta_{\sigma_i,\sigma'_i}\,\delta_{\sigma_{i+1},\sigma'_{i+1}} \exp\left(J_x \delta_{\sigma_i,\sigma_{i+1}}\right) = \prod_{i=1}^{N} v(i) \tag{8.47}$$

We doubled the Kronecker δ terms in the last equation for the sake of clarity. This is not a problem because $\delta^2 = \delta$. It is now apparent that $v(i)$ acts non-trivially only on the 2×2 spin space connecting the sites i and $i+1$ to i' and $i'+1$. The delta terms $\delta_{\sigma_i,\sigma'_i}\delta_{\sigma_{i+1},\sigma'_{i+1}}$ show that $v(i)$ is diagonal and we can write it as follows.

$$\begin{aligned} v(i) &= \mathbf{1} \otimes \cdots \otimes \overbrace{\mathrm{v}}^{i\otimes(i+1)} \otimes \cdots \otimes \mathbf{1} \\ \mathrm{v} &= \begin{pmatrix} e^{J_x} & 0 & 0 & 0 \\ 0 & 1 & 0 & 0 \\ 0 & 0 & 1 & 0 \\ 0 & 0 & 0 & e^{J_x} \end{pmatrix} \end{aligned} \tag{8.48}$$

We complement the matrices e_{2i-1} introduced earlier with

$$e_{2i} = \sqrt{p}\,\mathbf{1} \otimes \cdots \otimes \overbrace{\begin{pmatrix} 1&0&0&0 \\ 0&0&0&0 \\ 0&0&0&0 \\ 0&0&0&1 \end{pmatrix}}^{i\otimes(i+1)} \otimes \cdots \otimes \mathbf{1} \tag{8.49}$$

and we introduce the modified spatial coupling

$$\Upsilon_x = \frac{1}{\sqrt{p}}\left(e^{J_x} - 1\right) \tag{8.50}$$

A little manipulation shows that with $p = 2$ we have

$$v(i) = \left(\mathbf{1}_N + \Upsilon_x e_{2i}\right) \tag{8.51}$$

You have already noticed the obvious similarity between V and W obtained with this calculation. To pursue further we have to verify that the matrices e_i defined above define a *Temperley-Lieb algebra*

$$\begin{aligned} e_i^2 &= \beta e_i\,, & \beta &= \sqrt{p} \\ e_i e_j &= e_j e_i\,, & |i-j| &\geq 2 \\ e_i e_{i\pm1} e_i &= e_i \end{aligned} \tag{8.52}$$

The Temperley-Lieb algebra is not a usual algebra defined by commutation relations. It is a *word algebra*. The partition function can be written in terms of this algebra as

$$\mathcal{Z} = \left(\prod_{i=1}^{N} \sqrt{p}\left(\Upsilon_\tau \mathbf{1}_N + e_{2i-1}\right)\left(\mathbf{1}_N + \Upsilon_x e_{2i}\right)\right)^M \tag{8.53}$$

Obviously the algebra (8.52) does not depend on the parity of the indices of the generators. The fact that in Eq.(8.53) the odd indexed e_i are in the parts with Υ_τ while the even indexed e_i are associated with with Υ_x is of no importance because odd and even e are algebraically equivalent. In the large lattice size limit, the transformation $e_i \to e_{i+1}$ (permuting odd and even e) will not change the partition function. Suppose that we exchange odd and even e. We can extract a factor $(\Upsilon_\tau \Upsilon_x)^{NM}$ and write

$$\mathcal{Z}(\Upsilon_\tau, \Upsilon_x) = (\Upsilon_\tau \Upsilon_x)^{NM}\, \mathcal{Z}(1/\Upsilon_x, 1/\Upsilon_\tau) \tag{8.54}$$

Therefore the free energy per site

$$f = - \lim_{N,M\to\infty} \frac{1}{NM} \ln \mathcal{Z} \tag{8.55}$$

exhibits explicitly the duality relation

$$f(\Upsilon_\tau, \Upsilon_x) = -\ln(\Upsilon_\tau \Upsilon_x) + f(1/\Upsilon_x, 1/\Upsilon_\tau) \tag{8.56}$$

equivalent to eq. (1.54). On physical grounds, we expect the systems to have a disordered high temperature phase ($\Upsilon_{\tau,x}$ small) and an ordered low temperature phase ($\Upsilon_{\tau,x}$ large). The two parameter space of couplings must have a critical curve separating both regions. We deduce from eq. (8.56) that the non-analyticity of f along a curve $\Upsilon_\tau \Upsilon_x = \text{const} < 1$ automatically implies its non-analyticity along the dual curve $1/(\Upsilon_\tau \Upsilon_x) = 1/\text{const} > 1$. The simplest solution is that the critical curve satisfies the self-duality condition $\Upsilon_\tau \Upsilon_x = 1$. In the isotropic case where $J_\tau = J_x = J$, we recover the result of chapter 1 for the critical point: $e^{J_c} = p^{1/2} + 1$.

It is important to notice that Temperley-Lieb algebras exist for many exactly integrable models, whether they are critical or not. They reduce the complicated structure of a system to an algebraic problem, whose representation theory can be extensively discussed. A theorem shows that hermitian representations of the Temperley Lieb algebra are possible for the coupling $\beta < 2$ if and only if β of (8.52) takes one of the following values

$$\beta = 2 \cos\left(\frac{\pi}{m+1}\right) \tag{8.57}$$

We recognize for $m = 3, 5, \infty$ the values of β for the Potts models with $p = 2, 3, 4$. The formula (8.57) gives us the first sign that the p-state Potts models for $p > 4$ must be very different from the first three models. Actually, the parameter m can be identified with the one which labels the Virasoro central charge $c = 1 - 6/(m(m+1))$ and the partition function (8.53) describes the partition functions of the RSOS models with the parameter β of the Temperley-Lieb algebra given by eq. (8.57).

Temperley-Lieb algebras can be used to relate the spectra of the quantum Hamiltonians of quite distinct systems. In chapter 15, we shall map in this way the p-states Potts model onto the XXZ quantum chain.

8.6 Laudau-Ginzburg classification

We have seen earlier that the Ising model quantum Hamiltonian could be obtained from either an Ising spin system or a scalar ϕ^4 field theory. This universality suggests that there might be a series of field theory realizations of conformally invariant systems and we shall now briefly mention a correspondence between the A_m-RSOS models and a class of scalar field theories. Such a realization might for example be useful to make explicit contact with ε-expansion related studies.

Consider the Landau-Ginzburg-Wilson (LGW) action of the form

$$S = \int d^2x \left(\frac{1}{2} (\partial\phi)^2 + \sum_{n=1}^{2(m-1)} \lambda_n \phi^n \right) \tag{8.58}$$

For $\lambda_1 = \lambda_2 = \ldots = \lambda_{2m-3} = 0$, one has a multicritical point. These $2D$ critical systems turn out to have a central charge $c(m) = 1 - 6/(m(m+1))$ and the scaling dimensions of the fields are given by the Kac formula. The correspondence, due to A. Zamolodchikov [341], between the scaling fields ϕ^n and the conformal weigths $\Delta_{r,s}$ is illustrated in Table 6 for the cases $m = 3, 4, 5$ which already shows the general structure. The index r runs horizontally from 1 at the left to $m-1$ at the right and the index s run vertically from 1 below to m at the highest line. The conformal field $\phi_{1,1}$ always corresponds to the identity operator $\mathbf{1}$ and $\Delta_{1,1} = 0$. The LGW field ϕ^{2m-3} is redundant and can always be elimated form the action. To make the correspondence between the LGW fields ϕ^n and the

Table 6. Correspondence between conformal primary fields and the LGW fields ϕ^n.

$\frac{1}{2}$	0
$\frac{1}{16}$	ϕ
1	ϕ^2

$m = 3$

$\frac{3}{2}$	$\frac{7}{16}$	0
$\frac{3}{5}$	$\frac{3}{80}$	ϕ^2
$\frac{1}{10}$	ϕ	ϕ^4
1	ϕ^3	ϕ^6

$m = 4$

3	$\frac{7}{5}$	$\frac{2}{5}$	0
$\frac{13}{8}$	$\frac{21}{40}$	$\frac{1}{40}$	ϕ^3
$\frac{2}{3}$	$\frac{1}{15}$	ϕ^2	ϕ^6
$\frac{1}{8}$	ϕ	ϕ^5	
1	ϕ^4	ϕ^8	

$m = 5$

conformal primary fields $\phi_{r,s}$ explicit, recall the symmetry relation $\Delta_{r,s} = \Delta_{m-r,m+1-s}$. The LGW fields, reading the table from the lower left corner, correspond to the conformal weights descending from the upper right corner. In particular, since the LGW action is invariant under the Z_2-symmetry $\phi \to -\phi$, one expects these models to coincide with the A series in Table 13. This is the case.

In particular, the least relevant of the relevant operators corresponds to the conformal field $\phi_{1,3} \sim \phi^{2m-4}$, while the field $\phi_{3,1} \sim \phi^{2m-2}$ corresponds to the most relevant of the irrelevant primary fields.

The justification [341] of this comes from the operator product expansion. For a LGW field ϕ, it is trivial that $\phi^a \cdot \phi^b = \phi^{a+b}$ in the sense of the OPE. On the other hand, using the general OPE eq. (7.43) and the proposed identification of the conformal operators $\phi_{r,s}$ with the LGW fields ϕ^a, the consistency of the identification can be established. As a corrollary, the multipoint correlation functions of the composite LGW fields can be written down explicitly in terms of hypergeometric functions, following the techniques of chapter 7.

An alternative lattice construction of these models is provided by the A_m-RSOS models as introduced in chapter 1 whose critical point partition functions are given by the A series of Table 13.

9 Numerical Techniques

We now discuss the techniques necesssary for a numerical diagonalization of a given quantum Hamiltonian as obtained in chapter 8. First, the problem of diagonalizing H can be broken into the separate consideration of block-diagonal parts by using a few simple symmetry properties. In connection with conformal invariance, the systematic use of translation invariance allows to fix the normalization of H in a unique way. Numerical diagonalization and finite-lattice extrapolation algorithms will be described. In doing so, we shall merely mention a few selected techniques which have proved to work reliably in the context of critical quantum chains and do not attempt to give a systematic overview on the numerous numerical algorithms which exist in the literature. These methods can be applied independently of the conformal invariance of the model, but throughout this chapter, we shall take H to be defined on a quantum **chain** of N sites.

9.1 Simple properties of quantum Hamiltonians

For the purposes of this section, we shall use the example of the $2D$ p-states vector Potts model (see chapter 1), but the presentation is readily extended to any system with p states per site. The classical Hamiltonian is

$$\mathcal{H} = -\beta \sum_{r,t} \{ J_x \cos(\Theta_{r,t} - \Theta_{r+1,t}) + J_\tau \cos(\Theta_{r,t} - \Theta_{r,t+1}) \} \tag{9.1}$$

where the spin variables $\Theta = \frac{2\pi}{p} r, r = 0, \ldots, p-1$. The Hamiltonian limit of eq. (9.1) can be done in complete analogy with the calculations of chapter 8. For a quantum chain with N sites, the quantum Hamiltonian is in the limit $\beta J_x \to 0, \beta J_\tau \to \infty$ with λ fixed [306]

$$H = -2 \sum_{n=1}^{N} \left\{ \lambda \cos(\hat{\Theta}_n - \hat{\Theta}_{n+1}) + \cos \hat{p}_n \right\} \tag{9.2}$$

The limit taken implies that we are considering the geometry of an infinitely long strip of finite width $L = Na$, where a is the lattice constant. Expressed with variables

$$\sigma_n := \exp i\hat{\Theta}_n = \begin{pmatrix} 1 & & & & \\ & \omega & & & \\ & & \omega^2 & & \\ & & & \ddots & \\ & & & & \omega^{p-1} \end{pmatrix} \tag{9.3}$$

where $\omega = \exp(2\pi i/p)$ and

$$\Gamma_n := \exp i\hat{p}_n = \begin{pmatrix} 0 & & & & 1 \\ 1 & 0 & & & \\ & 1 & 0 & & \\ & & \ddots & \ddots & \\ & & & 1 & 0 \end{pmatrix} \tag{9.4}$$

the quantum Hamiltonian becomes

$$H = -\sum_{n=1}^{N} \{\lambda\left(\sigma_n \sigma_{n+1}^{+} + \sigma_n^{+} \sigma_{n+1}\right) + \Gamma_n + \Gamma_n^{+}\} \tag{9.5}$$

which has the dihedral group D_p as global symmetry. More general choices of the couplings are possible and lead to quantum Hamiltonians describing other universality classes. We now give a few properties of the $p^N \times p^N$ matrix H.

Charge symmetry. H commutes with the $\mathbf{Z}_p$ *charge operator*

$$\hat{Q} := \prod_{n=1}^{N} \Gamma_n \tag{9.6}$$

$\hat{Q}$ has the eigenvalues ω^Q. The corresponding eigenspaces are called *charge sectors*, $Q = 0, 1, \ldots, p-1$.

Parity. The *parity operator* $\hat{P}$ is defined by

$$\hat{P}\sigma_n \hat{P}^{+} = \sigma_{N-n+1} \quad , \quad \hat{P}\Gamma_n \hat{P}^{+} = \Gamma_{N-n+1} \tag{9.7}$$

and has eigenvalues $P = \pm 1$. The eigenspaces of $\hat{P}$ are called *parity sectors*.

Charge conjugation. The *charge conjugation* operator

$$\hat{C} := \prod_{n=1}^{N} \hat{c}_n \quad , \quad \hat{c}_n = \begin{pmatrix} 1 & & & & 0 \\ & 0 & & 0 & 1 \\ & & \ddots & 1 & \\ & 0 & 1 & \ddots & \\ 0 & 1 & & & 0 \end{pmatrix} \tag{9.8}$$

has the eigenvalues $C = \pm 1$. $\hat{C}$ transforms the charge sector $Q = q$ into $Q = p - q (\mathrm{mod}\, p)$ and vice versa. One has

$$\hat{C}\sigma_n \hat{C}^{+} = \sigma_n^{+} \quad , \quad \hat{C}\Gamma_n \hat{C}^{+} = \Gamma_n^{+} \tag{9.9}$$

Invariance under charge conjugation is a property of the Potts Hamiltonian eq. (9.5), but not necessarily of more general Hamiltonians with complex couplings. Consequently, H of eq. (9.5) has the same spectrum in the two charge sectors linked by $\hat{C}$. In the sector $Q = 0$ (and also $Q = p/2$ if p is even), the states can be labelled by the values of the C-*parity*.

Algebra of σ_n and Γ_m. The matrices σ_n and Γ_m satisfy the algebra

$$\begin{aligned} \sigma_n \Gamma_m &= \delta_{nm}\, \omega \Gamma_m \sigma_n \, + \, (1 - \delta_{nm})\, \Gamma_m \sigma_n \\ \sigma_n^p &= \Gamma_n^p = 1 \end{aligned} \tag{9.10}$$

This algebra is left invariant under the global transformation

$$\begin{aligned} \Gamma &\rightarrow U \Gamma U^{+} = \sigma \\ \sigma &\rightarrow U \sigma U^{+} = \Gamma^{+} \end{aligned} \tag{9.11}$$

Also, the matrix form (9.8) of $\hat{C}$ is kept invariant under this transformation. Sometimes, the application of (9.11) to the Hamiltonian eq. (9.5) is referred to as the change into the ***high-temperature basis.***

Duality. Another transformation, which leaves the algebra eq. (9.10) invariant, is the *duality transformation*

$$\begin{aligned} \sigma_n &\to \tilde{\sigma}_n := \Gamma_{n+1}\Gamma_n^+ \quad \text{if } n < N \\ \sigma_N &\to \tilde{\sigma}_N := \Gamma_N^+ \\ \Gamma_n &\to \tilde{\Gamma}_n := \prod_{l=1}^{n} \sigma_l \end{aligned} \tag{9.12}$$

Note the peculiar treatment of the boundary spin. Up to boundary terms (which means, in the $N \to \infty$ limit), the p-state Potts model satisfies the ***self-duality*** property

$$H(\lambda) = \lambda H(1/\lambda) \tag{9.13}$$

This is the analogue of self-duality as derived in chapters 1 and 8. To see this, one has to apply the Hamiltonian limit to the self-duality condition eq. (8.56). In order to have self-duality for N finite, special boundary conditions are needed. If the model has only one critical point, then by self-duality it has to be at $\lambda_c = 1$.

To summarize the contents of this section, consider the most general Z_p symmetric quantum Hamiltonian H of a model with p states per site and N sites

$$H = -\sum_{n=1}^{N}\sum_{k=1}^{p}\left[a_k(n)\sigma_n^k\sigma_{n+1}^{p-k} + b_k(n)\Gamma_n^k\right] \tag{9.14}$$

where $a_k(n)$ and $b_k(n)$ are (possibly site-dependent) couplings. Particular choices for the couplings reflect themselves in symmetry properties of H as shown in Table 7.

Table 7. Symmetry properties of a p-state Z_p symmetric quantum Hamiltonian.

Symmetry	Condition	
$H = H^+$	$a_k(n) = a_{p-k}^*(n)$	$b_k(n) = b_{p-k}^*(n)$
$[H, \hat{P}] = 0$	$a_k(n) = a_k(N-n+1)$	$b_k(n) = b_k(N-n+1)$
$[H, \hat{C}] = 0$	$a_k(n) = a_{p-k}(n)$	$b_k(n) = b_{p-k}(n)$
H self-dual	$a_k(n) = \lambda b_k(n)$	

9.2 Some further physical quantities and their critical exponents

We now define a few additional physical quantities. Since λ can be seen to correspond to a temperature-like variable, the analogon of the *specific heat* can be defined as

$$C = -\frac{1}{N}\frac{d^2E_0}{d\lambda^2} \sim (\lambda - \lambda_c)^{-\alpha} \tag{9.15}$$

To define an analogon of a *magnetic susceptibility*, we have to add a symmetry breaking term, e.g. $hH_1 = -h\sum_n(\sigma_n + \sigma_n^+)$, to H and to calculate the ground state energy $E_0(h)$. Then the susceptibility per spin is

$$\chi = -\frac{1}{N}\frac{d^2E_0}{dh^2}\bigg|_{h=0} \sim (\lambda - \lambda_c)^{-\gamma} \tag{9.16}$$

Turning to the *order parameter*, one faces the problem that a double limit: i) $N \to \infty$ with $T \neq T_c$ and only afterwards ii) $T \to T_c$; is needed for its definition. Carrying out the double limit is computationally difficult. A trick first used by Yang [333] to calculate the spontaneous magnetization in the Ising model can be used to cirvumvent this difficulty. Write the total Hamiltonian $H_{tot} = H + hH_1$. Close to criticality ($\lambda \to \lambda_c, h \to 0$), one has a very small energy gap $\xi^{-1} = E_1 - E_0 \to 0$. Now, suppose that even for finite N, ξ^{-1} is exponentially small and therefore treat E_1 and E_0 as degenerate. In first order degenerate perturbation theory in h, the magnetization per site M will be obtained from the smallest eigenvalue of the matrix [319, 164]

$$M = -\frac{\partial f^{sin}}{\partial h}\bigg|_{h=0} = \frac{1}{N}\begin{pmatrix} \langle 0|H_1|0\rangle & \langle 0|H_1|1\rangle & \dots \\ \langle 1|H_1|0\rangle & \langle 1|H_1|1\rangle & \dots \\ \vdots & \vdots & \dots \end{pmatrix} \tag{9.17}$$

where $|0\rangle, |1\rangle, \dots$ are the eigenstates of the smallest eigenvalues in the charge sectors $Q = 0, 1, \dots$. Since H_1 is connecting different charge sectors and breaks the global symmetry, its diagonal elements vanish. The order parameter defined this way can be used for either first order or second order transitions. In the latter case, it shows the correct finite-size scaling behaviour $M \sim N^{-\beta/\nu}$.

A similar problem arises in the definition of the *latent heat*. Here, we write

$$H_{tot} = H + \vartheta V \quad \text{with } \vartheta = \lambda - \lambda_c \tag{9.18}$$

V is proportional to the energy density and does not link states in different charge sectors. Denote by $|0\rangle, |\varepsilon\rangle$ the eigenstates with the lowest eigenvalues in the charge sector $Q = 0$. Assume these two states to be degenerate even for N finite. The ground state energy is then, to lowest order in ϑ, obtained from the eigenvalues of $\mathcal{E} = E_0 \cdot 1 + \vartheta\mathcal{E}_1$, where

$$\mathcal{E}_1 = \begin{pmatrix} \langle 0|V|0\rangle & \langle 0|V|\varepsilon\rangle \\ \langle \varepsilon|V|0\rangle & \langle \varepsilon|V|\varepsilon\rangle \end{pmatrix} \tag{9.19}$$

Let $e_\pm$ be the higher and lower eigenvalues of $\mathcal{E}_1$. Then the latent heat per site ℓ_h is [165]

$$\ell_h = \frac{e_+ - e_-}{N} = \frac{1}{N}\left[\left(\langle 0|V|0\rangle - \langle \varepsilon|V|\varepsilon\rangle\right)^2 + 4\langle 0|V|\varepsilon\rangle^2\right]^{1/2} \tag{9.20}$$

which can be used for both first order and second order transitions. While ℓ_h remains finite for a first order transition (as M does), for a second order transition one has $\ell_h \sim N^{-(1-\alpha)/\nu}$. For the p-state Potts model, it was checked [165] that ℓ_h reproduces the exactly known result (e.g. [G11]) for the latent heat.

These expressions, although they require the calculation of eigen**vectors**, provide direct access to the scaling dimensions of the order parameter (x_σ) and the energy density (x_ε)

$$x_\sigma = \beta/\nu \ , \ x_\varepsilon = (1-\alpha)/\nu \tag{9.21}$$

While this is the only known way to calculate β directly from finite-size data, α could also be obtained from the specific heat. If α is small, however, background terms may inhibit a precise determination and using ℓ_h becomes a useful alternative[8)].

Given some quantity X with a finite-size scaling behaviour $X_N \sim N^{-x/\nu}$, finite-size estimates for the exponent x/ν can be obtained by two ways, namely either from

$$\left(\frac{x}{\nu}\right)_N := -N\left(\frac{X_N}{X_{N-1}} - 1\right) \tag{9.22}$$

or else from

$$\left(\frac{x}{\nu}\right)_N := -\frac{\ln X_N - \ln X_{N-1}}{\ln N - \ln(N-1)} \tag{9.23}$$

Practical experience suggests that for most systems with power-law singularities, (9.23) produces smaller correction-to-scaling terms than (9.22), while the opposite seems to be true for Kosterlitz-Thouless-like systems.

9.3 Translation invariance

We go back to the quantum Hamiltonian eq. (9.5). Translation invariance, if permitted by the interactions contained in H and by the boundary conditions, is the most effective means to break H into smaller blocks which can be diagonalized separately. We shall here, in the context of p-state systems like the Potts model, consider *toroidal boundary conditions*

$$\sigma_{N+1} = \exp\left(\frac{2\pi i\tilde{Q}}{p}\right)\sigma_1 \tag{9.24}$$

with $\tilde{Q} = 0, 1, \ldots, p-1$. The special case $\tilde{Q} = 0$ gives periodic boundary conditions. (Note that these boundary conditions are not compatible with self-duality for N finite.) With these boundary conditions, H is broken into the blocks $H_Q^{(\tilde{Q})}$ which can be diagonalized separately (if terms breaking the global symmetry are absent). At the self-dual point $\lambda = 1$, one has

$$H_Q^{(\tilde{Q})} = H_{\tilde{Q}}^{(Q)} \tag{9.25}$$

Similarly, for the case of free boundary conditions, one has the sectors $H_Q^{(F)}$.

Translation invariance is expressed by the existence of some translation operator $T_{\tilde{Q}}$, commuting with H. Its eigenvalues are related in a well-known way to the discretized lattice momentum $\mathcal{P}$. It is therefore useful to write the quantum Hamiltonian not in the basis of individual spins spanned by the states $|\alpha_1, \ldots, \alpha_N\rangle$, where the α_k give the values of an individual spin variable at site k (with p possible values per site), but in a basis of states $|\mathcal{P}, Q, \ldots\rangle$ in which the translation operator $T_{\tilde{Q}}$ is diagonal. For example,

8) This situation already occurs in the $(2+1)D$ Ising model, where $\alpha \simeq 0.11$ [169].

if the global symmetry group of the quantum Hamiltonian is the abelian group Z_p, the eigenstates of H with a *lattice momentum* $\mathcal{P}$ on a chain of N sites are [142]

$$\begin{aligned}\aleph(N)^{1/2}|\mathcal{P},Q,\ldots\rangle &= |\alpha_1,\alpha_2,\ldots,\alpha_N\rangle \\ &+ \exp\left(\frac{2\pi i C_1}{pN}\right)|\alpha_2,\alpha_3,\ldots\alpha_N,\alpha_1\rangle \\ &+ \ldots \\ &+ \exp\left(\frac{2\pi i C_{N-1}}{pN}\right)|\alpha_N,\alpha_1,\alpha_2,\ldots,\alpha_{N-1}\rangle \end{aligned} \tag{9.26}$$

$$C_k = \tilde{Q}N(\alpha_1+\ldots+\alpha_k) - (Q\tilde{Q}+\mathcal{P}p)k \tag{9.27}$$

and $\aleph(N)$ is the number of states $|\alpha_1,\ldots,\alpha_N>$ appearing in the translation-invariant state
$|\mathcal{P},Q,\ldots\rangle$. Going to this basis amounts to a prediagonalization which reduces the matrices to be considered by a factor of order $\mathcal{O}(N)$.

Since N is finite, the eigenvalues of $T_{\tilde{Q}}$ will be quantized such that $\mathcal{P}$ is an integer. On the other hand, the Virasoro generator $L_0-\bar{L}_0$ measures the spin of a certain scaling field φ. If Δ and $\overline{\Delta}$ are the conformal weights of ϕ, the scaling dimension x_ϕ and the spin s_ϕ are given by

$$x_\phi = \Delta+\overline{\Delta}\ ,\quad s_\phi = \Delta-\overline{\Delta} \tag{9.28}$$

and the lattice momentum $\mathcal{P}$ can be read off as

$$\mathcal{P} = \Delta-\overline{\Delta}+\frac{1}{p}\tilde{Q} = L_0-\bar{L}_0+\frac{1}{p}\tilde{Q} \tag{9.29}$$

for either a primary or a secondary field.

In connection with eq. (8.37), this can be used to fix the normalization of the lattice Hamiltonian H. The momentum eigenvalues are $\frac{2\pi}{N}\mathcal{P}$. Thus the energy levels of a set of states belonging to the same *conformal tower*, should be integer spaced if $\mathcal{P}\to\mathcal{P}\pm 1$. If two states belong to different conformal towers, their energy spacing is not simply related. In any case, for a given value of $\mathcal{P}$, there will be minimal energy, in conformal units: $E_{min}(\mathcal{P})-E_0 = |\mathcal{P}|$. This means that all these states must be secondary to the unit operator. In particular, due to the presence of the energy-momentum tensor $T(z)$, there will be states with $\mathcal{P}=\pm 2$ and $E(\mathcal{P})-E_0 = 2$. This fixes the normalization of H. It follows that a linear dispersion relation is a necessary condition for conformal invariance.

9.4 Diagonalization

In order to diagonalize a lattice Hamiltonian, one of the best methods is provided by the *Lanczos algorithm*, see [205, 218, 295]. Let H be a real symmetric $n\times n$ matrix, but the algorithm is readily adapted to complex hermitian matrices as well. One wants to find a basis $v_1,\ldots,v_n$ such that H is tridiagonal. A convenient representation of the main loop which is easy to implement on a computer is

$$
\begin{array}{rcll}
 & & & [v_i, w] \\
 & & & \downarrow \\
w' & = & w + Hv_i & [v_i, w] \\
\alpha_i & = & v_i \cdot w' & \downarrow \\
w'' & = & w' - \alpha_i v_i & [v_i, w''] \\
\beta_i & = & (w'' \cdot w'')^{1/2} & \downarrow \\
v_{i+1} & = & w''/\beta_i & [v_i, v_{i+1}] \\
 & & & \downarrow \\
 & & \mathrm{swap}(v_i, v_{i+1}) & [v_{i+1}, v_i] \\
 & & & \downarrow \\
w & = & -\beta_i v_i & [v_{i+1}, w]
\end{array}
\tag{9.30}
$$

where v_i, w, w', w'' are n-component vectors. Having finished these steps, one jumps back to the beginning. The *initial conditions* are $w = 0$ and v_1 has to be chosen as a convenient, normalized *starting vector*. It should have a sizeable overlap with the true ground state eigenvector, but otherwise, the actual choice of v_1 is normally not very important. The square brackets indicate which vectors take the memory locations $[A, B]$ in storage. This algorithm is (possibly apart from the matrix multiplication Hv_i) vectorizable and parallelizable and one can make full use of the sparse structure of H.

In the basis of the v_i $(i = 1, \ldots, n)$, H takes the tridiagonal form

$$
H = \begin{pmatrix}
\alpha_1 & \beta_1 & & & & \\
\beta_1 & \alpha_2 & \beta_2 & & & \\
 & \ddots & \ddots & \ddots & & \\
 & & \ddots & \ddots & \ddots & \\
 & & & \ddots & \alpha_{n-1} & \beta_{n-1} \\
 & & & & \beta_{n-1} & \alpha_n
\end{pmatrix}
\quad , \quad \beta_n = 0
\tag{9.31}
$$

However, if one only computes a **truncated** tridiagonal matrix using only m steps of the Lanczos algorithm, the eigenvalues of the $m \times m$ principal submatrix of (9.31) converge extremely rapidly to those of H. It is this observation which makes the Lanczos algorithm useful for practical calculations. Typically, one has values of $n \sim 10^4 - 10^5$. Using $m \simeq 20 - 100$ Lanczos steps, one typically obtains the first few lowest eigenvalues of H to a precision of 8 digits or more (precise statements are of course model-dependent). These rates of convergence generically only depend very weakly on the model parameters and anything comparable to the critical slowing-down seen in Monte Carlo simulations is completely absent.

If one is only interested in the leading critical exponents, it is sufficient to obtain the lowest two eigenvalues of H, corresponding to states in different charge sectors. Then one never encounters a numerical instability of the Lanczos algorithm due to rounding errors which we discuss now.

Choosing a "good" starting vector v_1, one gets a sequence of estimates of the eigenvalues and the corresponding eigenvectors. While the second eigenvector is in principle orthogonal to the first, rounding errors will make it to get a significant overlap with the first eigenvector after sufficiently many iterations. When this occurs, the estimates for the second eigenvalue will cross over to the first, producing an artificial degeneracy. One can prevent this by re-orthogonalising the set of the v_i sufficiently often, but this may require much memory to keep all the vectors needed.

Practical experience suggests the following. Suppose that one is interested in the higher eigenvalues but not in the eigenvectors. Then one observes that the apparent convergence of the higher eigenvalue estimates increases after each cross-over of an eigenvector into the ground state. This cross-over usually occurs in a fairly regular fashion and stable estimates are reached again after just a few (up to about 5) iterations, then remaining stable for another $\sim 30-60$ iterations until the next cross-over. We illustrate this with an example. In Table 8 we give the first few eigenvalues of the Ising model quantum Hamiltonian

$$-H = \sum_{n=1}^{N} [t\sigma^z(n) + h\sigma^x(n) + \sigma^x(n)\sigma^x(n+1)] \tag{9.32}$$

with $t = 1$, $h = 0.3$, $N = 13$ sites and periodic boundary conditions as found for various sizes m of the Lanczos submatrix. We observe a systematic evolution of the eigenvalue estimates E_i with the number of iteration steps m. The first eigenvalue E_0 converges very rapidly. However, we see that already for E_1 there is some apparent convergence to some limit for moderate values of m, before the cross-over to another value (in this case E_0) mentioned above sets in, which is due to the numerical rounding errors. We also observe that for the very value of m where the cross-over takes place in E_1, also for E_2 a similar cross-over is seen. The estimates E_2 now converge to the true second eigenvalue, and similarly for the higher ones. In this particular example, after 60 iterations E_0, E_2, E_4 and E_5 will be accepted as genuine eigenvalues of $-H$, while E_1 and E_3 will be rejected. This trick allows a faster determination of the higher eigenvalues which are needed for

Table 8. Convergence of the Lanczos algorithm in the Ising model with m iterations.

m	E_0	E_1	E_2	E_3	E_4	E_5
30	19.795206293	16.961613590	15.253109231	14.289888075	13.579039982	12.929548392
33	19.795206293	16.961613590	15.253109251	14.289952969	13.581189795	13.232102549
36	19.795206293	16.961613590	15.253109251	14.289960211	13.582247409	13.333511487
39	19.795206293	16.961613590	15.253109251	14.289960793	13.582515210	13.357819874
42	19.795206293	16.961613590	15.253109251	14.289960816	13.582540980	13.360748774
45	19.795206293	19.721013310	16.961613590	15.253109251	14.289960815	13.582539177
48	19.795206293	19.795145787	16.961613590	15.253109251	14.289960815	13.582539476
51	19.795206293	19.795206176	16.961613590	15.253109251	14.289960815	13.582539482
54	19.795206293	19.795206293	16.961613590	15.253109251	14.289960815	13.582539483
57	19.795206293	19.795206293	16.961613590	15.253109251	14.289960815	13.961362436
60	19.795206293	19.795206293	16.961613590	16.949143430	15.253109251	14.289960815

conformal invariance applications. Some caution is in order, however. The procedure described yields artificial degeneracies not only for the ground state, but also for the higher states. The question whether an eigenvalue is degenerate or not must be answered by looking at the evolution of the eigenvalue estimates with the number of Lanczos steps m. The number of iterations necessary for a certain precision of some eigenvalue is model dependent, but usually does not exceed $m \sim \mathcal{O}(100)$ for the first few eigenvalues.

Obtaining eigenvectors with the Lanczos algorithm does require to reorthogonalise the v_i. An alternative method is provided by the BMKD algorithm [42] which converges somewhat slower, but does have the advantage of being numerically extremely stable. One writes, beginning from a normalised starting vector v_1 (we assume H to be real symmetric for notational simplicity)

$$Hv_1 = e_1 v_1 + d_1 v_2 \qquad \text{with } v_2 \cdot v_1 = 0 \quad \text{and } v_2 \cdot v_2 = 1 \tag{9.33}$$

from which the vector v_2 and e_1 and d_1 are calculated. Let $e_2 = v_2 H v_2$ and form the matrix of H in the subspace spanned by v_1, v_2

$$\check{h} := \begin{pmatrix} e_1 & d_1 \\ d_1 & e_2 \end{pmatrix} \tag{9.34}$$

If $d_1 \neq 0$, then one of the eigenvalues of $\check{h}$ is below and the other above e_1. To find the lowest eigenvalue of H, take the eigenvector of $\check{h}$ corresponding to its lower eigenvalue and let it be the v_1 in the next iteration step.

This gives a very reliable estimate for the lowest eigenvector. Combinations and hybridizations with the Lanczos algorithm are possible.

9.5 Extrapolation

We now turn to the important and difficult question of how to extrapolate a given sequence $\{f_N\}$ of finite-size estimates towards its limit f as $N \to \infty$ for critical temperatures, exponents or amplitudes. To give an example of what one would like to have, consider finite-size data for the exponent x_ε of the $(1+1)D$ Ising model as obtained from the universal finite-size scaling amplitude $S_\varepsilon(0,0) = 2\pi x_\varepsilon$ (see eq. (3.49)) and shown in Figure 25. For $N \to \infty$, one has the asymptotic behaviour (full line in Figure 25)

$$x_\varepsilon \simeq 1 - \frac{\pi^2}{24}\frac{1}{N^2} + \ldots \tag{9.35}$$

but it is another matter whether the **asymptotically valid** expressions can be used for N finite and small (usually, $N \leq 20$ for quantum chains). Any extrapolation procedure has therefore **by necessity to assume** that the asymptotic behaviour somehow applies to the values of N within reach. This is apparently satisfied for the Ising model example of Figure 25, even for the small lattices taken into account. However, this does by no means imply that a similarly rapid convergence is the rule.

A very clear and instructive warning comes from the $3D$ spherical model [242]. In Figure 26, we give finite-size estimates of the inverse critical temperature β_c as calculated from phenomenological renormalization. The asymptotic behaviour of $\beta_c(N)$ is, with $\beta_c = 0.25274\ldots$ [242]

$$\beta_c(N) - \beta_c(\infty) = 0.027807 \cdot N^{-2} + \ldots \tag{9.36}$$

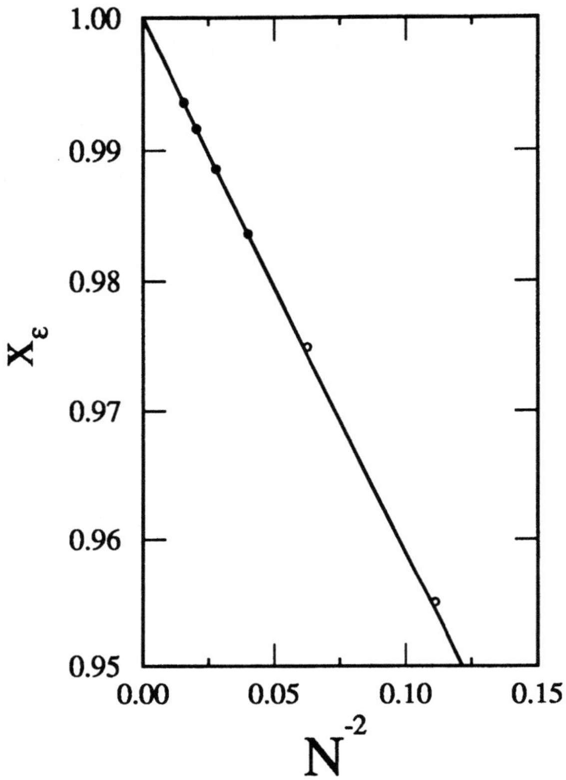

Fig. 25. Finite-size estimates of the exponent x_ϵ in the $(1+1)D$ Ising model.

and is given by the full line. It is clearly seen that up to lattice sizes $N \simeq 11$ (a cubic lattice of the shape $N \times N \times \infty$ with periodic boundary conditions was used) the behaviour of the finite-size data **has nothing to do** with the asymptotic expectation. Even the sign of the leading correction term is different ! We stress that this kind of phenomenon is not at all restricted to the spherical model. In order to illustrate the implications of this result, we remark that the largest square lattice of any model ever calculated exactly until today is the 6×6 lattice for the spin-1/2 quantum Heisenberg antiferromagnet [301]. Using the full space group and periodic boundary conditions, the size of the Hilbert space to be treated is 15804956 and there are $\sim 1.17 \cdot 10^9$ non-vanishing matrix elements. For the $(2+1)D$ Ising model, the 6×6 lattice was calculated using Monte Carlo methods [284].

Generally, turning to algorithms devised for the extrapolation versus $N \rightarrow \infty$ of a finite-lattice sequence, "... *the method should be able to represent the underlying singular*

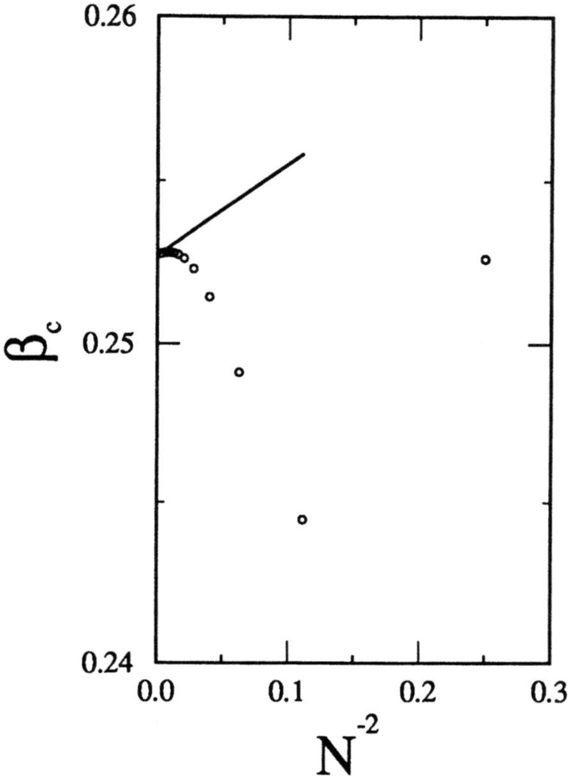

Fig. 26. Finite-size estimates for β_c in the $3D$ spherical model.

structure of the function under investigation" (Guttmann [159]). This implies that there is nothing like a straightforward and foolproof extrapolation method. The use of an extrapolation algorithm does require care and a considerable amount of judgement. By necessity, this implies that the final estimates for $N \to \infty$ are somewhat subjective. In particular, this holds for the estimated **accuracy** of the final results since the estimates eventually obtained will almost always contain systematic errors. If at all possible, one should therefore try to obtain a second estimate of the same quantity by a different method. The variation of the results thus obtained should give a fair idea on the lower bound of the extrapolation error.

After these remarks, we now turn to describe how finite-lattice extrapolations can be done in practice. We merely mention some of the most basic ideas and refer to the reviews of Guttmann [159] and of Weniger [324][9)] for more detailed information. We need some

[9)] These reviews deal mainly with the extrapolation of sequences obtained from series expansions.

terminology on the asymptotic convergence behaviour of finite-lattice sequences. Let f_N be a convergent sequence with $f = \lim_{N\to\infty} f_N$. Then consider

$$\rho := \lim_{N\to\infty} \frac{f_{N+1} - f}{f_N - f} \tag{9.37}$$

There are two principal situations to be distinguished:

1. **Linear convergence**. This situation is realized by definition if $0 < |\rho| < 1$. As an example, take for $N \to \infty$

$$f_N = f + a\exp(-bN) + \ldots \tag{9.38}$$

   where a and $b > 0$ are constants. For statistical systems, this corresponds to systems far away from criticality.
2. **Logarithmic convergence**. This case is defined by $\rho = 1$. Examples of this are given by, for $N \to \infty$

$$f_N = f + a_1 N^{-\omega_1} + a_2 N^{-\omega_2} + \ldots \tag{9.39}$$

$$f_N = f + a\exp(-bN^{\omega}) + \ldots \; ; \; \omega < 1 \tag{9.40}$$

   where a_1, a_2, a and $b, \omega_1, \omega_2, \omega > 0$ are constants. The first of these sequences is typically realized for critical systems and most of the determinations of critical temperature, exponents and amplitudes fall into this class. The cross-over between the two regimes of linear and logarithmic convergence may pose yet additional problems.

The Germain-Bonne theorem [146, 324] allows to construct extrapolation schemes which will accelerate the convergence of **any** linearly converging sequence, at least asymptotically for $N \to \infty$. On the other hand, one can show that there is nothing of the kind for logarithmically convergent sequences. Every algorithm which works apparently well on some types of logarithmically convergent sequences will fail on some other. For example, many extrapolation algorithms which improve the convergence of logarithmically converging sequences when $\omega_1, \omega_2, \ldots$ are positive **integers**, will cease to do so if non-integer values for the ω_i arise. We now present two widely used extrapolation algorithms designed for the extrapolation of logarithmically converging sequences, in particular for sequences of the type of eq. (9.39), which have been useful in the context of critical phenomena.

VBS algorithm. This algorithm was first proposed by van den Broeck and Schwartz [320] and introduced into the physics literature by Barber and Hamer [26, 163]. Consider a *consecutive sequence* of lattice sizes with constant increment between sizes such as $N = 1, 2, 3, 4, \ldots$ or $2, 4, 6, 8\ldots$, but avoid sequences like $2, 6, 3, \ldots$ or $1, 2, 3, 7, 8, 11, \ldots$.

These appear to have a structure slightly different from finite-size sequences which should play a role in the selection of a convenient algorithm.

Form approximants $[N, M]$, arranged as a triangular table

$$\begin{array}{lll} [0,0] & & \\ [1,0] & [1,1] & \\ [2,0] & [2,1] & [2,2] \\ [3,0] & [3,1] & \\ [4,0] & & \ddots \end{array} \tag{9.41}$$

where the $[N, M]$ are defined recursively

$$\begin{aligned} &[N,-1] = \infty \\ &[N,0] = f_N \\ &([N,M+1]-[N,M])^{-1} + \alpha\,([N,M-1]-[N,M])^{-1} \\ &= ([N+1,M]-[N,M])^{-1} + ([N-1,M]-[N,M])^{-1} \end{aligned} \tag{9.42}$$

The original sequence $f_N = [N, 0]$ is transformed into a new one, e.g. $[N, 1]$, which is expected to converge faster towards the desired limit f. For $\alpha = -1$, one can show that [26]

$$[N,2] = f + \mathcal{O}(N^{-\omega'}) \quad \text{where } \omega' = \min(\omega_1 + 2, \omega_2) \tag{9.43}$$

for a logarithmically convergent sequence of the type (9.39). The algorithm arises from a generalization of the Padé table, to which it reduces for $\alpha = 1$. The final estimate f should in principle be independent of the choice of α, but practically, performance may change considerably. A possible choice is $\alpha_M = -(1-(-1)^M)/2$. The consideration of a set of test sequences of the type (9.39), some of with non-integer ω_1, ω_2, suggests that this algorithm converges faster than many other conventional extrapolation algorithms (which only allow for ω_1, ω_2 to be integer or half-integer at most) [26].

The algorithm may converge to a spurious limit, if either the sequence to be extrapolated is non-monotonous or else the correction coefficients a_i have different signs. To rectify this, one might replace in the finite-size estimators eqs. (9.22) and (9.23) N by $N + \varepsilon$ where ε is a free parameter (*N-shift technique* [163]). The final result should be independent of ε, so one will look for the regions of stability with respect to it.

BST algorithm. A special case of this algorithm was proposed by Bulirsch and Stoer [62] as a variant of Romberg integration, but applications to critical phenomena are relatively recent [173]. Consider a sequence h_N ($N = 0, 1, 2, \ldots$) converging to zero as $N \to \infty$. In many applications, $h_N = 1/N$, but there is no need to have a consecutive sequence here. Form a table of extrapolants

$$\begin{array}{lll} T_0^{(0)} & & \\ & T_1^{(0)} & \\ T_0^{(1)} & & T_2^{(0)} \\ & T_1^{(1)} & \\ T_0^{(2)} & & \ddots \end{array} \tag{9.44}$$

where

$$T_{-1}^{(N)} = 0$$
$$T_0^{(N)} = T(h_N) = f_N \tag{9.45}$$
$$T_m^{(N)} = T_{m-1}^{(N+1)} + \left(T_{m-1}^{(N+1)} - T_{m-1}^{(N)}\right)\left[\left(\frac{h_N}{h_{N+m}}\right)^{\omega}\left(1 - \frac{T_{m-1}^{(N+1)} - T_{m-1}^{(N)}}{T_{m-1}^{(N+1)} - T_{m-2}^{(N+1)}}\right) - 1\right]^{-1}$$

with ω as a free parameter. This algorithm arises from approximating the function $T(h)$ by a sequence of rational functions in the variable h^{ω}. One expects that $T_1^{(N)}$ converges faster than $T_0^{(N)}$ to the desired limit $f = T(0)$ and so on.

For a logarithmically converging sequence of the type (9.39) and if one chooses $\omega = \omega_1$, one can show that [173]

$$T_1^{(N+1)} = f + \mathcal{O}(N^{-\omega'}) \quad \text{where } \omega' = \min(\omega_1, 2\omega_2) \tag{9.46}$$

Practically, one tries to choose ω such as to minimize simultaneously the differences $T_m^{(N+1)} - T_m^{(N)}$. It is also worth mentioning that the pseudoconverge of the VBS algorithm does not appear to show up for the BST algorithm and the N-shift technique is not really needed here. Furthermore, the BST algorithm appears to be much less sensible to rounding errors in the data than the VBS algorithm.

Comparing the two algorithms, we note that both tend to "absorb" the leading correction into the sequence. The VBS algorithm does so in two iteration steps, while the BST algorithm can do so (with a nearly optimal ω) already in one step. The algorithms will show their full strength only if they can be iterated several times on the sequence $\{f_N\}$. The application of both is best acquired by practical experience. Here, we just present one example how a relatively short sequence may be extrapolated. The data are estimates [178] for the critical point t_c of the following quantum Hamiltonian

$$H = -\frac{1}{2}\sum_{j=1}^{N}\left[t\sigma_j^z + \sigma_j^x\sigma_{j+1}^x + \sigma_j^y\sigma_{j+1}^y - \sigma_j^z\sigma_{j+1}^z + i\left((1-\sigma_j^z)\sigma_{j+1}^y + \sigma_j^y(1-\sigma_{j+1}^z)\right)\right] \tag{9.47}$$

which can be shown to be in the same universality class as directed percolation [72, 59, 219]. In Table 9, we display the iterated sequences as obtained with the VBS algorithm, where we have taken $\alpha = -1$. It is necessary to vary α and look for optimal convergence. In Table 10, the corresponding results for the BST algorithm are shown, where $\omega = 1.94$ was used. If the chosen values of α and ω were indeed the final ones, we would read off $t_c^{(VBS)} \simeq 2.606$ and $t_c^{(BST)} \simeq 2.60641$. From these tables, the reader should get some idea on how the iterated finite-size sequences behave. We recommend to investigate the structure of these extrapolation tables, while varying the parameters α and ω. For a detailed description of the many other extrapolation algorithms and further numerical examples, we refer to the literature, e.g. [159, 324, 27, 177] and references therein.

Table 9. Finite-lattice extrapolation with the VBS algorithm, $\alpha = -1$.

N				
8	2.6138388			
9	2.6125490	2.6081970		
10	2.6115541	2.6079750	2.6003815	
11	2.6107756	2.6077709	2.6059171	2.6059186
12	2.6101573	2.6075978	2.6059186	2.6059171
13	2.6096593	2.6074498	2.6068248	
14	2.6092529	2.6073364		
15	2.6089176			

Table 10. Finite-lattice extrapolation with the BST algorithm, $\omega = 1.94$.

2.6138388							
	2.6075368						
2.6125490		2.6063534					
	2.6071762		2.6064220				
2.6115541		2.6063865		2.6064067			
	2.6069493		2.6064116		2.6064131		
2.6107756		2.6063975		2.6064099		2.6064068	
	2.6068001		2.6064107		2.6064121		2.6064107
2.6101573		2.6064029		2.6064029		2.6064103	
	2.6066987		2.6064029		2.6064109		
2.6096593		2.6064029		2.6064029			
	2.6066275		2.6064321				
2.6092529		2.6064141					
	2.6065788						
2.6089176							

10 Conformal Invariance in the Ising Quantum Chain

At last, we shall give the first explicit example of the application of conformal invariance methods in the context of a specific lattice model. We choose the $2D$ Ising model as an example. Of course, the model was solved by Onsager back in 1944 and we shall not learn anything about this model which has not yet been deduced long ago by different means [G10, G11]. However, since all quantities of interest can be worked out explicitly, we get in this example absolute control of the critical behaviour of the model. This is useful to see how the techniques of conformal invariance work and one is not disturbed by convergence problems to reach the $N \to \infty$ limit. In particular, recalling the treatment of the same model in the language of continuum field theory in chapter 6, this example further illustrates the general machinery used for applying conformal invariance and useful insight will be obtained by comparing these two different approaches to the same problem.

We shall work with the following quantum Hamiltonian

$$H = -\frac{1}{\zeta}\sum_{n=1}^{N}\left[t\sigma^z(n) + \frac{1+\eta}{2}\sigma^x(n)\sigma^x(n+1) + \frac{1-\eta}{2}\sigma^y(n)\sigma^y(n+1)\right] \tag{10.1}$$

where N is the number of sites, $\sigma^x, \sigma^y, \sigma^z$ are Pauli matrices, t is the transverse field, η an irrelevant coupling introduced to check explicitly for universality and ζ is the normalization of H to be determined such that conformal invariance results can be applied. While this quantum Hamiltonian has been studied in detail for a long time (see e.g. [215, 258, 269]), the systematic application of conformal invariance techniques was done first in [138].

10.1 Exact diagonalization

General remarks. We begin by specifying the problem completely. The boundary conditions are

$$\sigma(N+1) = (-1)^{\tilde{Q}}\sigma(1) \tag{10.2}$$

where σ stands for σ^x, σ^y and $\tilde{Q} = 0, 1$. The charge operator, commuting with H, is

$$\hat{Q} = \frac{1}{2}\left(1 - \prod_{n=1}^{N}\sigma^z(n)\right) \tag{10.3}$$

and has eigenvalues $Q = 0, 1$. H is thus decomposed into the blocks $H_Q^{(\tilde{Q})}$. The global symmetry depends on the value of η. If $\eta = 0$, then H is invariant under a global $U(1)$ transformation

$$\begin{array}{ccccccc} \begin{pmatrix}\sigma^x\\ \sigma^y\end{pmatrix} & \to & \begin{pmatrix}\bar{\sigma}^x\\ \bar{\sigma}^y\end{pmatrix} & := & \begin{pmatrix}\cos\theta & -\sin\theta\\ \sin\theta & \cos\theta\end{pmatrix} & \begin{pmatrix}\sigma^x\\ \sigma^y\end{pmatrix} \\ \sigma^z & \to & \bar{\sigma}^z & := & \mathbf{1} & \sigma^z \end{array} \tag{10.4}$$

where θ is an arbitrary real number. If $\eta \neq 0$, one only has a global Z_2 invariance

$$\begin{array}{ccccc} \begin{pmatrix}\sigma^x\\ \sigma^y\end{pmatrix} & \to & \begin{pmatrix}\bar{\sigma}^x\\ \bar{\sigma}^y\end{pmatrix} & := - & \begin{pmatrix}\sigma^x\\ \sigma^y\end{pmatrix} \\ \sigma^z & \to & \bar{\sigma}^z & := & \sigma^z \end{array} \tag{10.5}$$

In the sequel, we shall only consider the case $\eta \neq 0$.

Jordan-Wigner transformation. Introduce the raising and lowering operators

$$\sigma^{\pm}(n) := \frac{1}{2}\left(\sigma^{x}(n) \pm i\sigma^{y}(n)\right) \tag{10.6}$$

The σ-variables commute on different sites and anticommute at the same site. This hybrid character of the variables is avoided by passing to truly fermionic variables c_n^+, c_n by the celebrated Jordan-Wigner transformation

$$\begin{aligned} \sigma^{+}(n) &= \exp\left(i\pi \sum_{\ell=1}^{n-1} c_\ell^+ c_\ell\right) c_n^+ \\ \sigma^{-}(n) &= c_n \exp\left(i\pi \sum_{\ell=1}^{n-1} c_\ell^+ c_\ell\right) \end{aligned} \tag{10.7}$$

The c's satisfy anticommutation relations

$$\left\{c_n^+, c_m\right\} = \delta_{n,m} \quad , \quad \left\{c_n, c_m\right\} = 0 \tag{10.8}$$

Let $\mathcal{N} := \sum_{\ell=1}^{N} c_\ell^+ c_\ell$ denote the *fermion number operator.* $\mathcal{N}$ is even in the sector $Q = 0$ and odd in the sector $Q = 1$. Then the Hamiltonian eq. (10.1) takes the form

$$\begin{aligned} H = -\frac{2}{\zeta}\Bigg\{ & \sum_{n=1}^{N}(-t)\left(c_n^+ c_n - \frac{1}{2}\right) \\ & + \frac{1}{2}\sum_{n=1}^{N-1}\left[c_n^+ c_{n+1} + c_n c_{n+1}^+ + \eta c_n^+ c_{n+1}^+ + \eta c_n c_{n+1}\right] \\ & - \frac{1}{2} e^{i\pi\mathcal{N}}\left[c_N^+ c_1 + c_N c_1^+ + \eta c_N^+ c_1^+ + \eta c_N c_1\right]\Bigg\} \end{aligned} \tag{10.9}$$

$$= -\frac{2}{\zeta}\sum_{n,m=1}^{N}\left[c_n^+ A_{nm} c_n + \frac{1}{2}\left(c_n^+ B_{nm} c_m^+ + c_n B_{nm} c_m\right)\right] + \text{ const.} \tag{10.10}$$

which, when restricted to the blocks $H_Q^{(\tilde{Q})}$, is a quadratic form in the c's and the integrability of the model is already clear. This result only depends on the nearest-neighbor nature of the interactions. The matrices $\hat{A}$ and $\hat{B}$ are, with $\mathcal{Q} := 1 - 2Q$ and for $\tilde{Q} = 0$

$$\hat{A} = \begin{pmatrix} -t & \frac{1}{2} & & & & -\frac{\mathcal{Q}}{2} \\ \frac{1}{2} & -t & \frac{1}{2} & & & \\ & \frac{1}{2} & -t & \ddots & & \\ & & \ddots & \ddots & \ddots & \\ & & & \ddots & -t & \frac{1}{2} \\ -\frac{\mathcal{Q}}{2} & & & & \frac{1}{2} & -t \end{pmatrix} , \quad \hat{B} = \frac{\eta}{2}\begin{pmatrix} 0 & 1 & & & & \mathcal{Q} \\ -1 & 0 & 1 & & & \\ & -1 & 0 & \ddots & & \\ & & \ddots & \ddots & \ddots & \\ & & & \ddots & 0 & 1 \\ -\mathcal{Q} & & & & -1 & 0 \end{pmatrix} \tag{10.11}$$

If $\tilde{Q} = 1$, replace $\mathcal{Q} \to -\mathcal{Q}$.

Diagonalization of a quadratic form. Equation (10.10) is most efficiently diagonalized by the well-known technique of Lieb, Schultz and Mattis [238]. In particular, the technique is directly applicable for quite general nearest-neighbor interactions, including for example quasiperiodic and random ones. One wants to rewrite the Hamiltonian in a diagonal form, using new fermionic variables a_k, a_k^+

$$H = \sum_k \Lambda_k \left(a_k^+ a_k - \frac{1}{2} \right) \tag{10.12}$$

where the Λ_k are the *one-fermion energies*. One makes the ansatz

$$\begin{aligned} a_k &= \sum_n \left(g_{kn} c_n + h_{kn} c_n^+ \right) \\ a_k^+ &= \sum_n \left(g_{kn} c_n^+ + h_{kn} c_n \right) \end{aligned} \tag{10.13}$$

where the g_{kn} and h_{kn} can be chosen to be real. In order that the a_k do satisfy the fermionic anticommutation relations, we have to require that

$$\begin{aligned} &\sum_n \left(g_{kn} g_{k'n} + h_{kn} h_{k'n} \right) = \delta_{kk'} \\ &\sum_n \left(g_{kn} h_{k'n} - g_{k'n} h_{kn} \right) = 0 \end{aligned} \tag{10.14}$$

Using the relation $[a_k, H] - \Lambda_k a_k = 0$, one finds

$$\begin{aligned} \Lambda_k g_{kn} &= \sum_m \left(g_{km} A_{mn} - h_{km} B_{mn} \right) \\ \Lambda_k h_{kn} &= \sum_m \left(g_{km} B_{mn} - h_{km} A_{mn} \right) \end{aligned} \tag{10.15}$$

From the matrices $\hat{g}$ and $\hat{h}$, build $2N$ vectors $\vec{\Phi}_k$ and $\vec{\Psi}_k$

$$\left(\vec{\Phi}_k \right)_i = g_{ki} + h_{ki} \;, \quad \left(\vec{\Psi}_k \right)_i = g_{ki} - h_{ki} \tag{10.16}$$

and we obtain

$$\vec{\Phi}_k \left(\hat{A} - \hat{B} \right) = \Lambda_k \vec{\Psi}_k \;, \quad \vec{\Psi}_k \left(\hat{A} + \hat{B} \right) = \Lambda_k \vec{\Phi}_k \tag{10.17}$$

Now, the one-fermion energies Λ_k are obtained from the eigenvalues of an $N \times N$ matrix

$$\widehat{\mathcal{M}} \vec{\Phi}_k := \left(\hat{A} + \hat{B} \right) \left(\hat{A} - \hat{B} \right) \vec{\Phi}_k = \Lambda_k^2 \vec{\Phi}_k \tag{10.18}$$

The achievement of the method is obvious: the problem of diagonalizing a $2^N \times 2^N$ matrix has been reduced to the eigenvalue problem of a $N \times N$ matrix. Even in cases where the eigenvalues of $\widehat{\mathcal{M}}$ cannot be written down analytically, numerical studies are considerably simplified. The price to pay is that one only obtains the one-particle energies directly. Multiparticle states have to be built up by linear superposition of the one-particle ones.

The matrix $\widehat{\mathcal{M}}$ takes the form (for $\tilde{Q} = 0$)

$$\widehat{\mathcal{M}} = \left(\frac{2}{\zeta}\right)^2 \begin{pmatrix} t^2 + \frac{1+\eta^2}{2} & -t & \frac{1-\eta^2}{4} & & & -\frac{1-\eta^2}{4}Q & tQ \\ -t & t^2 + \frac{1+\eta^2}{2} & -t & \ddots & & & -\frac{1-\eta^2}{4}Q \\ \frac{1-\eta^2}{4} & -t & \ddots & \ddots & \ddots & & \\ & \ddots & \ddots & \ddots & \ddots & \ddots & \\ & & \ddots & \ddots & \ddots & \ddots & \frac{1-\eta^2}{4} \\ -\frac{1-\eta^2}{4}Q & & & \ddots & \ddots & \ddots & -t \\ tQ & -\frac{1-\eta^2}{4}Q & & & \frac{1-\eta^2}{4} & -t & t^2 + \frac{1+\eta^2}{2} \end{pmatrix} \tag{10.19}$$

Since $\hat{A}$ is symmetric and $\hat{B}$ is antisymmetric, $\widehat{\mathcal{M}}$ is symmetric and positive semidefinite. Writing eq. (10.18) in components of $\vec{\Phi}$, we recover the discretized Schrödinger equation of a spinless electron in a potential (in our case just a constant one) given by the interactions of the Hamiltonian eq. (10.1).

Eigenvalue spectrum and normalization. The eigenvalue problem is now solved by the ansatz $(\vec{\Phi}_k)_r = e^{ikr}$ and gives, as first obtained by Katsura [215]

$$\Lambda_k^2 = \left(\frac{2}{\zeta}\right)^2 \left[(t - \cos k)^2 + \eta^2 \sin^2 k\right] \tag{10.20}$$

The critical point is determined by the vanishing of the lowest energy gap $\xi^{-1} = E_1 - E_0 = \Lambda_{k\to 0}$ in the $N \to \infty$ limit

$$\Lambda_{k\to 0} \simeq \frac{2}{\zeta}\left[(t-1)^2 + ((t-1)+\eta^2)k^2 + \mathcal{O}(k^4)\right]^{1/2} \tag{10.21}$$

from which we read off the longest wavelength mode ($k \to 0$) behaviour $\xi^{-1} \sim |t-1|$ which yields for the critical point t_c and the exponent ν

$$t_c = 1 \ , \ \ \nu = 1 \tag{10.22}$$

and we see that ν is independent of η, as expected from universality. We briefly outline the phase diagram [33] in the region $0 \leq \eta \leq 1$. For $t > 1$, the model is paramagnetic. If $t < 1$ and $t^2 + \eta^2 > 1$, we have an ordered ferromagnetic phase. This phase is separated by a *disorder line* at $t_{dis}^2 + \eta_{dis}^2 = 1$ from an ordered oscillating phase in the region $t^2 + \eta^2 < 1$. The correlation functions do not decay monotonously, but show a modulation. Finally, for $\eta = 0$, $t < 1$, one has an incommensurate phase. Note that for $\eta = 1$, t_c could have also been obtained from self-duality[10)].

Precisely at the critical point $t = t_c$, one has the dispersion relation $\Lambda_k \simeq 2\zeta^{-1}\eta|k|$. This is indeed a **linear** dispersion relation as required for conformal invariance. We now

---

10) Very recently, a generalization of self-duality valid for arbitray η was observed [191].

choose units of energy and momentum such that the speed of sound $v_{sound} = 2\zeta^{-1}\eta$ becomes unity in these units. This fixes the correct conformal normalization of H [138]

$$\zeta = 2\eta \tag{10.23}$$

Finally, we dicuss how the lattice momenta k should be discretized. The Jordan-Wigner transformation had introduced a non-local boundary term (see eq. (10.9)) which causes a mixture of periodic and antiperiodic boundary conditions in the fermionic language. This leads to the discretization

$$\begin{aligned} &\text{if } Q+\tilde{Q} = 0 \bmod 2 \quad k = \frac{2m+1}{N}\pi \; ; \; m = 0,1,\ldots,N-1 \\ &\text{if } Q+\tilde{Q} = 1 \bmod 2 \quad k = \frac{2m}{N}\pi \quad ; \; m = 0,1,\ldots,N-1 \end{aligned} \tag{10.24}$$

and the one-fermion energies are, with the above choice for k

$$\Lambda_k = \frac{1}{\eta}\left[(t-1)^2 + 4(t-1+\eta^2)\sin^2\frac{k}{2} - 4(\eta^2-1)\sin^4\frac{k}{2}\right]^{1/2} \tag{10.25}$$

This gives the complete solution for the spectrum of H, eq. (10.1).

10.2 Character functions

We now state explicitly the prediction of conformal invariance for the critical point spectrum, making use of the representation theory of the Virasoro algebra eq. (2.56) developed earlier. The basic objects of conformal theory are the **primary operators**. For unitary minimal models, their conformal weights were seen to be given by the Kac formula, see eqs. (4.48) and (4.49)

$$c = 1 - \frac{6}{m(m+1)} \quad ; \quad m = 3,4,5,\ldots \tag{10.26}$$

$$\Delta = \Delta_{r,s} = \Delta_{m-r,m+1-s} = \frac{[(m+1)r - ms]^2 - 1}{4m(m+1)} \quad ; \quad 1 \le r \le m-1 \; ; \; 1 \le s \le m$$

and we denote the primary operator corresponding to $\Delta_{r,s}$ by $\phi_{r,s}$. Higher states, corresponding to secondary operators, are obtained by acting with L_{-i} on the primary highest weight states $|\Delta_{r,s}\rangle$ (eq. (4.17))

$$L_{-i_1}L_{-i_2}\ldots|\Delta_{r,s}\rangle \quad i_1 \ge i_2 \ge \ldots > 0 \tag{10.27}$$

If $i_1 + i_2 + \ldots = I$, the number $d(\Delta_{r,s}, I)$ of excited states of *level* I is read from the *Virasoro character* [293], see eq. (4.52)

$$\begin{aligned} \chi_{r,s}(q) &= q^{-c/24}P(q)\sum_{k=-\infty}^{\infty}\left(q^{\Delta_{2mk+r,s}} - q^{\Delta_{2mk+r,-s}}\right) \\ &= q^{\Delta_{r,s}-c/24}\sum_{I=0}^{\infty} d(\Delta_{r,s}, I)q^I \end{aligned} \tag{10.28}$$

where

$$P(q) = \prod_{n=1}^{\infty}(1-q^n)^{-1} = \sum_{I=0}^{\infty} p(I)q^I \tag{10.29}$$

is the generating function for the number of partitions $p(I)$ of the integer I.

As an exercise, let us use the generating function $\chi_{r,s}(q)$ to calculate the first few $d(\Delta_{r,s}, I)$. Write $\chi_{r,s}(q) = \chi^+_{r,s}(q) + \chi^-_{r,s}(q)$ in an obvious notation. To each k in eq. (10.28) there are two new null states in the Verma module, one coming from $\chi^+_{r,s}$ an done coming from $\chi^-_{r,s}$. The corresponding Δ-values must be equal to $\Delta_{r,s}$ up to an integer

$$\Delta_{2mk+r,\pm s} = \Delta_{r,s} + k^2 m(m+1) + k\left[(m+1)r \mp ms\right] \tag{10.30}$$

We calculate the first few of these integers for the example $m = 3$. Then there are three primary operators, $\phi_{1,1}, \phi_{1,2}, \phi_{1,3}$. The integers involved in (10.30) are then up to the level $I = 20$

$$\begin{aligned} &\phi_{1,1} : 0^+, 1^-, 6^-, 11^+, 13^+, 20^-, \ldots \\ &\phi_{1,2} : 0^+, 2^-, 4^-, 10^+, 14^+, \ldots \\ &\phi_{1,3} : 0^+, 2^-, 3^-, 7^+, 17^+, \ldots \end{aligned} \tag{10.31}$$

where the index labels whether the corresponding term arises in $\chi^+_{r,s}$ or $\chi^-_{r,s}$, respectively. The *degeneracies* $d(\Delta_{r,s}, I)$ are then calculated by summing all the contributions for all values of k, where those terms coming from $\chi^+_{r,s}$ will make a positive and those coming from $\chi^-_{r,s}$ a negative contribution. This is illustrated in Table 11 for the operator $\phi_{1,1}$. The first contribution, labelled by the first integer coming from (10.30) by 0^+, is nothing else than the contribution of the generic character eq. (4.13), using the tabulated values [1] of the number of partitions $p(I)$ of I, obtained from $P(q)$. The second contribution, being negative since coming from $\chi^-_{1,1}$, repeats the same numbers $p(I)$, but shifted by one. This is nothing else than explicitly carrying out the Verma module construction described in chapter 4. Summing all contributions arising up to level $I = 15$ as indicated in Table 11, the final degeneracies are obtained. Similarly, the corresponding degeneracies

Table 11. Calculation of the degeneracies $d(0, I)$ of the operator $\phi_{1,1}$ for $m = 3$.

I	0	1	2	3	4	5	6	7	8	9	10	11	12	13	14	15
0^+	1	1	2	3	5	7	11	15	22	30	42	56	77	101	135	176
1^-		-1	-1	-2	-3	-5	-7	-11	-15	-22	-30	-42	-56	-77	-101	-135
6^-							-1	-1	-2	-3	-5	-7	-11	-15	-22	-30
11^+												1	1	2	3	5
13^+														1	1	2
$d(0,I)$	1	0	1	1	2	2	3	3	5	5	7	8	11	12	16	18

for the other primary operators are calculated. The results are collected in Table 12, and we recall that $\Delta_{1,1} = 0, \Delta_{1,2} = \Delta_{2,2} = 1/16, \Delta_{1,3} = \Delta_{2,1} = 1/2$.

Table 12. Virasoro characters of the primary operators $\phi_{r,s}$ for $m = 3$.

I	0	1	2	3	4	5	6	7	8	9	10	11	12	13	14	15	r	s
$d(0, I)$	1	0	1	1	2	2	3	3	5	5	7	8	11	12	16	18	1	1
$d(1/16, I)$	1	1	1	2	2	3	4	5	6	8	10	12	15	18	22	27	1	2
$d(1/2, I)$	1	1	1	1	2	2	3	4	5	6	8	9	12	14	17	20	1	3

From Table 12, recalling the relations between the spectra of the Hamiltonian ((8.37)) and of the momentum operator (eq. (9.29)) and the scaling dimensions and spin of the scaling fields, we can thus write down the conformal invariance prediction. We have two Virasoro algebras, the primary operators have **two real conformal weights** $\Delta_{r,s}, \overline{\Delta}_{\bar{r},\bar{s}}$ and the respective excitations are characterized by two integers $I, \bar{I}$. The states will occur at the scaled energy $\mathcal{E}$ and momentum $\mathcal{P}$ given by

$$\begin{array}{rcll} \mathcal{E}_Q^{(\tilde{Q})}(\mathcal{P}, \alpha; N) := \frac{N}{2\pi}(E_i - E_0) &=& (\Delta_{r,s} + I) + (\overline{\Delta}_{\bar{r},\bar{s}} + \bar{I}) & \\ \mathcal{P} &=& (\Delta_{r,s} + I) - (\overline{\Delta}_{\bar{r},\bar{s}} + \bar{I}) & + \mathcal{P}_{shift} \end{array} \tag{10.32}$$

with degeneracy $d(\Delta_{r,s}, I) \cdot d(\overline{\Delta}_{\bar{r},\bar{s}}, \bar{I})$ and α labels the eigenvalues. As discussed in chapter 9 in connection with translation invariance, the momentum shift is $\mathcal{P}_{shift} = \tilde{Q}/2$ for the Ising model, see eq. (9.29). This completes the statement of the critical point prediction for the Hamiltonian spectrum.

In the sequel, we shall refer to the characters $\chi_{r,s}(q)\bar{\chi}_{r',s'}(q)$ also by $(\Delta_{r,s}, \overline{\Delta}_{r',s'})$ and will give the numerical values of the $\Delta_{r,s}$. For example, the $m = 3$ character $\chi_{1,2}(q)\bar{\chi}_{1,2}(q)$ will be denoted $(1/16, 1/16)$.

10.3 Finite-size scaling analysis

We now turn to the explicit verification of conformal invariance in the Ising model at the critical point $t = t_c = 1$. At the same time, explicit expressions for the finite-size scaling functions and corrections to the leading finite-size scaling behaviour will be derived.

Ground state energy. We first have to find the central charge c. To do so, we need the universal finite-size scaling amplitude $Y(0,0)$ of the ground state energy E_0. It is given by (we take periodic boundary conditions)

$$\begin{aligned} E_0^{(0)} &= -\sum_{k=0}^{N-1} \left[\sin^2\left(\frac{\pi(2k+1)}{2N}\right) - \frac{\eta^2 - 1}{\eta^2} \sin^4\left(\frac{\pi(2k+1)}{2N}\right) \right]^{1/2} \\ &= -\left[\sin^2\left(\frac{\pi}{2N}\right) - \frac{\eta^2 - 1}{\eta^2} \sin^4\left(\frac{\pi}{2N}\right) \right]^{1/2} \\ &\quad - \sum_{k=1}^{N-1} \left[\sin^2\left(\frac{\pi(2k+1)}{2N}\right) - \frac{\eta^2 - 1}{\eta^2} \sin^4\left(\frac{\pi(2k+1)}{2N}\right) \right]^{1/2} \\ &=: G_1 + G_2 \end{aligned} \tag{10.33}$$

and

$$G_1 = -\frac{\pi}{2N} - \frac{\pi^3}{16N^3}\left(\frac{1}{\eta^2} - \frac{4}{3}\right) + \mathcal{O}(N^{-5})$$
$$G_2 = -\sum_{\ell=0}^{\infty}(-1)^\ell \binom{1/2}{\ell}\left(1 - \frac{1}{\eta^2}\right)^\ell \sum_{k=1}^{N-1}\left(\sin\frac{\pi(2k+1)}{2N}\right)^{2\ell+1} \tag{10.34}$$

We now analyse G_2. This is not merely a technical exercise, but also allows some insight in the nature of the finite-size terms arising. This will become important in chapter 13 when considering conformal perturbation theory. We use the identity [167]

$$\sum_{k=1}^{N-1}\left(\sin\frac{\pi(2k+1)}{2N}\right)^{2\ell+1} = \tag{10.35}$$
$$\left(-\frac{1}{4}\right)^\ell \sum_{k=0}^{\ell}(-1)^k\binom{2\ell+1}{k}\cos\left((l+\frac{1}{2}-k)\frac{\pi}{N}\right)\cot\left((l+\frac{1}{2}-k)\frac{\pi}{N}\right)$$

and expand in $1/N$. One then observes a clear separation of terms. The first term, of order $\mathcal{O}(N)$, receives contributions from all values of ℓ. This term is the **non-universal** bulk term and since each term in ℓ has a different η-dependence, it is not suprising to see that it will depend on all the finer details of the Hamiltonian. One might say that it is the presence of this term which complicates the exact solution of spin systems so much. The second term, of order $\mathcal{O}(1/N)$, only receives a contribution for $\ell = 0$, which is η-independent. We expect this term to be universal. The third term, of order $\mathcal{O}(1/N^3)$, receives contributions from $\ell = 0, 1$ and will be η-dependent. Since there is only **one** contribution depending on η, terms of this order will be the leading non-universal correction and their η-dependent coefficient can be identified with the non-universal metric factor of the leading correction operator.

These remarks are formally stated by the identities (see [167])

$$\sum_{k=0}^{\ell}(-1)^k\binom{2\ell+1}{k}\frac{1}{2\ell+1-2k} = (-1)^\ell 2^{4\ell}\frac{(\ell!)^2}{(2\ell+1)!}$$
$$\sum_{k=0}^{\ell}(-1)^k\binom{2\ell+1}{k}(2\ell+1-2k) = 0 \quad \text{if } \ell \geq 1$$
$$\sum_{k=0}^{\ell}(-1)^k\binom{2\ell+1}{k}(2\ell+1-2k)^3 = 0 \quad \text{if } \ell \geq 2 \tag{10.36}$$

Collecting terms, we find

$$G_2 = -\frac{N}{2}T_1(\eta) + \frac{5\pi}{12}\frac{1}{N} + \frac{67\pi^3}{960}\left(\frac{1}{\eta^2} - \frac{4}{3}\right)\frac{1}{N^3} + \mathcal{O}(N^{-5}) \tag{10.37}$$

where

$$T_1(\eta) = \frac{2}{\pi}\left(1 + \frac{\arccos\eta}{\eta\sqrt{1-\eta^2}}\right) \tag{10.38}$$

The critical free energy density $f = E_0^{(0)}/N$ is thus, in the strip geometry and with **periodic** boundary conditions [168]

$$f = -\frac{1}{2}T_1(\eta) - \frac{\pi}{12}\frac{1}{N^2} + \frac{7\pi^3}{960}\left(\frac{1}{\eta^2} - \frac{4}{3}\right)\frac{1}{N^4} + \mathcal{O}(N^{-6}) \tag{10.39}$$

We note the following:

1. The bulk free energy is η-dependent and thus non-universal. One can show by explicit calculation that it does not depend on the boundary condition $\tilde{Q} = 0, 1$, see eq. (10.42) below.
2. The leading **singular** term is $f^{sin} \sim N^{-2}$, in agreement with the finite-size scaling expectation. Further, we explicitly see that the finite-size scaling amplitude $Y(0,0) = -\pi/12$ is universal. Using eq. (3.54), we read off the desired value of the central charge
$$c = \frac{1}{2} \tag{10.40}$$
3. The leading correction term is η-dependent and thus non-universal.

The same techniques can also be applied for antiperiodic boundary conditions. Using the identity [167]

$$\sum_{k=1}^{N-1}\left(\sin\frac{\pi k}{N}\right)^{2\ell+1} = \left(-\frac{1}{4}\right)^{\ell}\sum_{k=0}^{\ell}(-1)^k\binom{2\ell+1}{k}\cot\left((\ell+\frac{1}{2}-k)\frac{\pi}{N}\right) \tag{10.41}$$

we obtain for the free energy density for **antiperiodic** boundary conditions in the strip geometry [168]

$$f_A = -\frac{1}{2}T_1(\eta) + \frac{\pi}{6}\frac{1}{N^2} - \frac{\pi^3}{120}\left(\frac{1}{\eta^2} - \frac{4}{3}\right)\frac{1}{N^4} + \mathcal{O}(N^{-6}) \tag{10.42}$$

We remark that the leading bulk term, although manifestly non-universal, is seen to be independent of the boundary conditions considered.

Operator content. Since $c = 1/2$, we can apply the $m = 3$ predictions from Table 12 to obtain the conformal predictions for the energy levels and their degeneracies. We still have to form the product of two such representations, since there are two Virasoro algebras, with generators $L_n, \bar{L}_n$. For the example of the primary operator $(0,0)$ with $m = 3$ (or $c = 1/2$), this leads to the degeneracy pattern

$$\begin{array}{ccccccccc} \vdots & & \vdots & & \vdots & & \vdots & & \vdots \\ 2 & & 0 & & 1 & & 0 & & 2 \\ & 1 & & 0 & & 0 & & 1 & \\ & & 1 & & 0 & & 1 & & \\ & & & 0 & & 0 & & & \\ & & & & 1 & & & & \end{array} \tag{10.43}$$

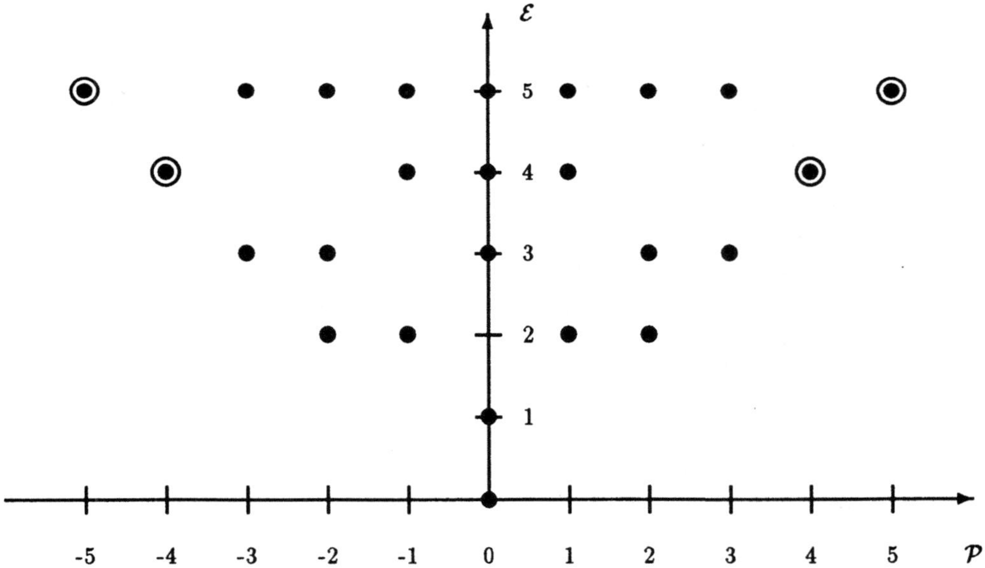

Fig. 27. Critical finite-size scaling spectrum for $H_0^{(0)}$ in the $2D$ Ising model.

which is beginning of the *conformal tower* of the operator $(0,0)$. From eq. (10.25), taking $t = t_c = 1$, the leading finite-size scaling amplitude for each multiparticle state can be easily obtained. The result for the block $H_0^{(0)}$ is shown in Figure 27. We first note that the boundary of the kinematically allowed region is indeed given by a linear dispersion relation $\mathcal{E} \sim \mathcal{P}$. Secondly, the energy spacings are indeed equidistant, as required from the structure of the Virasoro representations, and also integer-valued, which confirms the normalization of H from eq. (10.23). In particular, one recognizes that the degeneracies of the states with $\mathcal{E} = \pm\mathcal{P}$ are just those of a single Virasoro algebra and can be read from

Table 12. Turning to the quantitative comparison, we see that besides the (0,0)-operator, there is at least one further primary operator present with exponent $x = 1$. Given the list of allowed values for the conformal weights from the Kac formula eq. (10.26), this can be only $\Delta = \overline{\Delta} = 1/2$, which is the energy density $\varepsilon = (1/2, 1/2)$. Working out the degeneracy pattern of the higher states, the agreement is complete. Similarly, one sees that the *entire operator content* of the 2D Ising model is given by

$$\begin{aligned} H_0^{(0)} &= (0,0) + \left(\frac{1}{2},\frac{1}{2}\right) \\ H_1^{(0)} = H_0^{(1)} &= \left(\frac{1}{16},\frac{1}{16}\right) \\ H_1^{(1)} &= \left(0,\frac{1}{2}\right) + \left(\frac{1}{2},0\right) \end{aligned} \tag{10.44}$$

The interpretation is as follows. The block $H_0^{(0)}$ contains as primary operators the identity and the energy density (ε), both even under the Z_2 symmetry. The sectors $H_1^{(0)}$ and $H_0^{(1)}$ contain the order (σ) and disorder (μ) operators, respectively. These cannot be written in a local form with respect to neither the energy density nor the pair of conjugate Majorana spinors $(\psi, \bar{\psi})$ contained the sector $H_1^{(1)}$. In particular, from the operator (1/2, 1/2), we read off the exponent $x_\varepsilon = 1/2 + 1/2 = 1$ and similarly $x_\sigma = x_\mu = 1/8$ and $x_\psi = x_{\bar{\psi}} = 1/2$, as it should be.

It is of interest to compare these results with those obtained from the correlation functions using monodromy arguments. The various operator product algebras considered there appear when combining several of the sectors of the model. First, the sector $H_0^{(0)}$ by itself realizes the subalgebra generated by $\mathbf{1}$ and ε. In the fermionic language, this sector corresponds to the subspace of states generated by an even number of fermionic excitations from the vacuum. It is clear that a restriction to this subspace is consistent. Next, we combine the two sectors $H_0^{(0)}$ and $H_1^{(1)}$. The corresponding operator product algebra is now generated by $\mathbf{1}$, ε, ψ and $\bar{\psi}$. In the fermionic language, these two sectors do belong together, since they both have **antiperiodic** boundary conditions as can be seen from eq. (10.24). For periodic boundary conditions of the spin variables, we combine the sectors $H_0^{(0)}$ and $H_1^{(0)}$, which give rise to the operator product algebra spanned by $\mathbf{1}, \varepsilon$ and σ. Finally, the full operator product algebra is realized by taking all sectors of the model together.

For an alternative discussion of the continuum limit of the Ising model from the point of view of conformal invariance, see [25].

Finite-size corrections. Corrections to the leading finite-size scaling behaviour can be worked out easily. We give here just the results for the lowest levels, but similar expressions can be obtained [168] for any state desired without much effort, and show the same general characteristics.

We have [168]

$$\mathcal{E}_0^{(0)}(0,1;N) = 1 + \frac{\pi^2}{8}\left(\frac{1}{\eta^2} - \frac{4}{3}\right)\frac{1}{N^2}$$

$$+\frac{\pi^4}{16}\left[\frac{1}{8}\left(\frac{1}{\eta^2}-\frac{4}{3}\right)^2+\frac{1}{3}\left(\frac{1}{\eta^2}-\frac{16}{15}\right)\right]\frac{1}{N^4}+\mathcal{O}(N^{-6})$$

$$\mathcal{E}_1^{(0)}(0,1;N)=\frac{1}{8}-\frac{\pi^2}{128}\left(\frac{1}{\eta^2}-\frac{4}{3}\right)\frac{1}{N^2}+\mathcal{O}(N^{-4}) \tag{10.45}$$

and $\mathcal{E}_1^{(1)}(-1/2,1;N)=\mathcal{E}_1^{(1)}(1/2,1;N)=\frac{1}{2}\mathcal{E}_0^{(0)}(0,1;N)$. In the $N\to\infty$ limit, these scaled gaps converge towards the exponents x_ε, x_σ and x_ψ. We see again that the exponent estimates obtained for $N\to\infty$ are indeed universal. Note that only integer powers of $1/N$ appear. In addition, we observe that **all** leading finite-size correction terms (see eq. (10.39)) have the **same** η-dependent prefactor, to leading order in $1/N$, while the next terms appear to contain a second-order effect arising from the leading correction plus a new contribution. These observations will be explained in chapter 13.

Finite-size scaling functions. The scaled energies do depend on the finite-size scaling variable, in this case $z = N(t-1)$. The finite-size scaling functions for the inverse correlation lengths (see eq. (3.30)) are, again just for the lowest levels but easy to write down for the entire spectrum (we have also included the leading finite-size corrections) [168]

$$\frac{1}{2\pi}S_\varepsilon(z,0)=\left[1+\frac{z^2}{\pi^2\eta^2}\right]^{1/2}+\frac{z}{2\eta^2}\left[1+\frac{z^2}{\pi^2\eta^2}\right]^{-1/2}\frac{1}{N}+\mathcal{O}(N^{-2}) \tag{10.46}$$

$$\frac{1}{2\pi}S_\sigma(z,0)=\frac{1}{8}+\frac{1}{4\pi}\frac{z}{\eta}+\frac{\ln 2}{4\pi^2}\frac{z^2}{\eta^2}+\frac{1}{2}R_{1\frac{1}{2},0}\left(\frac{z^2}{4\pi^2\eta^2}\right)-\frac{1}{8}R_{1\frac{1}{2},0}\left(\frac{z^2}{\pi^2\eta^2}\right)+\mathcal{O}(N^{-1})$$

and $S_\psi = S_\varepsilon/2$ where

$$R_{1\frac{1}{2},0}(x):=-4\sum_{r=1}^{\infty}\left(\left(r^2+x\right)^{1/2}-r-\frac{x}{2r}\right) \tag{10.47}$$

is a particular *remnant function* [124]. These are special functions frequently arising in the study of finite lattice sums. We merely list the behaviour for small and large values of x [124]

$$R_{1\frac{1}{2},0}(x)\simeq\begin{cases}\frac{\zeta(3)}{2}x^2-\frac{\zeta(5)}{9}x^3+\mathcal{O}(x^4) & \text{if } x\to 0\\ x\ln x+2\left(C_E-\ln 2-\frac{1}{2}\right)x+2\sqrt{x}-\frac{1}{3}+\frac{2}{\pi}x^{-1/4}e^{-2\pi\sqrt{x}} & \text{if } x\to\infty\end{cases} \tag{10.48}$$

and $C_E = 0.5772\ldots$ is Euler's contant and ζ is the Riemann zeta function.

Considering these results, we note that the leading term in $1/N$ only depends on the combination z/η, in agreement with the two-scale factor universality expectation [278] discussed in chapter 3. We identify the non-universal metric constant $C_1 \sim 1/\eta$. The correction terms, however, do contain further η-dependent terms. One can also check explicitly that the results obtained for the finite-size scaling functions do not depend on taking the Hamitonian limit [65]. We can conclude that at least in this example, the expectations from finite-size scaling theory have been fully confirmed. This is remarkable, since the usual field-theoretic justification [58, G18] of finite-size scaling relies on the ε-expansion around the upper critical dimension $d^* = 4$, which is expected to break down in two dimensions. It will be shown later, using the n-point correlation functions obtained from conformal invariance, that the finite-size scaling functions and finite-size corrections can be reconstructed from knowing the operator content of a given lattice model.

10.4 The Virasoro generators

Finally, after having found the complete operator content, we complete the discussion of the Ising model by giving the explicit construction of the generators of the Virasoro algebra in terms of the fermionic oscillators a_k. This also further illustrates the field-theoretic discussion presented in chapter 6.

Since the discretization condition eq. (10.24) selects the k to be positive, we can adopt the convention that

$$a_{-k} = a_k^+ \quad , \quad k > 0 \tag{10.49}$$

and the anticommutation relations read simply

$$\{a_k, a_\ell\} = \delta_{k+\ell,0} \tag{10.50}$$

Further, in the sequel we shall always multiply k by $N/2\pi$, so they become integer or half-integer. Then the Virasoro generators can be written as (for the sectors $H_0^{(0)}$ and $H_1^{(1)}$, the two other sectors are treated similarly)

$$L_n := \frac{1}{2} \sum_{k \in Z + \frac{1}{2}} k : a_{n-k} a_k : \tag{10.51}$$

where the dots denote the fermionic **normal ordering** and it is understood that the a_k with a negative index are written to the left

$$a_k a_\ell =: a_k a_\ell : + \delta_{k+\ell,0}\, \Theta(k - \ell) \tag{10.52}$$

and

$$\Theta(x) := \begin{cases} 1 \ ; \ x > 0 \\ 0 \ ; \ x \le 0 \end{cases} \tag{10.53}$$

A few comments are necessary here. We have derived earlier the lattice dispersion relation eq. (10.20) and (10.24). In describing the critical behaviour, we are only interested in the lowest excitations of the systems. These lowest excitations come either from the smallest values of the (rescaled) k, or else from those values of k, which are close to the number of sites N. Since these two regimes are clearly distinct, we can treat them separately. In the construction of the L_n given above, we have just kept the linear term of the dispersion relation and only considered the small values of of k. Similarly, there is a second set of fermionic excitations, denoted by b_k, describing those excited states for which the k-values are close to N. The set of Virasoro generators constructed from the b_k are the conjugate generators $\bar{L}_n$. Since the two sets of fermionic operators anticommute, it follows immediately that $[L_n, \bar{L}_m] = 0$. Using eq. (10.49), it is an easy exercise to confirm that the L_n constructed indeed satisfy the hermiticity condition $L_n^+ = L_{-n}$. The commutator is

$$[L_n, L_m] = \frac{1}{4}\sum_{k,\ell} k\ell\,[a_{n-k}a_k, a_{m-\ell}a_\ell] \tag{10.54}$$

This can be evaluated using the identity

$$[AB, CD] = A\{B,C\}D - \{A,C\}BD + CA\{B,D\} - C\{A,D\}B \tag{10.55}$$

to yield

$$\begin{aligned}[L_n, L_m] = \frac{1}{4}\sum_{k,\ell} k\ell\,(&a_{n-k}a_\ell\delta_{k+m-\ell,0} - a_k a_\ell \delta_{n-k+m-\ell,0} \\ &+ a_{m-\ell}a_{n-k}\delta_{k+\ell,0} - a_{m-\ell}a_k\delta_{n-k+\ell,0})\end{aligned} \tag{10.56}$$

At this point one should be careful. Since this expression is singular, it is preferable not to perform an immediate translation on indices but first to reintroduce the normal ordering prescription via eq. (10.52). Using the relation $\sum_k : a_{n-k}a_k := 0$ leads to

$$\begin{aligned}[L_n, L_m] &= \frac{n-m}{2}\sum_k k : a_{n+m-k}a_k : \\ &+ \frac{1}{4}\sum_{k,\ell}\{k\ell\,\delta_{k+\ell,0}\delta_{n+m-k-\ell,0}(\Theta(k-\ell) - \Theta(m-n+k-\ell)) \\ &+ k\ell\,\delta_{k-\ell+m,0}\delta_{k-\ell-n,0}(\Theta(m-\ell-k) - \Theta(n-\ell-k))\} \\ &= (n-m)L_{n+m} + \frac{1}{24}\left(n^3 - n\right)\delta_{n+m,0}\end{aligned} \tag{10.57}$$

and we obtain again the expected central charge $c = 1/2$. Since the critical quantum Hamiltonian is $H = (N/2\pi)(L_0 + \bar{L}_0 - 1/24)$, we have an alternative proof that the critical two-dimensional Ising model is conformal invariant. We shall reuse this technique in chapter 15 in connection with surface critical phenomena.

10.5 Recapitulation

It may be helpful to sum up what the logical steps to be followed are. This is an algorithm which can be followed in a model-independent way.

1. One has to obtain finite-lattice data, thereby making the number of sites N large enough. The Ising model was so simple that these could be obtained from the exact solution. For more complex models, one often will have to be satisfied with numerical data.
2. Use these data to locate the critical point, for example by the phenomenological renormalization procedure described in chapter 3. If applicable, the critical point can also be found by analytic techniques, for example self-duality arguments.
3. Precisely at the critical point, the spectrum of the inverse correlation length finite-size scaling amplitudes should show a tower-like structure. The level spacings in the conformal tower(s) are integers. This can be used to fix the correct normalization factor ζ of the quantum Hamiltonian H. In this context, the conformal tower of the identity is particularly useful as its critical exponents are known beforehand.

4. Knowing the conformal normalization ζ, the central charge c can now be found from the ground state energy.
5. From the knowledge of the central charge c, at least for a minimal model, the finite list of possible primary operators and their scaling dimensions is now available. Using the conformal characters, the expected inverse correlation length spectrum can now be calculated in a straightforward way.
6. Starting from the lowest levels, identify the primary operators realized in the model. Compare with general results such as modular invariance to be discussed in the next chapter.

These calculations will rely on the capability to perform a precise numerical estimation of the $N \to \infty$ limit. This can be conveniently done with the extrapolation techniques described in chapter 9.

11 Modular Invariance

The result eq. (10.44) for the operator content of the $2D$ Ising model raises a question: in principle, having three irreducible representations of the Virasoro algebra for $m = 3$, one could combine in 9 possible ways the holomorphic and antiholomorphic parts to obtain the primary operators. However, just the 5 found can be realized. A part of the explanation comes from the locality requirement for the correlation functions discussed in chapters 5-7. A finer explanation for this selection comes from the requirement of *modular invariance* for the partition function. The presentation follows the work of Cardy [80, 81, G14].

11.1 The modular group

Consider a system defined on a $2D$ lattice with periodic boundary conditions and in both directions of finite extent which has the topology of a *torus*. Mathematically, a torus is conveniently characterized by a complex number τ with imaginary part $\mathrm{Im}\tau > 0$ (not to be confused with the lattice spacing in time direction). The finite periodic lattice can regarded as a rectangle in the complex plane with horizontal extent L_2 and vertical extent iL_1, which is a torus with $\tau = iL_1/L_2$. The partition function $\mathcal{Z}$ depends on τ.

The *modular group* $Sl(2,\mathbb{Z})$ is defined by the transformations

$$\tau \longrightarrow \frac{a\tau + b}{c\tau + d} \quad ; \quad ad - bc = 1 \tag{11.1}$$

where a, b, c, d are integers. It is generated by the two transformations T and S

$$T: \ \tau \longrightarrow \tau + 1 \ , \quad S: \ \tau \longrightarrow -\frac{1}{\tau} \tag{11.2}$$

whose effect on a torus characterized by τ in shown in Figure 28. They satisfy the defining relations $S^2 = 1$ and $(ST)^3 = 1$.

It is necessary to have the partition function $\mathcal{Z}(\tau)$ **invariant** under the modular group. To show this, it is sufficient to discuss the practical consequences of the two basic transformations T and S. Due to the periodic boundary conditions, invariance under T means that we could equivalently regard the same periodic lattice as having $iL_1 + L_2$ as "vertical" extent. We can also reverse our image of the lattice and look at the torus from the side, rather from underneath, to have a practical realization of S. The partition $\mathcal{Z}$ is insensitive to such arbitrary choices to describe the periodic lattice and is therefore modular invariant.

11.2 Implementation for minimal models

The requirement of modular invariance, although apparently trivial by itself, does become important when combined with the information obtained in the strip geometry. Let $q := \exp 2\pi i\tau$. In chapter 8 it was shown that (see eq. (8.37)) the quantum Hamiltonian

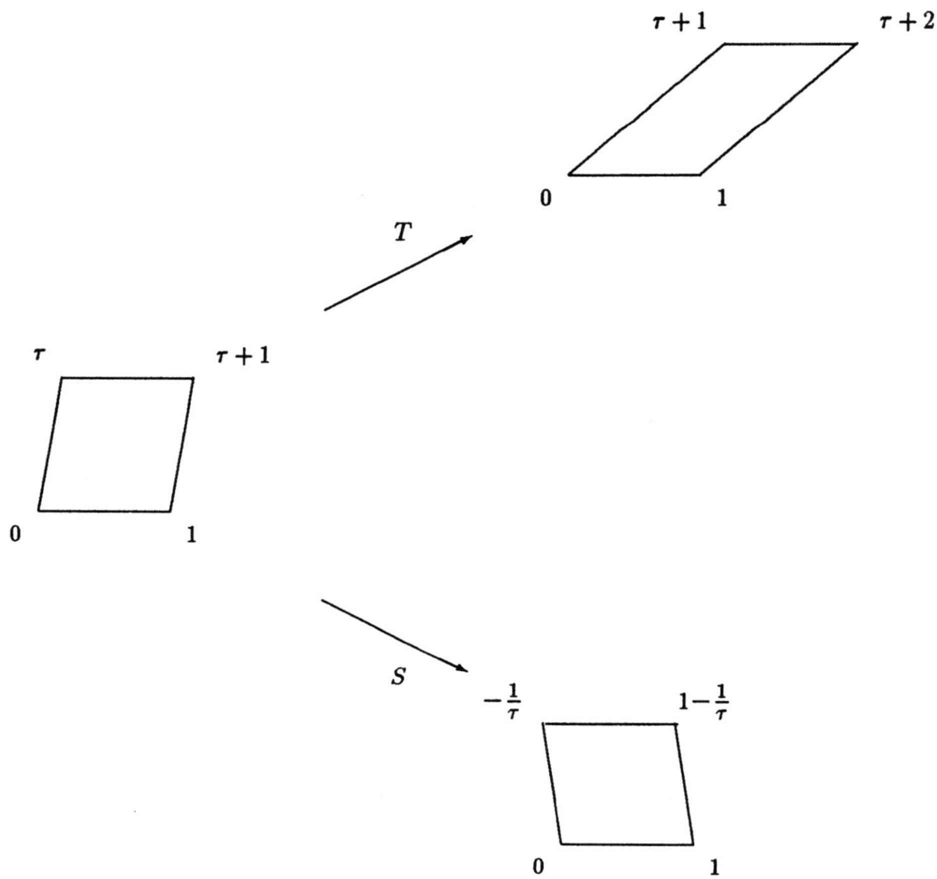

Fig. 28. The torus and the modular transformations T and S.

H and the momentum operator $\hat{\mathcal{P}}$ act as infinitesimal translation operators along and around the cylinder. Consequently

$$\begin{aligned}\mathcal{Z}(q) &= \mathrm{tr}\; e^{(-2\pi \mathrm{Im}\tau\, H + 2\pi i \mathrm{Re}\tau\, \hat{\mathcal{P}})} \\ &= e^{\pi c \mathrm{Im}\tau/6}\, \mathrm{tr}\left(e^{2\pi i\tau\, L_0} - 2\pi i\bar{\tau}\, \bar{L}_0\right) \\ &= q^{-c/24}\bar{q}^{-c/24}\, \mathrm{tr}\, q^{L_0}\bar{q}^{\bar{L}_0} \\ &= \sum_{r,s;r',s'} N_{r,s;r',s'}\, \chi_{r,s}(q)\chi_{r',s'}(\bar{q}) \end{aligned} \tag{11.3}$$

where the Virasoro algebra character functions eq. (10.28) are used and Re and Im denote the real and imaginary part, respectively. Note that $\chi_{r,s}(\bar{q}) = \bar{\chi}_{r,s}(q)$. The non-negative integers $N_{r,s;r',s'}$ give the number of times a certain primary field together with its conformal tower $(\Delta_{r,s}, \overline{\Delta}_{r',s'})$ appears in $\mathcal{Z}$. One also requires that $N_{1,1;1,1} = 1$, which means that there is a single vacuum. One needs the transformation of the characters under the modular group. These are given for the minimal models by $q \to \tilde{q} := \exp(-2\pi i/\tau)$ and [80, 81]

$$T: \quad \chi_{r,s}(q) \to \exp\left[2\pi i\left(\Delta_{r,s} - \frac{c}{24}\right)\right] \chi_{r,s}(q)$$

$$S: \quad \chi_{r,s}(q) \to \sum_{r',s'} S_{r,s}^{r',s'} \chi_{r',s'}(q) \tag{11.4}$$

where

$$S_{r,s}^{r',s'} = \sqrt{\frac{8}{m(m+1)}} (-1)^{(r+s)(r'+s')} \sin\left(\frac{\pi r r'}{m}\right) \sin\left(\frac{\pi s s'}{m+1}\right) \tag{11.5}$$

Below, we shall give an example how these modular transformations are done, but let us first examine the consequences of these transformation rules. Equation (11.4) implies that the $\chi_{r,s}$ transform under a finite-dimensional representation $\mathcal{R}_m$ of the modular group and that $\mathcal{Z}$ transforms under the product representation $\mathcal{R}_m^* \otimes \mathcal{R}_m$. Modular invariance is equivalent to the statement that this in general reducible product contains the identity representation.

Invariance under T requires that $N_{r,s;r',s'} = 0$ unless $\Delta_{r,s} - \overline{\Delta}_{r',s'}$ is an integer. This was shown in chapter 5 to imply the locality or singled-valuedness of the correlation functions. Exactly this same condition was met when discussing the monodromy of the multipoint correlation functions. Considering S shows the existence of at least one solution, since $S^2 = 1$, namely the diagonal one $N_{r,s;r',s'} = \delta_{r,r'}\delta_{s,s'}$. This solution contains all **scalar** operators permitted by the Kac formula. For a second solution, one notes the relations (if m is odd)

$$S_{r,s}^{r',s'} = (-1)^{s'-1} S_{r,s}^{r',m+1-s'} = (-1)^{s-1} S_{r,m+1-s}^{r',s'} \tag{11.6}$$

which implies that the space spanned by $\chi_{r,s} \pm \chi_{r,m+1-s}$ is left invariant. Similar expressions are found for m even. In chapter 7.4, we discussed the particular possible algebras containing integer (or half-integer) spin operators $(\Delta_{r,s}, \Delta_{r,m+1-s})$. The existence of modular invariant partition functions containing such non-scalar operators is a consequence of the very particular modular relationship between $\chi_{r,s}$ and $\chi_{r,m+1-s}$. A remarkable result, due to Cappelli, Itzykson and Zuber [70, 216], states that there are two infinite series of modular invariant partition functions and a finite number of known exceptions, which exhaust the classification of modular invariants. These are in a close correspondence with the *ADE* classification of the simply-laced simple Lie algebras. In Table 13, all the modular invariant critical partition functions are given. We restrict the presentation to the unitary case, the generalization to non-unitary minimal models being straightforward. The relation to the Lie algebra is particularly simple for the E-series. Either m or $m+1$ is the *Coxeter number* of the corresponding Lie algebra: 12, 18 and 30 for E_6, E_7 and E_8 respectively. Moreover the indices $(1,7,4,8,5,11)$, which appear in the E_6 cases, are the Betti numbers of E_6, i.e. if adding 1 to each of them , one obtains the orders of the invariant polynomials of the Lie algebra, the Casimir invariant. The multiplicities of the corresponding scalar operators also correspond to multiplicities of the

Table 13. Critical point partition functions as given by the A, D and E series of unitary models.

Condition	Partition function $\mathcal{Z}$	Class
$m \geq 3$	$\frac{1}{2}\sum_{r=1}^{m-1}\sum_{s=1}^{m}\lvert\chi_{r,s}\rvert^2$	(A_{m-1}, A_m)
$m=4\ell+1$	$\sum_{r=1}^{m-1}\sum_{a=0}^{\ell}\lvert\chi_{r,2a+1}\rvert^2+\sum_{r=1}^{m-1}\sum_{a=0}^{\ell-1}\chi_{r,2a+1}\bar{\chi}_{m-r,2a+1}$	(A_{m-1}, D_{2l+2})
$m=4\ell+2$	$\sum_{a=0}^{\ell}\sum_{s=1}^{m}\lvert\chi_{2a+1,s}\rvert^2+\sum_{a=0}^{\ell-1}\sum_{s=1}^{m}\chi_{2a+1,s}\bar{\chi}_{2a+1,m+1-s}$	$(D_{2\ell+2}, A_m)$
$m=4\ell+3$	$\sum_{r=1}^{m-1}\sum_{a=0}^{\ell}\lvert\chi_{r,2a+1}\rvert^2+\sum_{r=1}^{2\ell+1}\lvert\chi_{r,2\ell+2}\rvert^2$	$(A_{m-1}, D_{2\ell+3})$
	$+\sum_{r=1}^{m-1}\sum_{a=1}^{\ell}\chi_{r,2a}\bar{\chi}_{m-r,2a}$	
$m=4\ell+4$	$\sum_{a=0}^{\ell}\sum_{s=1}^{m}\lvert\chi_{2a+1,s}\rvert^2+\sum_{s=1}^{2\ell+2}\lvert\chi_{2\ell+2,s}\rvert^2$	$(A_{m-1}, D_{2\ell+3})$
	$+\sum_{a=1}^{\ell}\sum_{s=1}^{m}\chi_{2a,s}\bar{\chi}_{2a,m+1-s}$	
$m=11$	$\frac{1}{2}\sum_{r=1}^{10}[\lvert\chi_{r,1}+\chi_{r,7}\rvert^2+\lvert\chi_{r,4}+\chi_{r,8}\rvert^2+\lvert\chi_{r,5}+\chi_{r,11}\rvert^2]$	(A_{10}, E_6)
$m=12$	$\frac{1}{2}\sum_{s=1}^{12}[\lvert\chi_{1,s}+\chi_{7,s}\rvert^2+\lvert\chi_{4,s}+\chi_{8,s}\rvert^2+\lvert\chi_{5,s}+\chi_{11,s}\rvert^2]$	(E_6, A_{12})
$m=17$	$\frac{1}{2}\sum_{r=1}^{16}[\lvert\chi_{r,1}+\chi_{r,17}\rvert^2+\lvert\chi_{r,5}+\chi_{r,13}\rvert^2+\lvert\chi_{r,7}+\chi_{r,11}\rvert^2+\lvert\chi_{r,9}\rvert^2$	(A_{16}, E_7)
	$+(\chi_{r,3}+\chi_{r,15})^*\chi_{r,9}+\chi_{r,9}^*(\chi_{r,3}+\chi_{r,15})]$	
$m=18$	$\frac{1}{2}\sum_{s=1}^{18}[\lvert\chi_{1,s}+\chi_{17,s}\rvert^2+\lvert\chi_{5,s}+\chi_{13,s}\rvert^2+\lvert\chi_{7,s}+\chi_{11,s}\rvert^2+\lvert\chi_{9,s}\rvert^2$	(E_7, A_{18})
	$+(\chi_{3,s}+\chi_{15,s})^*\chi_{9,s}+\chi_{9,s}^*(\chi_{3,s}+\chi_{15,s})]$	
$m=29$	$\frac{1}{2}\sum_{r=1}^{28}[\lvert\chi_{r,1}+\chi_{r,11}+\chi_{r,19}+\chi_{r,29}\rvert^2$	(A_{28}, E_8)
	$+\lvert\chi_{r,7}+\chi_{r,13}+\chi_{r,17}+\chi_{r,23}\rvert^2]$	
$m=30$	$\frac{1}{2}\sum_{s=1}^{30}[\lvert\chi_{1,s}+\chi_{11,s}+\chi_{19,s}+\chi_{29,s}\rvert^2$	(E_8, A_{30})
	$+\lvert\chi_{7,s}+\chi_{13,s}+\chi_{17,s}+\chi_{23,s}\rvert^2]$	

Casimir invariants at the corresponding order. The same applies to all modular invariant partition functions. In the D-series, the indices $2a+1$ for the scalar operators correspond to the Betti number of some D_p Lie algebra. In this case, the doubling of some primary scalar operators corresponds to the existence of two Casimir invariants of the same order in the corresponding Lie algebra. In the A-series, all possible values for r, s are taken. Due to the two indices r, s of the characters, one can put in correspondence the values taken by each of them with a particular Lie algebra. This explains the pairing of Lie algebras which appears in Table 13. Notice that the A-Lie algebras are always present and that no (D, D), (D, E) or (E, E) pairing is possible.

We illustrate in an example the modular transformation generated by S [80]. For simplicity, let $\tau = i\delta$, so that the torus becomes a rectangle. Then under S, one has $\delta \to 1/\delta$ and $q \to \tilde{q} = e^{-2\pi/\delta}$. Consider the generating function of the partitions of the integers $P(q)$ (eq. (10.29))

$$P^{-1}(q) = \prod_{n=1}^{\infty} \left(1 - e^{-2\pi\delta n}\right) = \sum_{n=-\infty}^{\infty} \exp\left(-\pi\delta(3n^2 + n) + i\pi n\right) \tag{11.7}$$

by Euler's pentagonal theorem. One uses Poisson's resummation formula

$$\sum_{n=-\infty}^{\infty} f(n) = \sum_{p=-\infty}^{\infty} \int_{-\infty}^{\infty} dx\, f(x)\, e^{2\pi i p x} \tag{11.8}$$

and this yields, after having done the integration over x

$$P^{-1}(q) = \frac{1}{\sqrt{3\delta}} \exp\left[\frac{\pi}{12}\left(\delta - \delta^{-1}\right)\right] 2 \sum_{p=-\infty}^{\infty} \cos\left(\frac{(2p+1)\pi}{6}\right) \exp\left(-\frac{p(p+1)\pi}{3\delta}\right) \tag{11.9}$$

Now using that

$$\cos\frac{(2p+1)\pi}{6} = \begin{cases} (-1)^n \frac{1}{2}\sqrt{3} & \text{if } p = 3n, 3n-1 \\ 0 & \text{if } p = 3n+1 \end{cases} \tag{11.10}$$

one can replace the sum over p by two sums over n. This readily leads to the relation, written again in terms of τ and recalling the definition of $\tilde{q}$

$$P^{-1}(q) = (-i\tau)^{-1/2} \exp\left[-\frac{i\pi}{12}\left(\tau + \frac{1}{\tau}\right)\right] P^{-1}(\tilde{q}) \tag{11.11}$$

In particular, this implies that the partition function of a free Gaussian model with central charge $c = 1$

$$\mathcal{Z}_1(q, \bar{q}) := q^{-1/24}\bar{q}^{-1/24} P(q) P(\bar{q}) \left(\mathrm{Im}\tau\right)^{-1/2} \tag{11.12}$$

is modular invariant. It does differ from the naive expectation by the last factor, which can be understood [G12] in terms of a zero-mode subtraction from the otherwise ill-defined functional integral.

The discussion so far has ignored the possibility of internal symmetries and its applications are therefore restricted exclusively to lattice systems with **periodic** boundary conditions. However, if the model does have a global Z_2 symmetry, as it is possible for example in the A series (first line in Table 13), one can also write the partition function for antiperiodic boundary conditions [352]. The Z_2 symmetry allows the separation into a charge zero (integer spin) and a charge one (half-integer spin) sector. These charge sectors are exactly the eigenspaces of the lattice charge operator $\hat{Q}$. The corresponding partition functions (only for the A series) are listed in Table 14. They are not invariant under the full modular group, but only under some subgroup.

Table 14. Critical point partition function for Z_2 symmetric systems with antiperiodic boundary conditions in the A series.

	Charge	
m	0	1
$4\ell+1$	$\sum_{r=1}^{2\ell}\sum_{a=0}^{2\ell}\chi_{r,2a+1}\bar{\chi}_{r,m-2a}$	$\sum_{r=1}^{2\ell}\sum_{a=1}^{2\ell}\chi_{r,2a}\bar{\chi}_{r,m+1-2a}$
$4\ell+2$	$\sum_{a=0}^{2\ell+1}\sum_{s=1}^{2\ell}\chi_{2a+1,s}\bar{\chi}_{m-1-2a,s}$	$\sum_{a=1}^{2\ell}\sum_{s=1}^{2\ell+1}\chi_{2a,s}\bar{\chi}_{m-2a,s}$
$4\ell+3$	$\sum_{r=1}^{2\ell+1}\sum_{a=1}^{2\ell+1}\chi_{r,2a}\bar{\chi}_{r,m+1-2a}$	$\sum_{r=1}^{2\ell+1}\sum_{a=0}^{2\ell+1}\chi_{r,2a+1}\bar{\chi}_{r,m-2a}$
$4\ell+4$	$\sum_{a=1}^{2\ell+1}\sum_{s=1}^{2\ell+2}\chi_{2a,s}\bar{\chi}_{m-2a,s}$	$\sum_{a=0}^{2\ell+1}\sum_{s=1}^{2\ell+2}\chi_{2a+1,s}\bar{\chi}_{m-1-2a,s}$

To conclude, we again compare with the results found in the Ising model, eq. (10.44). Since $m = 3$, one has the A series as the only possibility. If one has periodic boundary conditions $\tilde{Q} = 0$, the partition function summed over both charge sectors is

$$\mathcal{Z} = (0,0) + \left(\frac{1}{2}, \frac{1}{2}\right) + \left(\frac{1}{16}, \frac{1}{16}\right) \tag{11.13}$$

while the results of Table 14 agree with the observations for antiperiodic boundary conditions, $\tilde{Q} = 1$. It should be stressed that the results presented in Table 13 do give a *classification of the critical point partition functions* of conformally and modularly invariant systems. They do not, however, specify how the primary fields contained in $\mathcal{Z}$ are distributed into the different charge sectors of the system. This leads to yet a further degree of freedom in constructing models. Although this possibility does not occur for the Ising case $m = 3$, it is present already for $m = 4$ (see chapter 12).

We finally mention a beautiful formula, due to E. Verlinde [321], which relates the formal operator product expansion (5.10) of primary operators

$$\phi_i \cdot \phi_j = N_{i,j}^{k}\, \phi_k \tag{11.14}$$

to the modular transformation S, which can be written

$$S: \qquad \chi_i(q) \quad \rightarrow \quad \sum_j S_i^j\, \chi_j(q) \tag{11.15}$$

to avoid a formalism with too many indices. It can be shown that

$$\sum_k S_k^\ell N_{i,j}^k = \frac{S_i^\ell S_j^\ell}{S_0^\ell} \tag{11.16}$$

where 0 stands for the identity operator $\phi_0 = \mathbf{1}$. This formula shows a direct link between the partition function and the OPE, which can be obtained without the knowledge of the correlation functions. We remark that this result was applied in a recent study of the Kondo effect [6].

11.3 Modular invariance at c = 1

Up to now, only systems with a **finite** number of primary fields as described by **minimal** models were studied. However, there exist plenty of models which do not fall into this category. We have already encountered the example of the free Gaussian field in chapter 7 described by the action

$$S = \int d^2x\,(\partial\phi)^2 = \int d^2z\,(\partial_{\bar{z}}\phi)(\partial_z\phi) \tag{11.17}$$

From its partition function eq. (11.12), one reads off the central charge $c = 1$. Primary operators for this action are the exponentials $e^{i\alpha\phi}$, also called *vertex operators*, which have the conformal weights $\Delta = \overline{\Delta} = \alpha^2$. Since the action is invariant under the transformations

$$\phi \to \phi + \text{ const.} \quad , \quad \phi \to -\phi \tag{11.18}$$

one may use these to restrict the configuration space. This leads to two possible compactifications, the *circle or Coulomb* models and the *orbifold* models.

For a Coulombic model, one identifies two configurations with fields ϕ and $\phi + 2\pi\rho$, where ρ is a constant and sometimes is called the *compactification radius.* Nothing changes for the correlation functions of the vertex operators, if one only takes $\alpha = n/\rho$. One example of these models is the **XY model** as defined in chapter 1. It is known that this model has a second-order transition described by the formation of *vortices* [229]. Besides the vortex operators one can also introduce operators corresponding to a *frustration line* such that as $\vec{r}$ circles around one of the end points (a vortex), ϕ increases by $2\pi m\rho$ where m is an integer. One would like to identify these operators within the framework of conformal invariance.

The implementation of modular invariance is done via the two compactification schemes mentioned above which we now describe in detail.

Circle or Coulomb models. Consider the model on the torus, characterized by the two periods ω_1 and ω_2 [130, 131]. The compactification condition is

$$\phi(z + k\omega_1 + k'\omega_2) - \phi(z) = 2\pi\rho(km + k'm') \tag{11.19}$$

where the dependence on $\bar{z}$ is suppressed. The fields satisfying this condition can be written as $\phi = \phi_{per} + \phi_{cl}$ where ϕ_{per} is a fluctuating periodic field and ϕ_{cl} is the solution of the equations of motion compatible with (11.19)

$$\phi_{cl}(z) = 2\pi\rho\left(\frac{z}{\omega_1}\frac{m\bar{\tau} - m'}{\bar{\tau} - \tau}\right) + \text{ c.c.} \tag{11.20}$$

The action is then $S = S_{per} + \pi\rho^2|m\tau - m'|^2/\mathrm{Im}\tau$. The model is thus decomposed into sectors described by the integers m, m'. The partition function of a sector is

$$\mathcal{Z}_{m',m}(\rho,\tau) := \mathcal{Z}_1(\tau)\exp\left(-\pi\rho^2\frac{|m\tau - m'|^2}{\mathrm{Im}\tau}\right) \tag{11.21}$$

where $\mathcal{Z}_1$ is defined in eq. (11.12). One can show that under a modular transformation, using the techniques of the previous section

$$\mathcal{Z}_{m',m}\left(\rho, \frac{a\tau+b}{c\tau+d}\right) = \mathcal{Z}_{am'+bm,cm'+dm}(\rho,\tau) \tag{11.22}$$

Consequently, there are two obvious modular invariants. One is simply $\mathcal{Z}_{0,0} = \mathcal{Z}_1$. The other one is obtained by summing over all values of m, m'

$$\begin{aligned} \mathcal{Z}(\rho,\tau) &:= \rho\sum_{m',m}\mathcal{Z}_{m',m}(\rho,\tau) \\ &= (q\bar{q})^{-1/24}P(q)P(\bar{q})\sum_{n,m}\exp\left[-2\pi i\mathrm{Re}\tau\cdot mn - 2\pi\mathrm{Im}\tau\frac{1}{2}\left(\rho^2m^2 + \frac{n^2}{\rho^2}\right)\right] \end{aligned} \tag{11.23}$$

after applying the Poisson formula eq. (11.8) to m'. Note that in this expression the factor $\mathrm{Im}\tau$ of eq. (11.12) has disappeared. Now let $q = \exp 2\pi i\tau$ and recalling eq. (11.3), one recognizes the scaling dimension $x_{n,m}$ and the spin $s_{n,m}$ of the primary operators $\mathcal{O}_{n,m}$ with

$$x_{n,m} = \frac{1}{2}\left(\rho^2m^2 + \frac{n^2}{\rho^2}\right) \quad , \quad s_{n,m} = nm \tag{11.24}$$

Finally, the Coulombic partition function becomes

$$\mathcal{Z}(\rho,\tau) = \mathcal{Z}(\rho^{-1},\tau) = (q\bar{q})^{-1/24}P(q)P(\bar{q})\sum_{n,m=-\infty}^{\infty} q^{\Delta_{n,m}}\bar{q}^{\overline{\Delta}_{n,m}} \tag{11.25}$$

with the conformal weights

$$\Delta_{n,m} = \frac{1}{4}\left(\frac{n}{\rho} + m\rho\right)^2 \quad , \quad \overline{\Delta}_{n,m} = \frac{1}{4}\left(\frac{n}{\rho} - m\rho\right)^2 \tag{11.26}$$

For $\rho > 1$, the quantum number n is said to describe *electric* fields, while the quantum number m is said to describe *magnetic* fields. It is apparent that all conformal weights are ρ-dependent, but the spins are always integer.

To make the connection with the XY model, one identifies the spin operator with the conformal operator $\mathcal{O}_{\pm 1,0}$ such that the exponent $\eta = 2x_{\pm 1,0} = 1/4$ is reproduced. This implies the choice $\rho = 2$. The corresponding magnetic or vortex operator is then $\mathcal{O}_{0,\pm 1}$, implying $x_{0,\pm 1} = 2$. This is satisfactory, since a **marginal** operator is needed to account for the temperature-dependent exponent $\eta(T)$ is the low-temperature phase of the XY model. In particular, if this identification is correct, this yields for the central charge of the XY model

$$c_{XY} = 1 \tag{11.27}$$

Orbifold models. An *orbifold* is defined as follows. Let $\mathcal{M}$ be a manifold and let G be a finite group acting on $\mathcal{M}$. Consider the ***fixed points*** under the action of G: $gx = x$, where g is an element of G. Define the equivalence relation $x \equiv gx$ for all elements g of G. Then the quotient space $\mathcal{M}/G$ is an orbifold. As an example, take the circle S^1 with points $x = x + 2\pi r \bmod 2\pi$ and the group Z_2 with the generator $gx = -x$. The fixed points are $x = 0, x = \pi r$. So topologically the orbifold S^1/Z_2 is a line segment (see Figure 29). It is distinguished from a manifold by the existence of a discrete set of singular points. In the

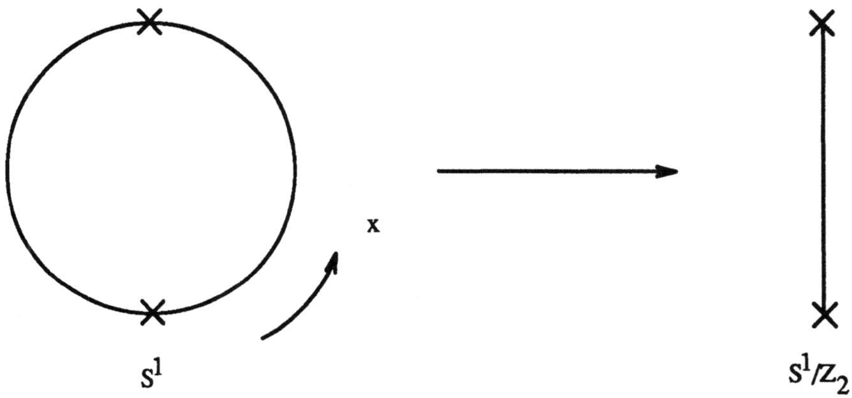

Fig. 29. The manifold S^1 and the orbifold S^1/Z_2.

context of conformal invariance, an orbifold construction on a modular invariant theory with some discrete symmetry group G amounts to forming a quotient of that theory with respect to G such that the resulting theory is also modular invariant.

Consider the Gaussian model. The symmetry group is Z_2 with the generator $g\phi = -\phi$. Consequently, on the torus one has four possible boundary conditions, obtained as combinations from periodic or antiperiodic boundary conditions ($\pm$) in either direction. Then the orbifold partition function is [147, G15]

$$\begin{aligned} \mathcal{Z}_{orb}(\rho,\tau) &:= (q\bar{q})^{-1/24}\left[\operatorname{tr}^{(+)}\frac{1}{2}(1+g)q^{L_0}\bar{q}^{\bar{L}_0} + \operatorname{tr}^{(-)}\frac{1}{2}(1+g)q^{L_0}\bar{q}^{\bar{L}_0}\right] \\ &= \frac{1}{2}\left(\mathcal{Z}(\rho,\tau) + 2\mathcal{Z}(2,\tau) - \mathcal{Z}(1,\tau)\right) \end{aligned} \tag{11.28}$$

where $\mathcal{Z}(\rho,\tau)$ is defined in eq. (11.25).

A few remarks are in order. First, one sees that $\mathcal{Z}_{orb}(1,\tau) = \mathcal{Z}(2,\tau)$ which is the XY model. Secondly, after some tedious algebra, one has $\mathcal{Z}_{orb}(\sqrt{2},\tau) = \mathcal{Z}^2_{Ising}$ and the model is equivalent to two decoupled Ising models. Third, by writing down the primary fields, some of the critical exponents do depend on the orbifold compactification radius ρ, while others do not. This will be made more explicit when discussing the Ashkin-Teller model below. Finally, one can write the partition functions of the p-state Potts model with $0 < p < 4$ and of the $O(n)$ model with $-2 \leq n \leq 2$ in terms of $\mathcal{Z}$. For the p-state

Potts model, take

$$\mathcal{Z}(p;\tau) = \frac{\rho}{2}\mathcal{Z}_1(\tau)\sum_{m',m} \exp\left[-\frac{\pi\rho^2}{4}\frac{|m\tau - m'|^2}{\mathrm{Im}\tau}\right]\cos[\pi e_0 < m, m' >]$$
$$+ \frac{1}{2}(p-1)\left[\mathcal{Z}(\rho,\tau) - \mathcal{Z}(\rho/2,\tau)\right] \qquad (11.29)$$

where $< m', m >$ is the largest common positive divisor of m' and m. Put

$$\rho^2 = \frac{4m}{m+1} \quad , \quad p = \left(2\cos\frac{\pi}{m+1}\right)^2 \quad , \quad e_0 = \frac{2}{m+1} \qquad (11.30)$$

where m is used to calculate the central charge from the unitary Kac formula eq. (10.26). For the $O(n)$ model, put

$$\rho^2 = \frac{m+1}{m} \quad , \quad n = \left(2\cos\frac{\pi}{m}\right)^2 \quad , \quad e_0 = \frac{1}{m} \qquad (11.31)$$

and the term proportional to $p-1$ in (11.29) should be dropped.

In Figure 30 the known conformal systems with central charge $c = 1$ are displayed. It is generally believed that this is the complete list of modular invariant systems with

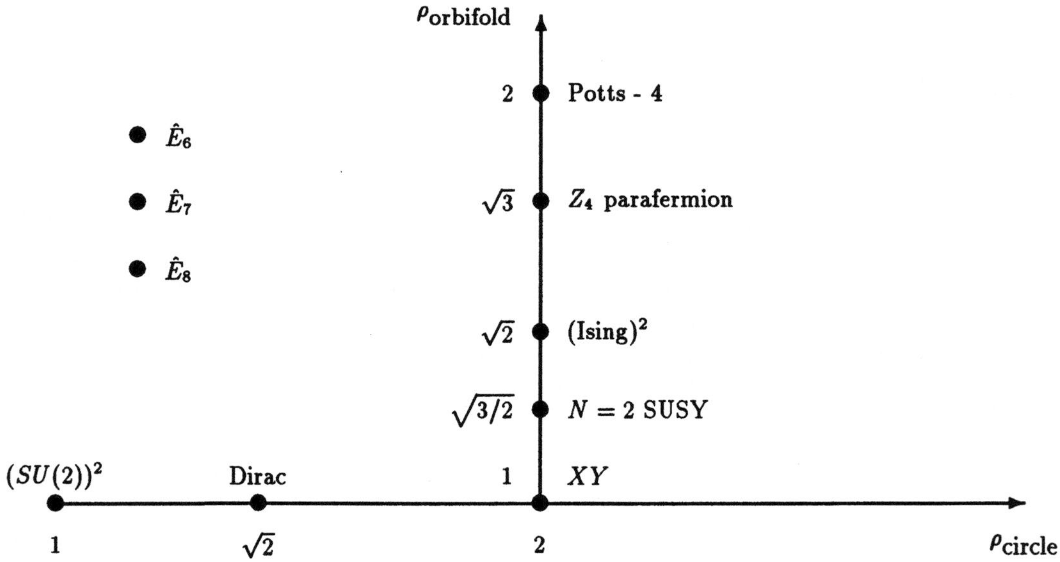

Fig. 30. Known conformal modular invariant systems with $c = 1$.

$c = 1$. We remark that the three isolated points correspond to systems which cannot be continuously connected to the Coulombic or orbifold models. Their partition functions were constructed by Pasquier [266] and read

$$\mathcal{Z}(\rho,\tau;\hat{E}_6) = \frac{1}{2}\left(2\mathcal{Z}(3,\tau) + \mathcal{Z}(2,\tau) - \mathcal{Z}(1,\tau)\right)$$

$$\mathcal{Z}(\rho,\tau;\hat{E}_7) = \frac{1}{2}\left(\mathcal{Z}(4,\tau) + \mathcal{Z}(3,\tau) + \mathcal{Z}(2,\tau) - \mathcal{Z}(1,\tau)\right)$$
$$\mathcal{Z}(\rho,\tau;\hat{E}_8) = \frac{1}{2}\left(\mathcal{Z}(5,\tau) + \mathcal{Z}(3,\tau) + \mathcal{Z}(2,\tau) - \mathcal{Z}(1,\tau)\right) \qquad (11.32)$$

Since the XY model appears at the meeting point of the two compactifications, one might speculate that this peculiar property might be related to the presence of the Kosterlitz-Thouless transition.

We have seen that the $c = 1$ models studied were classified according the Dynkin diagrams of the simple Lie algebras of the ADE series. It is well known that the subgroups of $SU(2)$ can be also classified according to them. It is then tempting to try to construct new integrable models from the Dynkin diagrams arising from the subgroups of larger algebras, like $SU(n)$. For a recent review on this line of research, see [132].

12 Further Developments and Applications

Having treated the Ising model in quite some detail, we now study several other systems to give an overview of some of the possible applications. Several examples will serve to illustrate the techniques developed so far and to introduce a few new concepts. Many of the results to be given were obtained by the numerical techniques described in chapter 9 and we shall mainly give a review on existing results. For computational details, we refer to the original papers.

12.1 Three-states Potts model

The quantum Hamiltonian of the three-states Potts model is, with the correct normalization for conformal invariance applications [140]

$$H = -\frac{2}{3\sqrt{3}} \sum_{n=1}^{N} \left\{ \sigma_n + \sigma_n^+ + \lambda \left(\Gamma_n^+ \Gamma_{n+1} + \Gamma_n \Gamma_{n+1}^+ \right) \right\} \tag{12.1}$$

where the 3×3 matrices σ and Γ were defined in chapter 9. The model has a single second-order transition at the self-duality point $\lambda_c = 1$. Consider the case of toroidal boundary conditions

$$\Gamma_{N+1} = \omega^{\tilde{Q}} \Gamma_1 \ ; \ \tilde{Q} = 0, 1, 2 \tag{12.2}$$

which gives the blocks $H_Q^{(\tilde{Q})} = H_{\tilde{Q}}^{(Q)}$ and also $H_1^{(\tilde{Q})} = H_2^{(\tilde{Q})}$ in the sense that the spectra are the same. One further has to replace $\mathcal{P}_{shift}$ from eq. (10.32) by $\mathcal{P}_{shift} = \tilde{Q}/3$.

From finite-size scaling calculations, the central charge $c = 4/5$ was found (for tables of finite-size data, see [140, 143]) and the operator content can thus be written in terms of the $m = 5$ Virasoro characters. The degeneracies following from the character functions of the primary fields are given in Table 15. This gives half of the entries of the Kac table. The second half of the conformal operators is obtained from

$$\phi_{r,s} = \phi_{m-r,m+1-s} \tag{12.3}$$

The operator content of the five independent sector is [140, 81]

$$\begin{aligned}
H_{0,+}^{(0)} &= (0,0) + \left(\frac{2}{5}, \frac{2}{5}\right) + \left(\frac{7}{5}, \frac{7}{5}\right) + (3,3) \\
H_{0,-}^{(0)} &= \left(\frac{2}{5}, \frac{7}{5}\right) + \left(\frac{7}{5}, \frac{2}{5}\right) + (3,0) + (0,3) \\
H_1^{(0)} &= \left(\frac{1}{15}, \frac{1}{15}\right) + \left(\frac{2}{3}, \frac{2}{3}\right) \\
H_1^{(1)} &= \left(\frac{2}{5}, \frac{1}{15}\right) + \left(0, \frac{2}{3}\right) + \left(\frac{7}{5}, \frac{1}{15}\right) + \left(3, \frac{2}{3}\right) \\
H_1^{(2)} &= \left(\frac{1}{15}, \frac{2}{5}\right) + \left(\frac{2}{3}, 0\right) + \left(\frac{1}{15}, \frac{7}{5}\right) + \left(\frac{2}{3}, 3\right)
\end{aligned} \tag{12.4}$$

Table 15. Virasoro characters of the primary fields $\phi_{r,s}$ for $m = 5$.

I	0	1	2	3	4	5	6	7	8	9	10	11	12	13	14	15	r	s
$d(0,I)$	1	0	1	1	2	2	4	4	7	8	12	14	21	24	34	41	1	1
$d(1/8,I)$	1	1	1	2	3	4	6	8	11	15	20	26	35	45	58	75	1	2
$d(2/3,I)$	1	1	2	2	4	5	8	10	15	19	27	34	46	58	77	96	1	3
$d(13/8,I)$	1	1	2	3	4	6	9	12	16	22	29	38	50	64	82	105	1	4
$d(2/5,I)$	1	1	1	2	3	4	6	8	11	15	20	26	35	45	58	74	2	1
$d(1/40,I)$	1	1	2	3	4	6	9	12	17	23	31	41	54	70	91	117	2	2
$d(1/15,I)$	1	1	2	3	5	7	10	14	20	26	36	47	63	81	106	135	2	3
$d(21/40,I)$	1	1	2	3	5	7	10	14	19	26	35	46	61	79	102	131	2	4
$d(7/5,I)$	1	1	2	2	4	5	8	10	15	19	26	33	45	56	74	92	3	1
$d(3,I)$	1	1	2	3	4	5	8	10	14	18	24	31	41	51	66	83	4	1

where the index $\pm$ refers to charge-conjugation eigenstates. Comparing with Table 6, we note that only a subset of the possible $m = 5$ primary fields is present and that some of the scaling operators occur twice in the full operator content. For periodic boundary conditions, one has two scalar relevant primary fields with exponents

$$x_\varepsilon = \frac{4}{5} \;, \quad x_\sigma = \frac{2}{15} \tag{12.5}$$

from which all other bulk exponents can be found by the familiar scaling relations eq. (9.21) and in agreement with earlier numerical studies, see [329, 247]. Constructing the partition function for periodic boundary conditions and comparing with Table 13, one sees that the model is the first member of the infinite D series. In contrast to the models of the A series, the three-states Potts model does contain also non-scalar fields, even for periodic boundary conditions. These correspond to **conserved currents** of spins 1 and 3, respectively and have negative charge conjugation.

These results can also be obtained analytically from a mapping onto the XXZ chain with boundary fields using the Temperley-Lieb algebra [12]. The simplest example of this correspondence occurs for free boundary conditions and will be described in chapter 15.

12.2 Tricritical Ising model

Operator content. As a further example, we consider the tricritical Ising model. There are quite a few known realizations of this universality class. One can start from a A_4 RSOS model or a ϕ^6 theory. Widely studied is the realization as an Ising model with vacancies, whose concentration is governed by a chemical potential. We have done so in chapter 1. This leads to the *Blume-Capel model* which has the quantum Hamiltonian [136, 10]

$$H = -\frac{1}{\zeta}\sum_{n=1}^{N}\left\{S^x(n)S^x(n+1) - \alpha\left(S^x(n)\right)^2 - \beta S^z(n) - \gamma\left(S^z(n)\right)^2\right\} \tag{12.6}$$

where

$$S^x = \frac{1}{\sqrt{2}}\begin{pmatrix} 0 & 1 & 0 \\ 1 & 0 & 1 \\ 0 & 1 & 0 \end{pmatrix} \;, \quad S^z = \begin{pmatrix} 1 & & \\ & 0 & \\ & & -1 \end{pmatrix} \tag{12.7}$$

and α, β, γ are free parameters and ζ is chosen to permit the application of conformal invariance. The charge operator is

$$\hat{Q} = \prod_{n=1}^{N} \left[1 - (S^z(n))^2\right] \tag{12.8}$$

Another possibility is to consider an *Ising metamagnet* with quantum Hamiltonian

$$H = -\frac{1}{\zeta} \sum_{n=1}^{N} \{t\sigma^z(n) - \sigma^x(n)\sigma^x(n+1) + \kappa\sigma^x(n)\sigma^x(n+2) + h\sigma^x(n)\} \tag{12.9}$$

Here the non-ordering field h plays the role of the chemical potential in the Blume-Capel model. The presence of the second term with $\kappa > 0$ is necessary to have a tricritical point. The non-ordering magnetic field h plays the role of the chemical potential $e^{-\mu}$ in the Blume-Capel model.

The phase diagram of these models (see Figure 5) contains lines of first-order and second-order transitions. The meeting point of these lines is the tricritical point. At this point, two ordered and one disorderd phase(s) become indistinguishable. The tricritical point is thus characterized by the fact that two lengths will diverge simultaneously. These are the correlation length ξ and the *persistence length* $\tilde{\xi}$. They can be calculated from the two lowest gaps in the energy spectrum using the notation of eq. (10.32)

$$\xi^{-1} = E_1^{(0)}(0,1;N) - E_0^{(0)} \quad , \quad \tilde{\xi}^{-1} = E_0^{(0)}(0,2;N) - E_0^{(0)} \tag{12.10}$$

and where $E_0^{(0)}$ is the ground state energy. The tricritical point is now found by solving simultaneously [291]

$$\begin{aligned} N\xi^{-1}(N;\alpha_t,\beta_t) &= (N+1)\xi^{-1}(N+1;\alpha_t,\beta_t) \\ N\tilde{\xi}^{-1}(N;\alpha_t,\beta_t) &= (N+1)\tilde{\xi}^{-1}(N+1;\alpha_t,\beta_t) \end{aligned} \tag{12.11}$$

where the parameter notation of the Blume-Capel model was used. Computationally, one has to be aware that the finite-size scaling region for tricritical behaviour, as opposed to ordinary second-order or first-order criticality, might become relatively small and a rather fine mesh of data points (α, β) or (t, h) may be necessary. The numerical results for the tricritical point are for the Blume-Capel quantum spin model with $\gamma = 0$ [144]

$$\alpha_t = 0.910207(4) \quad , \quad \beta_t = 0.415685(6) \quad , \quad \zeta_t = 0.56557(50) \tag{12.12}$$

and for the Ising metamagnet quantum chain [179]

$$\begin{aligned} \kappa = 1 &: t_t = 1.0940(9) \quad , \quad h_t = 1.8794(15) \quad , \quad \zeta_t = 4.36(4) \\ \kappa = \frac{1}{2} &: t_t = 0.5628(3) \quad , \quad h_t = 1.92776(5) \quad , \quad \zeta_t = 2.154(5) \end{aligned} \tag{12.13}$$

For estimates of the tricritical point of the isotropic systems, see [233, 234, 291, 36, 186].

The central charge is found to be $c = 7/10$ [24], which implies $m = 4$. The corresponding degeneracies of the primary fields are given in Table 16. The operator content

Table 16. Virasoro characters of the primary fields $\phi_{r,s}$ for $m = 4$.

I	0	1	2	3	4	5	6	7	8	9	10	11	12	13	14	15	r	s
$d(0,I)$	1	0	1	1	2	3	4	4	7	8	12	14	20	23	32	38	1	1
$d(1/10,I)$	1	1	1	2	3	4	6	8	11	14	19	24	32	40	51	64	1	2
$d(3/5,I)$	1	1	2	2	4	5	7	9	13	16	22	27	36	45	58	71	1	3
$d(7/16,I)$	1	1	1	2	3	4	6	8	10	14	18	23	30	38	48	61	2	1
$d(3/80,I)$	1	1	2	3	4	6	8	11	15	20	26	34	44	56	71	90	2	2
$d(3/2,I)$	1	1	2	2	3	4	6	7	10	13	17	21	28	34	44	54	3	1

is [24]

$$\begin{aligned} H_0^{(0)} &= (0,0) + \left(\frac{1}{10},\frac{1}{10}\right) + \left(\frac{3}{5},\frac{3}{5}\right) + \left(\frac{3}{2},\frac{3}{2}\right) \\ H_0^{(1)} = H_1^{(0)} &= \left(\frac{3}{80},\frac{3}{80}\right) + \left(\frac{7}{16},\frac{7}{16}\right) \\ H_1^{(1)} &= \left(0,\frac{3}{2}\right) + \left(\frac{3}{2},0\right) + \left(\frac{1}{10},\frac{3}{5}\right) + \left(\frac{3}{5},\frac{1}{10}\right) \end{aligned} \tag{12.14}$$

and one sees that the partition function for periodic boundary conditions

$$\mathcal{Z}(A_4) := H_0^{(0)} + H_1^{(0)} \tag{12.15}$$

is a member of the modular invariant A series of Table 13. In particular, one reads off the tricritical bulk exponents

$$x_\varepsilon = \frac{1}{5} \quad , \quad x_{\varepsilon'} = \frac{6}{5} \quad , \quad x_\sigma = \frac{3}{40} \quad , \quad x_{\sigma'} = \frac{7}{8} \tag{12.16}$$

in execellent agreement with the more traditional approaches,see [237].

However, this is **not** the only model with a modular invariant partition function and $m = 4$. A second model can be obtained by projecting out the corresponding states from the XXZ quantum chain to be described later. The model obtained has among others the sectors [14]

$$\begin{aligned} D_0^{(1)} &= (0,0) + \left(\frac{7}{16},\frac{7}{16}\right) + \left(\frac{3}{2},\frac{3}{2}\right) \\ D_0^{(2)} &= \left(\frac{3}{80},\frac{3}{80}\right) + \left(\frac{1}{10},\frac{1}{10}\right) + \left(\frac{3}{5},\frac{3}{5}\right) \end{aligned} \tag{12.17}$$

and the partition function

$$\tilde{\mathcal{Z}} := D_0^{(1)} + D_0^{(2)} = \mathcal{Z}(A_4) \tag{12.18}$$

but the *assignment of the primary fields to the sectors* of the model is different. In particular, one can also show that the operator product expansions of the sector $D_0^{(1)}$ closes. One reads off the exponents of the relevant fields

$$x_\varepsilon = \frac{7}{8} \;,\; x_\sigma = \frac{3}{40} \;,\; x_{\sigma'} = \frac{1}{5} \;,\; x_{\sigma''} = \frac{6}{5} \tag{12.19}$$

This identification is based on taking $(0,0)$ as the identity primary field.

This example shows that the ADE classification of the modular invariant partition functions $\mathcal{Z}$ of unitary minimal models is **not** equivalent to a classification of the universality classes of the $2D$ conformally and modularly invariant systems. Also, it has become evident that one cannot characterize an universality class just by giving a few critical exponents. The specification of the critical point partition function and the distribution of the primary fields into all the sectors $H_Q^{(\bar{Q})}$ of the model is required.

Supersymmetry and superconformal invariance. Sometimes it is possible to put several irreducible representations of some algebra together into a **single** irreducible representation of a larger algebra. This happens in the tricritical Ising model with *superconformal invariance.* The concept of *extended symmetries* is an important one with arises in many systems and we present superconformal invariance just as one example standing for a wealth of others.

We first recall briefly the notion of *supersymmetry.* For a detailed mathematical introduction, see [298] and for supersymmetric field theory, see e.g. [325]. A *Lie superalgebra* is spanned by so-called *even* generators E_i and so-called *odd* generators V_α, where the indices i and α run over the values $i = 1, \ldots N$ and $\alpha = 1, \ldots M$. The generators satisfy the commutation and anticommutation relations

$$[E_i, E_j] = C_{i,j}^k \, E_k \;,\; [E_i, V_\alpha] = D_{i,\alpha}^\beta \, V_\beta \;,\; \{V_\alpha, V_\beta\} = F_{\alpha,\beta}^i \, E_i \tag{12.20}$$

where C, D, F are structure constants such that the *Jacobi identies*

$$\begin{aligned} [E_i,[E_j,E_k]] + [E_j,[E_k,E_i]] + [E_k,[E_i,E_j]] &= 0 \\ [V_\alpha,[E_i,E_j]] + [E_i,[E_j,V_\alpha]] + [E_j,[V_\alpha,E_i]] &= 0 \\ [E_i,\{V_\alpha,V_\beta\}] + \{V_\alpha,[V_\beta,E_i]\} - \{V_\beta,[E_i,V_\alpha]\} &= 0 \\ [V_\alpha,\{V_\beta,V_\gamma\}] + [V_\beta,\{V_\gamma,V_\alpha\}] + [V_\gamma,\{V_\alpha,V_\beta\}] &= 0 \end{aligned} \tag{12.21}$$

are satisfied. The commutator $[,]$ and the anticommutator $\{,\}$ can be defined in terms of products of matrices

$$[A,B] := AB - BA \;,\; \{A,B\} := AB + BA \tag{12.22}$$

To motivate a physicist's interest in this apparently rather abstract definition, we note that supersymmetries already show up for very well-known systems. For example, the quantum mechanic harmonic oscillator as described by the Hamiltonian $H = (\hat{x}^2 + \hat{p}^2)/2$, where $\hat{x}$ and $\hat{p}$ are the canonical coordinate and momentum operators, has a dynamic supersymmetry [97]. The even generators are S^z, S^+, S^- where

$$S^z = H = \frac{1}{2}\{b^-, b^+\} \;,\; S^- = \frac{1}{2}\{b^-, b^-\} \;,\; S^+ = \frac{1}{2}\{b^+, b^+\} \tag{12.23}$$

and the odd generators are the bosonic creation and annihilation operators b^+, b^- built in the familiar way from $\hat{x}, \hat{p}$ via $b^{\pm} = (\hat{x} \mp i\hat{p})/\sqrt{2}$. These five generators generate a Lie superalgebra which is known as ***orthosympletic*** superalgebra $Osp(1;2)$. Since the Hamiltonian appears as one of the generators of the algebra, representation theory can be used to calculate the energy spectrum and transition rates. Because the harmonic oscillator spectrum can be described by a **single** irreducible representation of $Osp(1;2)$, this Lie superalgebra gives the **maximal** dynamical symmetry for that system [97]. The familiar unitary ($U(1)$) or symplectic ($Sp(2)$) dynamic symmetries of the harmonic oscillator are Lie subalgebras of the orthosymplectic Lie superalgebra [97].

The ***superconformal algebra*** [134, 46, 116] contains besides the "even" Virasoro generators L_n also "odd" generators G_n. The commutation and anticommutation relations are

$$\begin{aligned} [L_n, L_m] &= (n-m)L_{n+m} + \frac{c}{12}\left(n^3 - n\right)\delta_{n+m,0} \\ [L_n, G_m] &= \left(\frac{n}{2} - m\right) G_{n+m} \\ \{G_n, G_m\} &= 2L_{n+m} + \frac{c}{3}\left(m^2 - \frac{1}{4}\right)\delta_{n+m,0} \end{aligned} \tag{12.24}$$

While the L_n have integer indices, for the G_n one can have a half-integer index (Neveu-Schwarz) or else an integer index (Ramond). This implies that in the strip geometry, the Neveu-Schwarz "fermionic" fields have antiperiodic boundary conditions, while the Ramond "fermionic" fields have periodic ones. In the Ramond sector, G_0 acts as a supersymmetry charge, since $G_0^2 = L_0 - c/24$. Thus in the Ramond sector, the heighest weight states come in pairs $|\Delta^{\pm}\rangle$, with $|\Delta^-\rangle = G_0|\Delta^+\rangle$. Supersymmetry is unbroken if $G_0|\Delta^+\rangle = 0$, which is possible if $\Delta^+ = c/24$.

The unitary irreducible representations are described by a super Kac formula, if $c < 3/2$

$$\begin{aligned} c &= \frac{3}{2}\left(1 - \frac{8}{M(M+2)}\right) \quad ; \quad M = 2,3,4,\ldots \\ \Delta_{r,s} &= \frac{[(M+2)r - Ms]^2 - 4}{8M(M+2)} + \frac{1}{32}\left(1 - (-1)^{r-s}\right) \end{aligned} \tag{12.25}$$

For the Neveu-Schwarz sector, one has $r - s$ even and for the Ramond sector, one has $r - s$ odd. The character functions [150, 256] are for the Neveu-Schwarz sector

$$\begin{aligned} \chi_{r,s}(q) &:= q^{-c/24} \prod_{n=1}^{\infty} \left(\frac{1+q^{n-1/2}}{1-q^n}\right) \sum_{k=-\infty}^{\infty} \left(q^{\Delta_{2kM+r,s}} - q^{\Delta_{2kM+r,-s}}\right) \\ \tilde{\chi}_{r,s}(q) &:= q^{-c/24} \prod_{n=1}^{\infty} \left(\frac{1-q^{n-1/2}}{1-q^n}\right) \sum_{k=-\infty}^{\infty} (-1)^{Mk} \left(q^{\Delta_{2kM+r,s}} - q^{\Delta_{2kM+r,-s}}\right) \end{aligned} \tag{12.26}$$

and for the Ramond sector

$$\chi_{r,s}(q) := q^{-c/24+1/16} \prod_{n=1}^{\infty} \left(\frac{1+q^n}{1-q^n}\right) \sum_{k=-\infty}^{\infty} \left(q^{\Delta_{2kM+r,s}} - q^{\Delta_{2kM+r,-s}}\right) \tag{12.27}$$

and there does exist a classification of the modular invariant partition functions [214, 251, 71]. If M is odd, then $\Delta_{M/2,1+M/2} = c/24$ and supersymmetry can be preserved in the Ramond sector. The techniques to calculate the operator product expansion coefficients and the n-point correlation functions have been extended to superconformal invariance as well [220, 18].

Returning to the tricritical Ising model, the central charge $c = 7/10$ corresponds to $M = 3$ and one has the superconformal representations in the Neveu-Schwarz sector (the index Vir labels the Virasoro characters)

$$(0) = (0)_{Vir} + \left(\frac{3}{2}\right)_{Vir} , \quad \left(\frac{1}{10}\right) = \left(\frac{1}{10}\right)_{Vir} + \left(\frac{3}{5}\right)_{Vir} \tag{12.28}$$

and one has the decomposition of the superconformal characters, e.g.

$$\chi_{0,Vir} \pm \chi_{3/2,Vir} = \begin{cases} \chi_0 \\ \tilde{\chi_0} \end{cases} \tag{12.29}$$

For the Ramond sector, one has

$$\left(\frac{3}{80}\right) = \left(\frac{3}{80}\right)_{Vir} , \quad \left(\frac{7}{16}\right) = \left(\frac{7}{16}\right)_{Vir} \tag{12.30}$$

We remark that also for the second $m = 4$ model eq. (12.17) those Virasoro fields which combine into a super Virasoro field are in the same sector $D_a^{(b)}$ of the model.

12.3 Yang-Lee edge singularity

The Yang-Lee singularity had been introduced in chapter 1. It may be described by an Ising model in an imaginary magnetic field, for which the quantum Hamiltonian is [145]

$$H = -\frac{1}{2\zeta} \sum_{n=1}^{N} [\sigma^z(n) + \lambda\sigma^x(n)\sigma^x(n+1) + ih\sigma^x(n)] \tag{12.31}$$

Although the critical behaviour is signalled by the divergence of some correlation length, given by the lowest gap of H, there is no apparent symmetry of the model. A few values for the location of the Yang-Lee edge singularity in the Ising quantum chain are given in Table 17 [145]. Since the transition mechanism is given by the merging of two non-

Table 17. Location of the Yang-Lee singularity and conformal normalization.

λ	0.002	0.10	0.20	0.40	0.60	0.80	1.0
$h_c(\lambda)$	0.970715(1)	0.636640(3)	0.458498(1)	0.23202(1)	0.09807(2)	0.02468(2)	0
$\zeta(\lambda)$	0.097	0.376	0.49	0.657	0.773	0.890	1
$B(\lambda)$	0.02975(2)	0.4741(2)	0.8500(3)	1.714(6)	2.97(1)	5.50(6)	$\frac{\pi}{4}$

degenerate levels and the subsequent development of non-vanishing imaginary parts of the eigenvalues, one can show [145] that the lowest energy gap takes the form, for μ not too large

$$\xi^{-1}(\lambda, h; N) = \Delta E(\lambda, h; N) \simeq \frac{2\pi x_1}{N}\sqrt{1 + \frac{\mu}{B(\lambda)}} + \ldots \qquad (12.32)$$

where $\mu := (h_c(\lambda) - h)N^{2-x_{inf}}$ is the scaling variable and x_{inf} the scaling dimension of the lowest primary field. The constant $B(\lambda)$ is also included in Table 17.

Since the Hamiltonian is not hermitian, there is no reason to demand that the Virasoro generators L_n come from a unitary representation. In fact, there exists a class of **minimal**, albeit non-unitary theories which contain a **finite** number of primary fields. Their structure is quite similar to the unitary theories. Minimal models, denoted by $\mathcal{M}_{p,p'}$, are characterized by two positive coprime integers p and p' with $p > p'$. One has the Kac formula

$$c = 1 - \frac{6(p - p')^2}{pp'} \qquad (12.33)$$

$$\Delta_{r,s} = \Delta_{p'-r,p-s} = \frac{(rp - sp')^2 - (p - p')^2}{4pp'} \quad ; \quad 1 \leq r \leq p' - 1 \ ; \ 1 \leq s \leq p - 1$$

and the unitary case is recovered when $p - 1 = p' = m$ with $m \geq 2$. In particular, the Rocha-Caridi formula eq. (10.28) can be used for the calculation of the characters. The considerations of modular invariance can be repeated and a classification of the critical modular invariant partition functions can be obtained.

The simplest example of the non-unitary minimal models [77] is the one with just one primary field besides the identity. Since the Kac table will contain each primary field twice, one needs a table with four entries. This is achieved by choosing $p' = 2, p = 5$. Then one finds for the central charge and the conformal weights

$$c = -\frac{22}{5} \ , \ \Delta = 0, -\frac{1}{5} \qquad (12.34)$$

Due to the non-unitarity, both the central charge and some conformal weights are negative in this example. In Table 18 we give the first few degeneracies $d(\Delta, I)$.

Table 18. Virasoro characters of the primary fields $\phi_{r,s}$ for $p = 5, p' = 2$.

I	0	1	2	3	4	5	6	7	8	9	10	11	12	13	14	15	r	s
$d(0,I)$	1	0	1	1	1	1	2	2	3	3	4	4	6	6	8	9	1	1
$d(-1/5,I)$	1	1	1	1	2	2	3	4	5	6	8	10	13	15	20	25	1	2

However, since the lowest Δ is negative, some extra care is needed when interpreting the finite-size data. Usually, one assumes that the lowest eigenstate of H corresponds to the identity operator with zero conformal weight. Here, one has a case where the lowest eigenstate does have a non-vanishing scaling dimension, sometimes referred to as the *anomalous dimension of the vacuum* x^*. For example, the two-point function is now [127]

$$< \phi(\vec{r_1})\phi(\vec{r_2}) >= |r_{12}|^{-2x+x^*} \tag{12.35}$$

Consequently, when just calculating the eigenvalues of H, one will find **effective** values for the central charge and the exponents, see eqs. (3.49,3.54)

$$\begin{aligned} f &= f_b - \frac{\pi c_{eff}}{6}\frac{1}{N^2} + \ldots, \; c_{\text{eff}} = c - 12x_{\text{inf}} \\ \xi_i^{-1} &= 2\pi\, x_{\text{eff},i}\, N^{-1} + \ldots, \; x_{eff,i} = x_i - x_{\text{inf}} \end{aligned} \tag{12.36}$$

where $x_{inf} = \Delta_{inf} + \bar{\Delta}_{inf} = x^*/2$ is the scaling dimension of the lowest state.

Returning to the Yang-Lee singularity, numerical finite-size calculations have given evidence that [202]

$$c_{eff} = \frac{2}{5}\;,\; x_{eff,1} = \frac{2}{5} \tag{12.37}$$

from which (12.36) implies

$$c = -\frac{22}{5}\;,\; x_1 = -\frac{2}{5} \tag{12.38}$$

which means that the Yang-Lee singularity of the $2D$ Ising model is described by the minimal model $\mathcal{M}_{5,2}$ [77, 202]. The operator content is for periodic boundary conditions

$$H = (0,0) + \left(-\frac{1}{5}, -\frac{1}{5}\right) \tag{12.39}$$

which is modular invariant. In particular, since $\eta_{YL} = 2x_1 = -4/5$, this implies, using the scaling relations of chapter 1 that indeed $\delta_{YL} = -6$ in agreement with earlier conclusions [125, 77].

One might get the impression that models with a non-vanishing x^* are a peculiarity of non-unitary models. This is not the case. As an example, consider the three-states Potts quantum chain eq. (12.1) with the boundary conditions

$$\Gamma_{N+1} = \omega^{\tilde{Q}}\Gamma_1^+ \tag{12.40}$$

This model has only a global Z_2 symmetry. The Hamiltonian spectrum can be decomposed into the sectors $K_Q^{(\tilde{Q})}$, with $Q = 0,1$ and $\tilde{Q} = 0,1,2$. One has $K_Q^{(0)} = K_Q^{(1)} = K_Q^{(2)}$. The operator content is [81, 14]

$$\begin{aligned} K_0^{(0)} &= \left(\frac{1}{40}, \frac{1}{40}\right) + \left(\frac{1}{8}, \frac{1}{8}\right) + \left(\frac{21}{40}, \frac{21}{40}\right) + \left(\frac{13}{8}, \frac{13}{8}\right) \\ K_1^{(0)} &= \left(\frac{21}{40}, \frac{1}{40}\right) + \left(\frac{1}{40}, \frac{21}{40}\right) + \left(\frac{1}{8}, \frac{13}{8}\right) + \left(\frac{13}{8}, \frac{1}{8}\right) \end{aligned} \tag{12.41}$$

where the energies are measured with respect to the lowest state in the sector $H_{0,+}^{(0)}$, see eq. (12.4). Consequently, already the lowest state of $K_0^{(0)}$ has a non-vanishing scaling dimension. We also remark that the unitary $m = 5$ primary fields present in the K sectors are exactly those absent from the ordinary Potts model.

12.4 Ashkin-Teller model

We now introduce a further spin system not already defined in chapter 1. This is the Ashkin-Teller model. It describes the coupling of two sets of Ising spin variables placed on the same lattice and is defined by the classical spin Hamiltonian

$$\mathcal{H} = -J_2 \sum_{(i,j)} (\sigma_i \sigma_j + \tau_i \tau_j) - J_4 \sum_{(i,j)} \sigma_i \sigma_j \tau_i \tau_j \tag{12.42}$$

where $\sigma_i = \pm 1$ and $\tau_i = \pm 1$ are the Ising spin variables. The coupling J_4 measures the coupling between the two Ising models described by the σ or τ spins alone.

This example might give some idea on the complexities of the conformal description of more general models than the simple spin systems considered so far. Furthermore, the $2D$ Askin-Teller model turns out to contain a section of the orbifold compactification line of Figure 30. The global symmetry of the model (pending a precise statement of the boundary conditions) is the dihedral group D_4, which is generated by the transformations $\sigma \to -\sigma$, $\tau \to -\tau$ and $\sigma \leftrightarrow \tau$. Already at this stage, there is an apparent relationship with the orbifold partition function. $\mathcal{Z}_{orb}$ contains two spin fields $\hat{\sigma}_{1,2}$ with conformal weights $(1/16, 1/16)$, which arise in the $(-)$ sector of the traces, eq. (11.28). This implies that the statistical model having $\mathcal{Z}_{orb}$ as partition function should have as a global symmetry at least the group generated by $\hat{\sigma}_1 \to -\hat{\sigma}_1$, $\hat{\sigma}_2 \to -\hat{\sigma}_2$, $\hat{\sigma}_1 \leftrightarrow \hat{\sigma}_2$, which is nothing else than D_4.

To obtain the quantum Hamiltonian, it is convenient to rewrite the model in terms of a four-state variable Θ via

$$2\cos(\Theta_i - \Theta_j) = \sigma_i\sigma_j + \tau_i\tau_j \quad , \quad \cos(2\Theta_i - 2\Theta_j) = \sigma_i\sigma_j\tau_i\tau_j \tag{12.43}$$

Now one proceeds as in chapter 8. In the extreme anisotropic limit, where $\tau \to 0$ and

$$\beta J_{2,r} = \tau t \; , \; \beta J_{4,r} = \tau t \varepsilon \tag{12.44}$$
$$\beta J_{2,\tau} = \frac{1}{4}\left(\ln \tau^{-1} - \ln t\right) \; , \; \beta J_{4,\tau} = \frac{1}{4}\left(\ln \tau^{-1} + \ln t\right)$$

the quantum Hamiltonian becomes [228]

$$H = -\frac{1}{\zeta} \sum_{n=1}^{N} \left\{ \left(\Gamma_n + \Gamma_n^3 + \varepsilon\, \Gamma_n^2\right) + t\left(\sigma_n \sigma_{n+1}^3 + \sigma_n^3 \sigma_{n+1} + \varepsilon\, \sigma_n^2 \sigma_{n+1}^2\right)\right\} \tag{12.45}$$

where

$$\zeta^{-1} = \frac{1 - 4g}{4\sqrt{t}g \sin(\pi/4g)} \quad , \quad g = \frac{\pi}{4 \arccos(-\varepsilon)} \tag{12.46}$$

and σ and Γ are 4×4 matrices as defined in chapter 9 and ζ is already chosen in view for conformal invariance applications. In this case, the charge conjugation invariance corresponds to the exchange $\sigma_i \leftrightarrow \tau_i$ in the isotropic Ashkin-Teller model. The self-duality of H is evident.

Relation with the XXZ quantum chain. There exists a mapping onto a staggered XXZ chain with $2N$ sites. By making first a duality transformation on the τ spins alone and then a second duality transformation on all spins, one finds, up to boundary terms [228]

$$H = -\frac{1}{\zeta}\sum_{n=1}^{2N}\left(\frac{1+t}{2} + (-1)^n\frac{1-t}{2}\right)\left\{\sigma_n^x\sigma_{n+1}^x + \sigma_n^y\sigma_{n+1}^y - \varepsilon\,\sigma_n^z\sigma_{n+1}^z\right\} \tag{12.47}$$

where the $\sigma^x, \sigma^y, \sigma^z$ are Pauli matrices. This quantum Hamiltonian can also be obtained from the *six-vertex model* [G11]. For the critical point $t = 1$, it can be solved via Bethe ansatz techniques. If the boundary conditions are taken into account, the relationship becomes more complicated. In Table 19 [12] the correspondence is given for periodic boundary conditions in the Ashkin-Teller model with N sites. For each sector Q of the

Table 19. Sector equivalence of the XXZ chain with the Ashkin-Teller chain.

Q	XXZ boundary conditions		
0	$\sigma_{2N+1}^x = \sigma_1^x$	$\sigma_{2N+1}^y = \sigma_1^y$	$\sigma_{2N+1}^z = \sigma_1^z$
1, 3	$\sigma_{2N+1}^x = \sigma_1^x$	$\sigma_{2N+1}^y = -\sigma_1^y$	$\sigma_{2N+1}^z = \sigma_1^z$
2	$\sigma_{2N+1}^x \pm i\sigma_{2N+1}^y = -(\sigma_1^x \pm i\sigma_1^y)$		$\sigma_{2N+1}^z = \sigma_1^z$

Ashkin-Teller model, the corresponding boundary conditions of the XXZ chain are given. This result follows [12] since the link variables in both the Ashkin-Teller chain and the XXZ chain can be shown to satisfy the same algebra. This relationship has recently lead to an exact solution of the model on a three-dimensional submanifold of the full six-dimensional space of the thermodynamic variables [268].

Global symmetry and boundary conditions. The global symmetry of H is given by the transformation [20]

$$(\sigma_n)^a \to \left(\sigma_n^{'}\right)^a := A^{a,b}\,(\sigma_n)^b \tag{12.48}$$

where the 3×3 matrices $\hat{A}$ form a representation of the group G. For the Ashkin-Teller model, $\hat{A}$ can be taken to be one of the matrices $\Sigma^k C^\ell$, with $k = 0,1,2,3$, $\ell = 0,1$ and

$$\Sigma = \begin{pmatrix} i & & \\ & -1 & \\ & & -i \end{pmatrix} \;,\; C = \begin{pmatrix} & & 1 \\ & 1 & \\ 1 & & \end{pmatrix} \tag{12.49}$$

which form a representation of the finite group D_4. The symmetry sectors are given by the *irreducible representations* of D_4. There is a two-dimensional one

$$\mathcal{R}_Q(\Sigma) = \begin{pmatrix} i^Q & \\ & i^{-Q} \end{pmatrix} \;,\; \mathcal{R}_Q(C) = \begin{pmatrix} 0 & 1 \\ 1 & 0 \end{pmatrix} \tag{12.50}$$

and four one-dimensional ones

$$\begin{aligned}\mathcal{R}_{0,+}(\Sigma) = 1\,, \mathcal{R}_{0,+}(C) = 1 \\ \mathcal{R}_{0,-}(\Sigma) = 1\,, \mathcal{R}_{0,-}(C) = -1 \\ \mathcal{R}_{2,+}(\Sigma) = -1\,, \mathcal{R}_{2,+}(C) = 1 \\ \mathcal{R}_{2,-}(\Sigma) = -1\,, \mathcal{R}_{2,-}(C) = -1\end{aligned} \tag{12.51}$$

The presence of the two-dimensional irreducible representation of D_4 implies that in the charge sectors $Q = 1$ and $Q = 3$ one has the same spectrum. The one-dimensional irreducible representations provide the decomposition with respect to the charge conjugation C in the sectors $Q = 0$ and $Q = 2$. The *conjugacy classes* of D_4 are

$$\{\Sigma^0\}\,,\{\Sigma^2\}\,,\{\Sigma,\Sigma^3\}\,,\{C,\Sigma^2 C\}\,,\{\Sigma C,\Sigma^3 C\} \tag{12.52}$$

Toroidal boundary conditions are defined by

$$(\sigma_{N+1})^a = B^{a,b}\,(\sigma_1)^b \tag{12.53}$$

where $\hat{B}$ is any of the matrices of (12.52). Two Hamiltonians with boundary conditions specified by matrices $\hat{B}_1, \hat{B}_2$ have the same spectrum if $\hat{B}_1$ and $\hat{B}_2$ are in the same conjugacy class. The global symmetry depends on the choice of $\hat{B}$ as illustrated in Table 20 [20]. For further details and the generalization to n-state quantum chains, we refer to the review [292]. These symmetries are generically valid. For peculiar choices of ε, the

Table 20. Symmetries of the Ashkin-Teller model in dependence of the boundary conditions.

Boundary condition	Group	number of elements	Elements[a]
Σ^0	D_4	8	$\Sigma^k C^\ell$
$\Sigma^{\tilde{Q}}$	Z_4	4	Σ^k
Σ^2	D_4	8	$\Sigma^k C^\ell$
$\Sigma^R C$	$Z_2 \otimes Z_2$	4	$\Sigma^0, \Sigma^2, \Sigma^R C, \Sigma^{R+2} C$

[a] The indices are $k = 0, 1, 2, 3$ and $\ell = 0, 1$.

global symmetry can be higher. An obvious case is $\varepsilon = 1$, when $G = S_4$, the permutation group. This case corresponds to the four-state Potts model.

Phase diagram. The various phases described by the quantum Hamiltonian eq. (12.45) are displayed in Figure 31 [228]. One distinguishes five phases.

I A paramagnetic phase, both σ_i and τ_i are disordered.
II A partially ordered phase, with one of the σ_i or τ_i ordered and the other disordered.
III A fully ordered ferromagnetic phase.
IV A *"critical fan"*, where the order parameters of σ_i, τ_i are both zero, but their correlation functions decay algebraically.

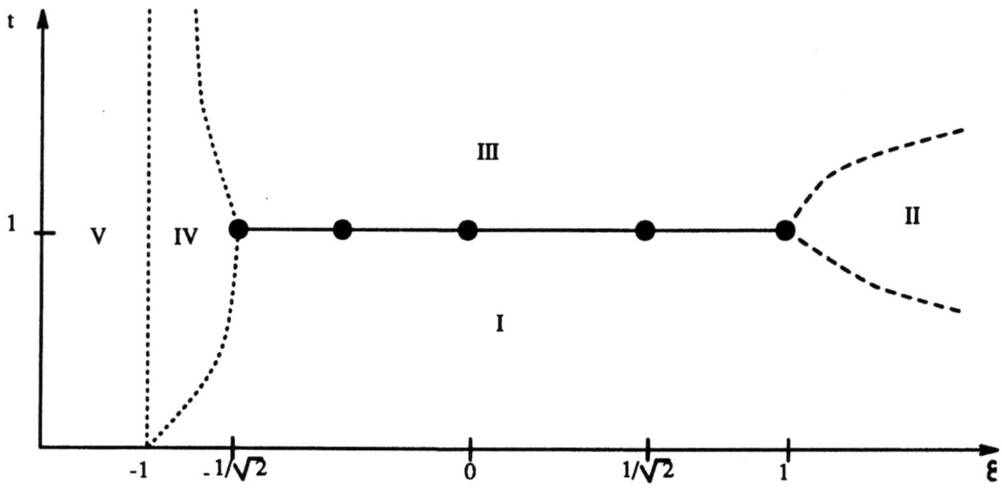

Fig. 31. Phase diagram of the $2D$ Ashkin-Teller quantum chain.

V An antiferromagnetically frozen phase.

The transitions between these phase are the following [228, 142]

1. Along the self-dual line $t = 1$, critical exponents change continuously as a function of the *four-spin coupling* ε (full line). The central charge is $c = 1$. The section between $\varepsilon = -1/\sqrt{2}$ and $\varepsilon = 1$ of this line corresponds to the orbifold compactification line of Figure 30 with ρ between 1 and 2. The case $\varepsilon = -1/\sqrt{2}$ corresponds to the meeting point of the two compactification lines and is in the Kosterlitz-Thouless universality class.
2. Two lines which are in the $2D$ Ising universality class (dashed lines). These transition values of t are dual to each other. It was checked that indeed $c = 1/2$ [142].
3. The borders of the critical fan have a Kosterlitz-Thouless transition (dotted lines).

It should be noted that the *isotropic* Ashkin-Teller model eq. (12.42) does not contain the critical fan, although the antiferromagnetic frozen phase is again present. The critical fan corresponds to a region with negative Boltzmann weights in the isotropic system. To have positive Boltzmann weights, one needs that $\varepsilon > -1/2$ [228].

Operator content on the $c = 1$ line. Since the actual expressions become quite lengthy, we shall only write here the operator content for those boundary conditions which preserve the full D_4 symmetry, see Table 20. The operator content is known for all toroidal boundary conditions and is given in [20]. One now needs quite a few notations. Let

$$A = (\{0\},\{0\}) + (\{1\},\{1\}) + A_1$$
$$B = (\{0\},\{1\}) + (\{1\},\{0\}) + A_1$$
$$C = R(4,2;4,2|g) + R(4,2;-4,-2|g)$$

$$\begin{aligned}
F(g) &= \sum_{n\geq 0}\left(\frac{(2n+1)^2}{16},\frac{(2n+1)^2}{16}\right) + R(4,2;4,4|g) + R(4,2;-4,-4|g)\\
G(g) &= F(1/g)\\
H &= (\{1/16\},\{1/16\}) + (\{9/16\},\{9/16\})\\
K &= (\{1/16\},\{9/16\}) + (\{9/16\},\{1/16\})
\end{aligned} \tag{12.54}$$

with the abbreviations

$$\begin{aligned}
R(p,q;r,s|g) &= \sum_{m,n\geq 0}\left(\frac{pm+q+(rn+s)g)^2}{16g},\frac{pm+q-(rn+s)g)^2}{16g}\right)\\
A_1 &= \sum_{n\geq 0}((n+1)^2 g,(n+1)^2 g) + \left(\frac{(n+1)^2}{g},\frac{(n+1)^2}{g}\right)\\
&\quad + R(4,4;4,4|g) + R(4,4;-4,-4|g)
\end{aligned} \tag{12.55}$$

and

$$\begin{aligned}
\left\{\frac{1}{16}\right\} &= \sum_{m=-\infty}^{\infty}\left(\frac{(8m+1)^2}{16}\right),\ \left\{\frac{9}{16}\right\} = \sum_{m=-\infty}^{\infty}\left(\frac{(8m+3)^2}{16}\right)\\
\{0\} &= \sum_{m=0}^{\infty}(4m^2)\ ,\ \{1\} = \sum_{m=0}^{\infty}((2m+1)^2)
\end{aligned} \tag{12.56}$$

where (Δ) stands everywhere for a $c=1$ Virasoro character. These are given by

$$\chi_\Delta(q) = q^{-1/24}\begin{cases} q^{n^2/4}\left(1-q^{n+1}\right)P(q) & \text{if } \Delta = \frac{n^2}{4}\\ q^\Delta P(q) & \text{if } \Delta \neq \frac{n^2}{4}\end{cases} \tag{12.57}$$

where n is an integer and $P(q)$ is the generating function of the partitions of the integers, see eq. (10.29). The particular form when $\Delta = n^2/4$ is important for the calculation of the secondary states. In Table 21 [20] the correspondence between the sectors of the model and the operator content eq. (12.54) is stated. Note that since $\mathcal{R}_1$ is two-dimensional, the corresponding primary fields appear twice. One can use the charge conjugation quantum number to distinguish between them. We also observe that only the one-dimensional irreducible representations of D_4 are related to g-dependent primary fields while the fields coming from the two-dimensional irreducible representations are g-independent.

In order to illustrate these results, we list as an example the relevant (or marginal) scalar primary fields for periodic boundary conditions as shown in Table 22, extending earlier results of [210]. The correspondence is given in terms of the primary fields $\hat{\sigma}, \hat{\varepsilon}_\sigma$ and $\hat{\tau}, \hat{\varepsilon}_\tau$ of the two Ising models into which the Ashkin-Teller chain decouples at $\varepsilon = 0$. The $\pm$ refers to the charge conjugation eigenvalue, or C-parity. For $\varepsilon \neq 0$, this simple decomposition no longer applies, but the fields are continuously deformed when changing ε. Since there are no level crossings, the fields retain their identity, however. The fields in the first three columns are often referred to as *magnetization*, *polarization* and *energy density*, respectively. Their bulk critical exponents can be read off

$$x_\sigma = \frac{1}{8}\ ,\ x_P = \frac{g}{2}\ ,\ x_\varepsilon = 2g \tag{12.58}$$

Table 21. Operator content of the Ashkin-Teller model for the D_4-invariant boundary conditions Σ^0, Σ^2.

Sector	Σ^0	Σ^2
$\mathcal{R}_{0,+}$	A	G
$\mathcal{R}_{0,-}$	C	F
$\mathcal{R}_{2,+}$	G	B
$\mathcal{R}_{2,-}$	F	C
$\mathcal{R}_1$	H	K

Table 22. Scalar primary fields for periodic boundary conditions of the Ashkin-Teller quantum chain.

$(0,0)$	$\left(\frac{1}{16},\frac{1}{16}\right)\times 2$	$\left(\frac{g}{4},\frac{g}{4}\right)$	(g,g)	$\left(\frac{9}{16},\frac{9}{16}\right)\times 2$	$\left(\frac{1}{4g},\frac{1}{4g}\right)$	$(1,1)$
$\mathbf{1}$	$\hat{\sigma}\pm\hat{\tau}$	$\hat{\sigma}\hat{\tau}$	$\hat{\varepsilon}_\sigma+\hat{\varepsilon}_\tau$	$\hat{\sigma}\hat{\varepsilon}_\tau\pm\hat{\tau}\hat{\varepsilon}_\sigma$	$\hat{\varepsilon}_\sigma-\hat{\varepsilon}_\tau$	$\hat{\varepsilon}_\sigma\hat{\varepsilon}_\tau$

Examining the operator content, one notes a few special values of ε where the model has a particular symmetry. A few cases are listed below, where the first column refers to the coupling, the second is the symmetry and the third identifies the model

$$\begin{array}{lccl}
\varepsilon = & 1 & S_4 & \text{four-states Potts model} \\
\varepsilon = & 1/\sqrt{2} & Z_4 & \text{parafermion} \\
\varepsilon = & 0 & Z_2\otimes Z_2 & \text{two decoupled Ising models} \\
\varepsilon = & -\sqrt{2-\sqrt{2}}\Big/2 & & N=2 \text{ twisted SUSY} \\
\varepsilon = & -1/\sqrt{2} & & \text{Kosterlitz-Thouless}
\end{array} \tag{12.59}$$

These findings, some of which were obtained long before the construction of the modular invariant partition functions, are in agreement with the conformal invariance results of Figure 30.

One aspect completely left out here is the reorganisation of the operator content into irreducible representations of larger algebraic structures which are not necessarily Lie algebras any more. This is treated in detail in the references [20, 130, 335, 336].

12.5 XY model

Previously, the XY model eq. (1.58) was tentatively identified within the framework of modular invariant systems with central charge $c = 1$. We now describe, following the original work of Allton and Hamer [17], a finite lattice study to confirm this identification. The quantum Hamiltonian of the XY model is

$$H = \frac{1}{\zeta}\sum_{n=1}^{N}\left[J_3(n) - \frac{\lambda}{2}\left(J_+(n)J_-(n+1) + J_-(n)J_+(n+1)\right) - \frac{h}{2}\left(J_+(n)+J_-(n)\right)\right] \tag{12.60}$$

where the $J_\pm$, J_3 are angular momentum operators with commutators

$$[J_3(n), J_\pm(n')] = \pm J_\pm(n)\delta_{n,n'} \tag{12.61}$$

while λ is the thermal variable and h a magnetic field. The Hamiltonian commutes with the total spin

$$\hat{S} = \sum_{n=1}^{N} J_3(n) \tag{12.62}$$

Since the model is expected to have not a power-law divergence of the correlation length and other physical quantities but rather a singularity of the form

$$\xi \sim \exp b \left(\frac{T_K}{T - T_K}\right)^{\sigma} \tag{12.63}$$

finite-size techniques must be adapted accordingly. Consider the Callan-Symanzik ***beta function***

$$\beta(\lambda) := \frac{\xi^{-1}(\lambda)}{\xi^{-1}(\lambda) - 2\lambda(\xi^{-1})'(\lambda)} \tag{12.64}$$

where the prime denotes the derivative. For a singularity of the form (12.63) one has $\beta(\lambda) \sim (\lambda - \lambda_c)^{1+\sigma}$, while for an ordinary second-order transition, the beta function would be linear in the vicinity of λ_c. This allows to distingusish these two situations and to find the exponent σ. A very useful tool to compute the beta-function from finite-lattice data is provided by the ***Roomany-Wyld approximant*** [294]

$$\begin{aligned} \beta_N^{RW} := &\left[1 + \ln\left(\frac{\xi^{-1}(\lambda;N)}{\xi^{-1}(\lambda;N-1)}\right) \Big/ \ln\frac{N}{N-1}\right] \\ &\cdot \left[1 - \lambda\left(\frac{\partial\xi^{-1}(\lambda;N)/\partial\lambda}{\xi^{-1}(\lambda;N)} + \frac{\partial\xi^{-1}(\lambda;N-1)/\partial\lambda}{\xi^{-1}(\lambda;N-1)}\right)\right]^{-1} \end{aligned} \tag{12.65}$$

Then finite-size scaling yields

$$\lambda_c = 2.06(4) \ , \quad \sigma = 0.501(5) \tag{12.66}$$

in excellent agreement with the Kosterlitz-Thouless prediction $\sigma = 1/2$. On the other hand, various determinations of λ_c using different methods have yielded results which scatter between 1.9 and 2.1.

Due to the conservation of $\hat{S}$, one can consider its sectors separetely. Allton and Hamer [17] considered the two lowest states in the sector $S = 0$, denoted by $|0\rangle$ and $|2\rangle$, and in the sector $S = 1$ giving $|1\rangle$ and $|3\rangle$. Now, using eq. (3.49), the energy gaps are at the critical point

$$E_i - E_j = 2\pi x_{i,j}\, N^{-1} \tag{12.67}$$

with $i, j = 0, 1, 2, 3$. One expects that $x_{0,1} = x_{2,3} = \eta/2$ and $x_{0,2} = x_{1,3} = 2$. In writing this, the normalization factor ζ was determined from the ground state energy and eq. (3.54) under the assumption that $c = 1$. The determination of the exponent η

was checked by also computing it from the magnetization and the susceptibility per spin. These are computed as outlined in chapter 9

$$\chi = \frac{1}{N^2} \sum_{n,m=1}^{N} \langle 0|J_+(n)J_-(m) + J_-(n)J_+(m)|0\rangle$$
$$M = \frac{1}{N} \sum_{n=1}^{N} \langle 1|J_+(n) + J_-(n)|0\rangle \qquad (12.68)$$

Since finite-size scaling was formulated in terms of the correlation length, see chapter 3, one can calculate the exponent η from

$$\chi_N \sim \xi^{2-\eta} \sim N^{2-\eta} \ , \ \ M_N \sim \xi^{-\eta/2} \sim N^{-\eta/2} \qquad (12.69)$$

In Table 23, some of the results of [17] for η as a function of λ are displayed. The indices refer to the results obtained either from the susceptibility χ, the magnetization M, the gap $2x_{0,1}$ or the leading weak-coupling result $\eta_{weak} = 1/(\pi\sqrt{2\lambda})$ for $\lambda \to \infty$. The inferred conformal normalization ζ is also given. The results are seen to be in good

Table 23. Exponent $\eta(\lambda)$ for the $(1+1)D$ XY model.

λ	2.0	2.5	3.0	3.5	4.0
η_χ	0.205	0.173	0.154	0.141	0.132
η_M	0.204	0.171	0.152	0.137	0.127
$2x_{0,1}$	0.205	0.173	0.154	0.140	0.129
η_{weak}			0.130	0.120	0.113
ζ	1.62	1.86	2.08	2.28	2.47

agreement with each other. This is consistent with the central charge being equal to unity. On the other hand, the data do not reproduce very precisely the expected value $\eta = 1/4$ at the critical point. This is ascribed to the presence of additional logarithmic factors, see eq. (1.60), which are known to severely affect the convergence of finite-lattice data. The same difficulty also appears in high-temperature series expansions. In fact, the precise determination of the critical behaviour in the two-dimensional XY model remains a challenging problem. For a recent Monte Carlo simulation using lattices up to 1200 x 1200 sites, see [204].

12.6 XXZ quantum chain

It was already mentioned above that the partition functions for the minimal models are contained in those of the compactified Coulomb gas partition functions, provided that certain frustrations lines at the boundaries were introduced. In the language of quantum Hamiltonians, the same can be achieved by projecting out the spectra of the minimal models from the one of the spin-1/2 XXZ quantum chain. The presentation follows the original work of Alcaraz, Grimm and Rittenberg [14].

The XXZ Hamiltonian is

$$H = -\frac{\gamma}{2\pi \sin\gamma} \sum_{n=1}^{N} [\sigma^x(n)\sigma^x(n+1) + \sigma^y(n)\sigma^y(n+1) - \cos\gamma\, \sigma^z(n)\sigma^z(n+1)] \quad (12.70)$$

where the σ are Pauli matrices. This Hamiltonian can also be obtained by taking the Hamiltonian limit in the eight-vertex model [G11]. Writing

$$g = \frac{1}{4}(1-\gamma/\pi)^{-1} \quad (12.71)$$

$g = \rho^2/2$ is related to the bosonic compactification radius. Consider the toroidal boundary conditions, characterized by the parameter ℓ

$$\sigma^x(N+1) + i\sigma^y(N+1) = e^{2\pi i \ell}(\sigma^x(1) + i\sigma^y(1)) \quad (12.72)$$

with $0 \leq \ell < 1$. The charge operator is

$$\hat{Q} = \frac{1}{2}\sum_{n=1}^{N} \sigma^z(n) \quad (12.73)$$

Denote by $E^{\ell}_{Q,i}(P;N)$ the ith level in the sector Q, with lattice momentum P, boundary condition ℓ and for N sites. Define the *scaled gaps*

$$\mathcal{E}^{\ell}_{Q,i} := \frac{N}{2\pi}\left(E^{\ell}_{Q,i}(P;N) - E^{0}_{0,0}(0;N)\right) \quad (12.74)$$

The operator content [13] is given by the $N \to \infty$ limit of the partition functions

$$\begin{aligned} \mathcal{Z}^{\ell}_{Q}(q,\bar{q}) &:= \sum_{i,P} q^{(\mathcal{E}^{\ell}_{Q,i}(P;N)+P)/2}\, \bar{q}^{(\mathcal{E}^{\ell}_{Q,i}(P;N)-P)/2} \\ &= P(q)P(\bar{q}) \sum_{k=-\infty}^{\infty} q^{[Q+4g(\ell+k)]^2/16g}\, \bar{q}^{[Q-4g(\ell+k)]^2/16g} \end{aligned} \quad (12.75)$$

In order to give the projection rules for the unitary minimal models, a few preparations are necessary. First, discretize the boundary conditions

$$\ell = \frac{k}{n} \;;\; k = 0,1,\ldots,n-1 \quad (12.76)$$

where the integer n will be fixed later. Second, take $E^{1/n}_{0,0}(0;N)$ as the new ground state and define the new scaled gaps

$$\tilde{\mathcal{F}}^{k}_{Q,i}(P;N) := \frac{N}{2\pi}\left(E^{k/n}_{Q,i}(P;N) - E^{1/n}_{0,0}(0;N)\right) \quad (12.77)$$

Since the boundary conditions are in the finite group Z_n, take the charges Q to be in Z_n by summing up all charges $Q = R \bmod n$ and obtain

$$\mathcal{F}^k_{R,i}(P;N) := \frac{N}{2\pi}\left(E^{k/n}_{R,i}(P;N) - E^{1/n}_{0,0}(0;N)\right) \tag{12.78}$$

where $R = 0, 1, \ldots, n-1$ for N even and $2R = 1, 3, \ldots, 2n-1$ for N odd. Finally, choose

$$g = \frac{n^2}{4m(m+1)} \tag{12.79}$$

where m gives the central charge via the Kac formula eq. (4.48). Then the partition functions are, where again the leading factor $(q\bar{q})^{-1/24}$ is suppressed

$$\mathcal{Y}^k_R(q,\bar{q}) := \mathcal{Y}^{n-k}_{n-R}(q,\bar{q}) = \sum_{i,P} q^{(\mathcal{F}^k_{R,i}(P;N)+P)/2}\,\bar{q}^{(\mathcal{F}^k_{R,i}(P;N)-P)/2} \tag{12.80}$$

One now specifies two classes of systems, labelled by a positive integer M. The systems are called M_L and M_R and are defined through the angles $\varphi_{M,L}$ and $\varphi_{M,R}$

$$\begin{aligned} \varphi_{M,L} &= \pi\left(\frac{1}{M^2 g} - 1\right) \quad ; \quad 0 \le \varphi_{M,R} \le \frac{\pi}{M} \\ \varphi_{M,R} &= \pi\left(1 - \frac{1}{M^2 g}\right) \quad ; \quad 0 \le \varphi_{M,R} \le \frac{\pi}{M} \end{aligned} \tag{12.81}$$

One can now recognise a few known models. These are given by the values $M = 1, 2$ as follows [14].

1. *The low-temperature $O(p)$ model* with $p = 2\cos\varphi_{1,R}$ and $g \ge 1$.
2. *The $O(p)$ model* with $p = 2\cos\varphi_{1,L}$ and $\frac{1}{2} \le g \le 1$.
3. *The p-states Potts model* with $\sqrt{p} = 2\cos\varphi_{2,R}$ and $\frac{1}{4} \le g \le \frac{1}{2}$.
4. *The tricritical p-states Potts model* with $\sqrt{p} = 2\cos\varphi_{2,L}$ and $\frac{1}{6} \le g \le \frac{1}{4}$.

The values of n and g are chosen as

$$\begin{aligned} M_R: \quad n &= \frac{2(m+1)}{M}, \quad g = \frac{m+1}{mM^2} \quad ; \quad m \ge M-1 \\ M_L: \quad n &= \frac{2m}{M}, \quad g = \frac{m+1}{(m+1)M^2} \quad ; \quad m \ge M \end{aligned} \tag{12.82}$$

Together with eqs. (12.71) and (12.79), the models are completely specified.

Now the projection to the unitary minimal models can be stated. We do so here for the case $M = 2$ and refer to [14] for the other cases. For the 2_R system, consider the quantities

$$\begin{aligned} \mathcal{X}^k_R = \mathcal{X}^{m+1-k}_{m+1-R} &:= \mathcal{Y}^k_R - \mathcal{Y}^R_k \\ &= \sum_{r=1}^{m-1} \chi_{r,k-R}\bar{\chi}_{r,k+R} = \sum_{r=1}^{m-1} \chi_{r,R-k}\bar{\chi}_{r,-k-R} \end{aligned} \tag{12.83}$$

where the $\chi_{r,s}$ are the Virasoro character functions of the unitary model with central charge $c = 1 - 6/(m(m+1))$. As an example, consider $m = 3$. The sectors are

$$\begin{aligned} \mathcal{X}_0^1 &= \mathcal{X}_0^3 = (0,0) + \left(\frac{1}{2}, \frac{1}{2}\right) \\ \mathcal{X}_0^2 &= 2\left(\frac{1}{16}, \frac{1}{16}\right) \\ \mathcal{X}_1^2 &= \mathcal{X}_3^2 = \left(0, \frac{1}{2}\right) + \left(\frac{1}{2}, 0\right) \end{aligned} \tag{12.84}$$

Combining all sectors, one obtains the partition function $\mathcal{Z}_T = 2Z$, where Z is called the ***half-partition function.*** One can check that the sectors of Z are identical to the ones of the Ising model given in eq. (10.44). In general, it is the half-partition function which should be considered. As a second example, we note that for $m = 4$ one finds the system with exponents given in eq. (12.17). In general, the thermal sector is always $\mathcal{X}_0^1$ and the magnetic sectors are $\mathcal{X}_0^k$ with $k > 1$. The corresponding scaling dimensions are

$$\begin{aligned} x_{\varepsilon,v} &= \frac{v^2(m+1) + 2v}{2m} \quad ; \quad 1 \leq v \leq m-2 \\ x_{\sigma,k} &= \frac{k^2 - 1}{2m(m+1)} \quad ; \quad k > 1 \end{aligned} \tag{12.85}$$

For the 2_L system, one has

$$\begin{aligned} \mathcal{X}_R^k = \mathcal{X}_{m-R}^{m-k} &:= \mathcal{Y}_R^k - \mathcal{Y}_k^R \\ &= \sum_{s=1}^{m} \chi_{k+R,s}\bar{\chi}_{k-R,s} = \sum_{s=1}^{m} \chi_{-k-R,s}\bar{\chi}_{R-k,s} \end{aligned} \tag{12.86}$$

For $m = 4$, one recovers the tricritical Ising model. The thermal and magnetic scaling dimensions are

$$\begin{aligned} x_{\varepsilon,v} &= \frac{v^2 m - 2v}{2(m+1)} \quad ; \quad 1 \leq v \leq m-1 \\ x_{\sigma,k} &= \frac{k^2 - 1}{2m(m+1)} \quad ; \quad k > 1 \end{aligned} \tag{12.87}$$

We leave it as an exercise to find the operator contents for $m = 5$.

Also, it was observed that the projection mechnism outlined in these cases and valid for $N \to \infty$, is still working even for finite systems [14]. The energy levels remaining in $\mathcal{X}_R^k$ will be those which do **not** occur simultaneously in both $\mathcal{Y}_R^k$ and $\mathcal{Y}_{R'}^{k'}$, where R' and k' depend on the class of models studied.

Let us finish this model with an experimental application [5, E1]. Consider the isotropic antiferromagnetic spin-s chain

$$H = \sum_n \left(S_n^x S_{n+1}^x + S_n^y S_{n+1}^y + S_n^z S_{n+1}^z\right) \tag{12.88}$$

This model can be solved by mapping it onto a $SU(2)$ Wess-Zumino-Witten σ model [5, G16], with Kac-Moody central charge $k = 2s$ (see chapter 7). The exact zero-temperature susceptibility can be calculated and is found to be

$$\chi v_s = \frac{k}{2\pi} \tag{12.89}$$

where v_s is the spin-wave velocity. This agrees with experiments on Cu $Cl_2 \cdot 2NC_5 H_5$ (with $s = 1/2$) and $(CH_3)_4$ N Mn Cl_3 (with $s = 5/2$). Similar results are obtained for the specific heat at low temperature T, where $C \sim AT$. The constant A is proportional to the central charge c. This follows from the finite-size scaling result eq. (3.52), where the lattice size L is reinterpreted as an inverse temperature, see chapter 1.

To close this chapter, we stress again that the examples provided are merely meant as an illustration for the applications of two-dimensional conformal invariance and were selected in order to have the main lines usually occuring in applications illustrated by an example. We do not pretend to have covered the complete range of possible applications, and the field is by no means complete at present. Further information must be sought from the literature.

12.7 A few remarks on 3*D* systems

Given the success of combined conformal and modular invariance to describe the operator content of a given model in detail, one might ask whether at least some of the techniques could be extended into higher dimensions as well. It is an open question whether such an extension is possible. Here we merely sketch proposals made towards this possibility.

Cardy [78, 86] proposed to consider a $3D$ space of the topology $\Sigma^3 = S^2 \otimes S^1$ with the metric

$$ds^2 = R^2 \left(d\theta^2 + \sin^2\theta d\phi^2\right) + R'^2 d\tau^2 \tag{12.90}$$

where R and R' are the radii of the 2-sphere S^2 and the 1-sphere S^1, respectively. Interpreting τ as imaginary time, the quantum Hamiltonian H in this formulation will be defined on the 2-sphere S^2 and by the transformation $\tau \longrightarrow (R/R')\ln r$, H is the generator of scale transformations $r \longrightarrow \lambda r$ for the space $R^3\backslash\{0\}$ with the flat metric. Then the partition function is, up to anomaly terms [86]

$$\mathcal{Z} = \sum_x q^x \tag{12.91}$$

where x runs over the list of scaling dimensions in the model and $q = e^{-2\pi R'/R}$ is the modular parameter. If the eigenvalues of H are taken to be in units of R' one has for the critical point correlation length [78]

$$\xi_i^{-1} = R^{-1}\, 2\pi x_i \tag{12.92}$$

in close analogy to the $2D$ situation (see eq. (3.47)). Unfortunately, there is at present no confirmation from any model beyond free field theory. For a numerical study in the Ising model, see [11]. Cardy has further shown that for $3D$ free fields, modular invariance of the partition function (12.91) can be obtained if **antiperiodic** boundary conditions on the 1-sphere are taken [86]. At present, it is not known whether this result remains true for interacting fields, at least in some asymptotic sense.

On the other hand, some model-dependent information is available. Consider the quantum Hamiltonian of the $(2+1)D$ Ising model [169]

$$H = -t\sum_{n}\sigma^z(n) - \frac{1}{2}\sum_{(n,n')}[(1+\eta)\sigma^x(n)\sigma^x(n') + (1-\eta)\sigma^y(n)\sigma^y(n')] \tag{12.93}$$

with nearest neighbor interactions and on a square $N \times N$ lattice, with the topology $S^1 \otimes S^1$. As in the $(1+1)D$ case eq. (10.1), η is an irrelevant parameter introduced to test universality explicitly. The inverse correlation lengths are given by the energy spectrum $\xi_i^{-1} = E_i - E_0$. In Table 24 [170], we give the location of the Ising transition $t_c(\eta)$ and for both periodic (P) and antiperiodic (A) boundary conditions the amplitude ratio

$$\Xi := S_\varepsilon(0,0)/S_\sigma(0,0) \tag{12.94}$$

as obtained from a finite-size scaling calculation. We observe that Ξ is universal for

Table 24. Critical point t_c and amplitude ratio Ξ in the $(2+1)D$ Ising model.

η	1.0	0.9	0.7	0.5	0.3
t_c	3.047(1)	2.938(1)	2.720(1)	2.500(2)	2.30(1)
Ξ_P	3.64	3.64	3.66	3.67	3.5
Ξ_A	2.76	2.78	2.70	2.80	2.77

both periodic (P) and antiperiodic (A) boundary conditions, as expected from finite-size scaling. The apparent variation of Ξ with η gives a realistic idea on the numerical precision of the data. Motivated from the simple $2D$ result eq. (3.47) for $S_i(0,0)$ we compare with the exponent ratio $x_\varepsilon/x_\sigma \simeq 2.76(2)$ (e.g. [G18]) for the $(2+1)D$ Ising model. This yields evidence for a **linear amplitude-exponent relation** in the $(2+1)D$ Ising model with **antiperiodic** boundary conditions on a square lattice

$$\xi_i = NAx_i^{-1} \tag{12.95}$$

as checked for $i = \sigma, \varepsilon$. The constant A cannot be determined from the quantum Hamiltonian (12.93), since its correct normalization is unknown. The value of A was provided by a Monte Carlo calculation [326] of the correlation functions in the classical $3D$ Ising model which also checked that Ξ does not depend[11)] on taking the Hamiltonian limit. Further, the correlation length amplitudes were calculated exactly in the $3D$ spherical model. The result $\Xi = x_\varepsilon/x_\sigma = 2$ is again consistent with eq. (12.95) and for **antiperiodic** boundary conditions [183]

$$A \simeq \begin{cases} 0.136\ldots & \text{spherical model} \\ 0.12 & \text{Ising model} \end{cases} \tag{12.96}$$

11) $\Xi_P \simeq 3.7, \Xi_A \simeq 2.7$ [326].

We see that A depends at least only weakly on the universality class and one might be tempted to conclude that A were a model-independent constant [12]. This is a purely empirical remark, however, and it is not understood why antiperiodic boundary conditions are important, or whether for other systems some "twisting" boundary conditions should be used. A recent study of a $3D$ dimer model with both periodic and antiperiodic boundary conditions [56] apparently fails to reproduce the linear relation (12.95) in $3D$, although it does so with periodic boundary conditions in $2D$, but it is not clear yet whether for some different boundary conditions (12.95) might work again.

Another attempt tries to calculate the entire set of scaling dimensions for some three-dimensional system. This program has been carried through for the $O(n)$ vector model via a $1/n$ expansion up to second order [235]. The available results point towards the existence of analogues of conformal families.

[12] A Gaussian free field theory gives $A = 1/(\sqrt{2}2\pi)$, which illustrates the important role of the interactions.

13 Conformal Perturbation Theory

In the last chapters we have obtained numerous results for statistical systems exactly at the critical point. Finite-size corrections were eliminated by the finite-lattice extrapolation procedures described and the finite-size scaling variables were set to zero. We now turn towards an investigation of these and shall show how finite-size corrections and finite-size scaling functions can be derived from the known operator content of a given model. These techniques do not require the integrability of the system under consideration.

13.1 Correlation functions in the strip geometry

We recall notation used with the logarithmic transformation Figure 17. The complex coordinate in the strip is $w = u + iv = \frac{N}{2\pi}\ln z$, where the lattice constant was put to one. The critical two-point correlation function of a primary operator is in the strip (see eq. (3.45))

$$
\begin{aligned}
&< \varphi(w_1, \bar{w}_1)\varphi(w_2, \bar{w}_2) >_w \\
&= \left(\frac{\pi}{N}\right)^{2x} \left(\sinh \frac{\pi}{N}(w_1 - w_2)\right)^{-2\Delta} \left(\sinh \frac{\pi}{N}(\bar{w}_1 - \bar{w}_2)\right)^{-2\overline{\Delta}} \\
&= \left(\frac{2\pi}{N}\right)^{2x} \sum_{m=0}^{\infty}\sum_{n=0}^{\infty} a_m \bar{a}_{\bar{m}} \exp\left[-\frac{2\pi}{N}(x + m + \bar{m})(u_1 - u_2)\right] \\
&\cdot \exp\left[\frac{2\pi i}{N}(s + m - \bar{m})(v_1 - v_2)\right]
\end{aligned}
\tag{13.1}
$$

where $x = \Delta + \overline{\Delta}$, $s = \Delta - \overline{\Delta}$ are the scaling dimension and the spin, respectively and

$$
a_m = \frac{\Gamma(2\Delta + m)}{\Gamma(2\Delta)\, m!} \quad , \quad \bar{a}_{\bar{m}} = \frac{\Gamma(2\overline{\Delta} + \bar{m})}{\Gamma(2\overline{\Delta})\, \bar{m}!}
\tag{13.2}
$$

This follows from the binomial theorem and the identities

$$
\begin{aligned}
\binom{-2\Delta}{k} &= (-1)^k \frac{\Gamma(2\Delta + k)}{\Gamma(2\Delta)k!} \\
(\sinh W)^{-2\Delta} &= 2^{2\Delta} \sum_{k=0}^{\infty} \frac{\Gamma(2\Delta + k)}{\Gamma(2\Delta)k!} (-1)^{-2\Delta} e^{W(2\Delta+2k)}
\end{aligned}
\tag{13.3}
$$

On the other hand, rewrite the two-point function while going from the "Heisenberg picture" with fields $\varphi(u, v)$ to the "Schrödinger picture" with fields $\varphi(v)$

$$
\langle \varphi(u_1, v_1)\varphi(u_2, v_2)\rangle = \sum_{n,k} \langle 0|\varphi(v_1)|n, k\rangle e^{-(E_n - E_0)(u_1 - u_2)} \langle n, k|\varphi(v_2)|0\rangle
\tag{13.4}
$$

where $|n, k\rangle$ is a complete set of intermediate states with energy E_n and momentum k, the lowset one of which is just the primary state $|\varphi\rangle$. Comparison with eq. (13.1) yields, see [80, G9]

$$
\langle 0|\varphi(v)|\varphi\rangle = \left(\frac{2\pi}{N}\right)^{x}
\tag{13.5}
$$

When considering a lattice operator, the coordinate v will correspond to the sites along the strip. This gives a means to fix the normalization of some lattice operator.

Similarly, the critical three-point function of primary operators is in the plane

$$< \varphi_1(z_1, \bar{z}_1)\varphi_2(z_2, \bar{z}_2)\varphi_3(z_3, \bar{z}_3) >= \mathbf{C}_{123}\, z_{12}^{-\Delta_{12}} z_{23}^{-\Delta_{23}} z_{31}^{-\Delta_{31}} \bar{z}_{12}^{-\overline{\Delta}_{12}} \bar{z}_{23}^{-\overline{\Delta}_{23}} \bar{z}_{31}^{-\overline{\Delta}_{31}} \tag{13.6}$$

where $\mathbf{C}_{123}$ is the *operator product expansion coefficient* and $\Delta_{12} := \Delta_1 + \Delta_2 - \Delta_3$ and $z_{12} := z_1 - z_2$ and so on. To be more precise, the $\mathbf{C}_{123}$ as written here must be written as a product of two OPE coefficients, each of which is determined separately from one of the (left or right) conformal algebras. In the strip geometry, for $u_1 \gg u_2 \gg u_3$

$$\begin{aligned} < \varphi_1\varphi_2\varphi_3 > \simeq & \left(\frac{2\pi}{N}\right)^{x_1+x_2+x_3} \mathbf{C}_{123} \exp\left[-\frac{2\pi}{N}\left(x_1(u_1-u_2)+x_3(u_2-u_3)\right)\right] \\ & \cdot \exp\left[\frac{2\pi i}{N}\left(s_1(v_1-v_2)+s_3(v_2-v_3)\right)\right] \end{aligned} \tag{13.7}$$

Rewriting in the Schrödinger picture

$$\begin{aligned} < \varphi_1\varphi_2\varphi_3 > = & \langle 0|\varphi_1(v_1)|\varphi_1\rangle e^{-\frac{2\pi}{N}x_1(u_1-u_2)}\langle\varphi_1|\varphi_2(v_2)|\varphi_3\rangle \\ & \cdot e^{\frac{2\pi}{N}x_3(u_2-u_3)}\langle\varphi_3|\varphi_3(v_3)|0\rangle \end{aligned} \tag{13.8}$$

one obtains [80, G9]

$$\langle\varphi_1|\varphi_2(v)|\varphi_3\rangle = \left(\frac{2\pi}{N}\right)^{x_2} \mathbf{C}_{123} \exp\left[-\frac{2\pi i}{N}(s_1 - s_3)v\right] \tag{13.9}$$

This can be used either to calculate some $\mathbf{C}_{123}$ from finite-lattice data or else to obtain some matrix elements from a known operator product. As shown in chapter 7, the knowledge of x_1, x_2, x_3 and s_1, s_2, s_3 is enough to find $\mathbf{C}_{123}$. The explicit expression was given in eq. (7.44). If the lattice operator happens to be not yet normalized, eq. (13.5) can be used to fix the correct conformal normalization.

13.2 General remarks on corrections to the critical behaviour

In general, any lattice Hamiltonian will contain correction terms to the critical Hamiltonian H_c

$$H = H_c + \sum_v \sum_k a_k \varphi_k(v) \tag{13.10}$$

where the a_k are (non-universal) parameters. In the sequel, we assume that the critical spectrum is non-degenerate for technical simplicity. So the following discussion only applies to the few lowest eigenvalues of H, but the generalization, if needed, is straightforward. The eigenvalues of H are then, if $E_{n,c}$ are the critical eigenvalues

$$\begin{aligned} E_n = & E_{n,c} + \sum_k a_k \sum_v \langle n|\varphi_k(v)|n\rangle \\ & + \sum_{k,\ell} a_k a_\ell \sum_{v,v'}\sum_{i\neq n} \frac{\langle n|\varphi_k(v)|i\rangle\langle i|\varphi_\ell(v')|n\rangle}{E_{n,c}-E_{i,c}} \\ & + \text{ higher order terms} \end{aligned} \tag{13.11}$$

Consider the first-order corrections for the correlation lenghts. One has, taking for simplicity just one perturbing field

$$\begin{aligned}\xi_n^{-1} = E_n - E_0 &= \frac{2\pi}{N}x_n + \sum_v a_k \langle n|\varphi_k(v)|n\rangle \\ &= \frac{2\pi}{N}x_n + a_k\left(\frac{2\pi}{N}\right)^{x_k} \mathrm{C}_{nkn}N \\ &= \frac{2\pi}{N}\left[x_n + 2\pi a_k \mathrm{C}_{nkn}\left(\frac{2\pi}{N}\right)^{x_k-2}\right] \end{aligned} \tag{13.12}$$

One now has to distinguish two cases.

1. **Irrelevant fields**. One has $y_k = 2 - x_k < 0$. These fields give rise to *finite-size corrections* to the critical behaviour. The possible finite-size correction exponents, as defined in chapter 2, can be read off from the operator content of H_c. We illustrate this in a few examples. One source of corrections is the conformal tower of the identity operator. These fields are present in **any** conformal invariant system. Possible correction terms appearing in the Hamiltonian must satisfy a few requirements. They should be quasiprimary and if they shall contribute in first order, they should be invariant under the global symmetry of the model. In principle, the following operators appear in the conformal tower of the identity

$$L_{-2},\ L_{-3},\ L_{-4},\ L_{-2}^2,\ \ldots \tag{13.13}$$

   and their combinations with the $\bar{L}_n$. It is usually sufficient to consider symmetric combinations like $L_{-2} + \bar{L}_{-2}$. This particular operator only changes the normalization ζ. The antisymmetric combination $L_{-2} - \bar{L}_{-2}$ is not consistent with parity conservation $E(\mathcal{P}) = E(-\mathcal{P})$. Terms involving L_{-3} alone are not quasiprimary. The first terms allowed are $L_{-2}\bar{L}_{-2}$ and $L_{-2}^2 - \frac{3}{5}L_{-4} + \bar{L}_{-2}^2 - \frac{3}{5}\bar{L}_{-4}$. The second one describes the breaking of rotational symmetry by a square lattice and is a tensor operator under the dihedral group D_4. We shall see in the examples of the Ising and the three-states Potts model that only this operator has a non-vanishing coupling. These two operators have a scaling dimension $x = 4$. They thus produce finite-size corrections of order $1/N^2$. Corrections of this size will therefore be always present (unless the coefficient vanishes). For the Ising model, corrections of this size are in fact the only ones present (see below). This is not so in other systems. For the tricritical Ising model with periodic boundary conditions, the operator content eq. (12.14) shows an irrelevant field with $\Delta = \overline{\Delta} = 3/2$ in the even sector. This field generates correction terms $1/N$ in leading order. Finally, for the three-states Potts model, eq. (12.4) has in the sector $Q = 0$ a field with $\Delta = \overline{\Delta} = 7/5$. The leading finite-size correction due to this field is of order $1/N^{0.8}$.
   In chapter 12, we remarked that for the tricritical Ising model the correction to scaling terms might be larger than for the Ising model. This numerical observation is confirmed by the above result on the leading correction terms.
2. **Relevant fields**. For these, $y_k > 0$. Relevant fields describe the cross-over to another universality class. We shall describe this in more detail in the next chapter. Here we

shall limit ourselves to a study of the **finite-size scaling limit** and define **finite-size scaling variables** $z_k := a_k N^{y_k}$ which are kept finite in the **finite-size scaling limit** $a_k \to 0$ and $N \to \infty$ (see chapter 3). We then obtain a power series expansion of the finite-size scaling functions. We stress that this can be obtained without having to know how to solve a given model exactly on the lattice!

After these general remarks about the order of the leading finite-size corrections, we now turn towards the calculation of the correction amplitudes, following [289]. For an alternative approach, see [297].

13.3 Finite-size corrections

The Hamiltonian on the strip of width N is

$$H = H_c + \sum_k a_k \int_{-N/2}^{N/2} dv\, \varphi_k(0, v) \tag{13.14}$$

where a_k is a non-universal constant and φ_k an irrelevant conformal field with scaling dimension

$$x_k = \Delta_k + r_k + \overline{\Delta}_k + \bar{r}_k \tag{13.15}$$

where $\Delta_k, \overline{\Delta}_k$ are the conformal weights of the corresponding primary field. Note that N does not appear as a coupling constant. To calculate the matrix element of φ_k consider the two-point function

$$\begin{aligned}
&< \varphi_{\Delta,\overline{\Delta}}(u_1, v_1)\varphi_{\Delta,\overline{\Delta}}(u_2, v_2) > \\
&= \sum_{r,\bar{r};\alpha} \langle 0,0|\varphi_{\Delta,\overline{\Delta}}(0,0)|\Delta + r, \overline{\Delta} + \bar{r}; \alpha\rangle\langle \Delta + r, \overline{\Delta} + \bar{r}; \alpha|\varphi_{\Delta,\overline{\Delta}}(0,0)|0,0\rangle \\
&\cdot\ (p_1 p_2)^{\Delta + r}\, (\bar{p}_1 \bar{p}_2)^{\overline{\Delta} + \bar{r}} \\
&= \sum_{r,\bar{r}} b(r, \bar{r})\, (p_1 p_2)^{\Delta + r}\, (\bar{p}_1 \bar{p}_2)^{\overline{\Delta} + \bar{r}}
\end{aligned} \tag{13.16}$$

with

$$p_k := \exp\left\{\frac{2\pi}{N}\left[(u_k - u_{k+1}) + i(v_k - v_{k+1})\right]\right\} \tag{13.17}$$

which defines $b(r, \bar{r})$. The three-point function is

$$\begin{aligned}
&< \varphi_{\Delta,\overline{\Delta}}(u_1, v_1)\varphi_k(u_2, v_2)\varphi_{\Delta,\overline{\Delta}}(u_3, v_3) > \\
&= \sum_{r_1,r_2,\bar{r}_1,\bar{r}_2} \sum_{\alpha,\beta} \langle 0,0|\varphi_{\Delta,\overline{\Delta}}(0,0)|\Delta + r_1, \overline{\Delta} + \bar{r}_1; \alpha\rangle \\
&\cdot\ \langle \Delta + r_1, \overline{\Delta} + \bar{r}_1; \alpha|\varphi_k(0,0)|\Delta + r_2, \overline{\Delta} + \bar{r}_2; \beta\rangle \\
&\cdot\ \langle \Delta + r_2, \overline{\Delta} + \bar{r}_2; \beta|\varphi_{\Delta,\overline{\Delta}}|0,0\rangle p_1^{\Delta + r_1} p_2^{\Delta + r_2} \bar{p}_1^{\overline{\Delta} + \bar{r}_1} \bar{p}_2^{\overline{\Delta} + \bar{r}_2} \\
&=: \sum_{r_1,r_2,\bar{r}_1,\bar{r}_2} a(r_1, r_2; \bar{r}_1, \bar{r}_2)\, p_1^{\Delta + r_1} p_2^{\Delta + r_2} \bar{p}_1^{\overline{\Delta} + \bar{r}_1} \bar{p}_2^{\overline{\Delta} + \bar{r}_2}
\end{aligned} \tag{13.18}$$

defing $a(r_1, r_2; \bar{r}_1, \bar{r}_2)$. Comparing, one has the desired matrix element, taking translation invariance into account

$$\langle \Delta + r, \overline{\Delta} + \bar{r}; \alpha | \varphi_k(0,0) | \Delta + r, \overline{\Delta} + \bar{r}; \alpha \rangle = \frac{a(r, r; \bar{r}, \bar{r})}{b(r, \bar{r})} \tag{13.19}$$

Tower of the identity. This is always present and thus studied first. As shown by Reinicke [289], the operators $\int_{-N/2}^{N/2} dv L_{-k}(w)$ with $k \geq 3$ do not give a correction in any order of perturbation theory since the corresponding matrix elements vanish. To leading order there remain two possibilites

$$\begin{aligned} \varphi_1(w, \bar{w}) &:= L_{-2}(w) \bar{L}_{-2}(\bar{w}) \\ \varphi_2(w, \bar{w}) &:= L_{-2}^2(w) + \bar{L}_{-2}^2(\bar{w}) \end{aligned} \tag{13.20}$$

One now has to calculate the correlation functions of these non-primary fields. In the plane, these are known and can be obtained by applying the differential operator $\mathcal{L}_k$ of eq. (4.38) to correlation functions of primary fields. These in turn were shown in chapter 4 to satisfy linear partial differential equations which can be solved. This was demonstrated for the Ising model example in chapter 6. Here, it is sufficient to recall that under the conformal transformation $z = \exp \frac{2\pi}{N} w$ the stress-energy tensor transforms as

$$T(w) \rightarrow \left(\frac{2\pi}{N}\right)^2 \left(z^2 T(z) - \frac{c}{24}\right) \tag{13.21}$$

and that

$$L_{-k}(w) = \oint dw' \frac{T(w')}{(w' - w)^{k-1}} \tag{13.22}$$

It then follows

$$< L_{-2}(w_1) L_{-2}(w_2) >= \left(\frac{2\pi}{N}\right)^4 \left\{ \frac{c}{2} \frac{(p_1 p_2)^2}{(1 - p_1 p_2)^4} + \left(\frac{c}{24}\right)^2 \right\} \tag{13.23}$$

and with a similar expression for the three-point function one obtains [289]

$$\begin{aligned} \langle \Delta + r | L_{-2}(0,0) | \Delta + r \rangle &= \left(\frac{2\pi}{N}\right)^2 \left(\Delta + r - \frac{c}{24}\right) \\ \langle \Delta + r | L_{-2}^2(0,0) | \Delta + r \rangle &= \left(\frac{2\pi}{N}\right)^4 \left[\left(\frac{c}{24}\right)^2 + \frac{11}{1440} c + A(\Delta, r)\right] \end{aligned} \tag{13.24}$$

where $r = 0, 1, 2, 3$ for $\Delta \neq 0$ and $r = 0, 2, 3$ for $\Delta = 0$ and

$$\begin{aligned} A(0, r) &= \left(\frac{11}{30} + \frac{c}{12}\right) r(2r^2 - 3) \\ A(\Delta, r) &= (\Delta + r) \left[\left(\Delta - \frac{2 + c}{12}\right) + \frac{r(2\Delta + r)(5\Delta + 1)}{(\Delta + 1)(2\Delta + 1)}\right] \quad ; \quad \Delta \neq 0 \end{aligned} \tag{13.25}$$

Summarising, the leading finite-size corrections coming from the tower of the identity are described by

$$H = H_c + a_1 \int_{-N/2}^{N/2} dv\,(L_{-2}\bar{L}_{-2}(0,v)) + a_2 \int_{-N/2}^{N/2} dv\,(L_{-2}^2(0,v) + \bar{L}_{-2}^2(0,\bar{v})) \quad (13.26)$$

and lead to the follwing results for the singular free energy density f^{sin} and the scaled and non-degenerate gaps $\mathcal{E}$ [289] (see eq. (10.32))

$$\begin{aligned}(2\pi)^{-1} f^{sin} &= -\frac{c}{12} + \frac{\pi^3}{N^2}\left[a_1\frac{c^2}{72} + a_2\left(\frac{c^2}{36} + \frac{11}{90}c\right)\right] + \mathcal{O}(N^{-4}) \\ \mathcal{E}(\Delta + r, \overline{\Delta} + \bar{r}) &= \Delta + r + \overline{\Delta} + \bar{r} \\ &+ \frac{8\pi^3}{N^2}\left\{a_1\left[\left(\Delta + r - \frac{c}{24}\right)\left(\overline{\Delta} + \bar{r} - \frac{c}{24}\right) - \left(\frac{c}{24}\right)^2\right]\right. \\ &+ \left. a_2\left[A(\Delta, r) + A(\overline{\Delta}, \bar{r})\right]\right\} + \mathcal{O}(N^{-4}) \end{aligned} \quad (13.27)$$

The only model-dependent parameters in these expressions are the values of the central charge c and the conformal weights $\Delta, \overline{\Delta}$ of the primary fields. For the values of r one has the same restrictions as above.

We now compare these results with explicit model calculations. Besides illustrating the general expressions with some specific numbers, these applications also point towards some further subleties involved in calculating finite-size corrections.

Application to the Ising model. Finite-size corrections for the Ising model were already calculated for the entire spectrum and the lowest levels are given in eqs. (10.39) and (10.45). Comparison with the general results of eq. (13.27) gives complete consistency if [289]

$$a_1 = 0 \;\; , \;\; a_2 = \frac{3}{56\pi}\left(\frac{1}{\eta^2} - \frac{4}{3}\right) \quad (13.28)$$

This results identifies the ubiquitous non-universal correction amplitude a_2 in terms of the Ising model parameters. This is one of the still very few examples where scaling and conformal predictions can be confirmed in full detail.

However, the $2D$ Ising model has the peculiarity that there is a further irrelevant field in the conformal tower of the energy density ε, namely $\varphi_3 := L_{-1}\bar{L}_{-1}\varepsilon$ which also gives corrections of order $1/N^2$. Since its scaling dimension $x_3 = 3$, the first order, if present, would give rise to terms of order $1/N$. These terms vanish since φ_3 is a derivative field, however. Similarly, the field $L_{-2}\varepsilon = 3/4L_{-1}^2\varepsilon$ does not contribute in first order ($\sim 1/N^2$). Second-order terms involving φ_3 do contribute, however. If a_1 and a_2 were both vanishing, consitency were achieved if

$$a_3^2 = \frac{1}{16\pi^2}\left(\frac{1}{\eta^2} - \frac{4}{3}\right) \quad (13.29)$$

To have H hermitian, one needs the a_k to be real. For $a_1 = a_2 = 0$, this is not possible if $\sqrt{3}/2 < \eta \leq 1$. Consequently, it is not enough to have a_3 alone non-vanishing. The leading finite-size corrections of the Ising model are thus described by the Hamiltonian [289]

$$H = H_c + a \int_{-N/2}^{N/2} dv \, (L_{-1}\bar{L}_{-1}\varepsilon) + \frac{3}{56\pi}\left[\frac{1}{\eta^2} - \frac{4}{3} - (4\pi a)^2\right] \int_{-N/2}^{N/2} dv \, \left(L_{-2}^2(0,v) + \bar{L}_{-2}^2(0,\bar{v})\right) \quad (13.30)$$

where a is a free parameter. There is no known criterion to fix its value.

Application to the three-states Potts model. The three-states Potts model serves to illustrate the case of **non-analytic** corrections arising from an additional irrelevant primary field. The finite-size corrections of this model were studied in great detail in [143]. The numerical techniques needed were described in chapter 9. The numerical results are

$$\begin{aligned}(2\pi)^{-1} f^{sin} &= -\frac{1}{15} + 0.0280(5)N^{-2} + \dots \\ \mathcal{E}(\Delta + r, \overline{\Delta} + \bar{r}) &= x + c_1 N^{-0.8} + c_2 N^{-2} + \dots \end{aligned} \quad (13.31)$$

where c_1 and c_2 are level-dependent constants. Their values are given in Table 25 [143].

Table 25. Finite-size correction coefficients of eq. (13.31) for the three-states Potts model.

$(\Delta + r, \overline{\Delta} + \bar{r})$	c_1	c_2
$(0+2, 0)$		-5.810(1)
$(1/15, 1/15)$	0.00657(2)	0.03238(2)
$(1/15+1, 1/15)$	0.034(2)	-1.6(1)
$(2/5, 2/5)$	0.2364(2)	-0.328(2)
$(2/5, 1/15)$	-0.0395(3)	-0.15(3)

We now discuss how these data can be explained in terms of conformal finite-size corrections.

1. For terms of order $N^{-0.8}$, write the perturbed Hamiltonian

$$H = H_c + a \int_{-N/2}^{N/2} dv \, \varphi_{7/5,7/5}(0,v) \quad (13.32)$$

and take $a = 0.009237(7)$ to reproduce the values of the c_1 as given in Table 25. Since the corresponding operator product expansion coefficient vanishes, there is no contribution to f^{sin} in this order.

2. For the order N^{-2}, take the constants a_1, a_2 as in eq. (13.27). A fit to the numerical data yields $a_1 = -0.002(2)$ and $a_2 = -0.0056(3)$. Since this is consistent with $a_1 = 0$, the fit was repeated with a_2 alone and yields [289]

$$a_1 = 0 \ , \ a_2 = -0.0056(6) \tag{13.33}$$

From the operator content of the Potts model it is clear that there are no further contributions at this order. It is non-trivial that a single parameter is enough to reproduce all values of c_2 in Table 25.

3. There are second-order contributions from $\varphi_{7/5,7/5}$, of order $N^{-1.6}$. Generally, the contribution to f^{sin} from a second-order term of an irrelevant scalar primary field is

$$(2\pi)^{-1} f^{sin} + \frac{c}{12} = -\left[2\pi a \left(\frac{2\pi}{N}\right)^{x-2}\right]^2 \frac{1}{2}\int_0^1 dv\, v^{x/2-1}{}_2F_1(x,x,1;v) \tag{13.34}$$

with $x = 2\Delta > 2$ and ${}_2F_1$ is the hypergeometric function. The integral is only convergent for $0 < \mathrm{Re}x < 1$, but is needed for $x > 2$. It can be made finite by analytical continuation which is achieved by $2k$-fold partial integration [289]

$$\begin{aligned}
&\int_0^1 dv\, v^{x/2-1}{}_2F_1(x,x,1;v) \\
&= -\sum_{m=1}^{k} \frac{\Gamma^2(1-x)\Gamma^2(m-x/2)\Gamma(2m-2x)}{\Gamma^2(1-x/2)\Gamma^3(1+m-x)\Gamma(m-x)}(3m-2x) \\
&\quad + \frac{\Gamma^2(k+1-x/2)\Gamma^2(1-x)}{\Gamma^2(1-x/2)\Gamma^2(k+1-x)}\int_0^1 dv\, v^{x/2-1}{}_2F_1(x-k,x-k,1;v)
\end{aligned} \tag{13.35}$$

where the right-hand side is now well-defined for $0 < \mathrm{Re}x < k+1$ and x non-integer. One then finds for the Potts model

$$\begin{aligned}
(2\pi)^{-1} f^{sin} &= -\frac{1}{15} - 0.001721(3)N^{-1.6} + \ldots \\
\mathcal{E}(2,0) &= 2 - 0.01093(2)N^{-1.6} + \ldots
\end{aligned} \tag{13.36}$$

Comparing these with the results of Table 25, one sees that these second order contributions are at least an order of magnitude smaller than those terms of order N^{-2}, at least for the largest lattice of size $N = 12$ used in the numerical calculations reported. This finding nicely illustrates a general remark that for small lattices the effective correction exponent can be substantially affected by relatively large or small values of the correction amplitudes, see [277].

The same techniques were also applied to study the finite-size corrections of the order parameter [290], where the finite-lattice order parameter is defined in eq. (9.17). Finite-size corrections have been calculated in a huge variety of integrable systems. For a recent review, see [312].

Checking the operator content from finite-size corrections. We have seen how the known operator content was used to predict the leading finite-size corrections of critical conformal invariant systems. Blöte and den Nijs [51] have asked conversely whether these corrections are always the **only** ones appearing. This question is well motivated, since, as we have seen in chapter 6, we came across solutions for the values of the conformal weights Δ which did not fit into the Kac table. In particular, in the context of the Ising model, there is an operator which would correspond to the conformal operator $\phi_{3,1}$, see eq. (6.22), but which is not in the Kac table for $m = 3$. This operator, if it **were** present, would represent a *vacancy operator* with scaling dimension $x_{vac} = 10/3$, producing finite-size corrections of order $N^{-4/3}$. The presence of such an operator would still be consistent with conformal invariance but **not** with the closed operator product algebra used to classify the possible primary fields of unitary models. This vacancy operator is absent in the integrable realisation of the $2D$ Ising universality class, as it is evident from the exact results for the finite-size correction terms of chapter 10. One might argue, however, that in an antiferromagnetic Ising model with next to nearest neighbor interactions an eventual coupling constant of the vacancy operator to the rest of the model might be enhanced. Numerical calculations [51] of the correlation length, however, clearly indicate the leading finite-size corrections to be of order N^{-2}, which implies that the vacancy operator cannot be present.

It does occur, however, in the random-cluster representation of the $p = 2 + \varepsilon$ state Potts model. The amplitude of this operator vanishes in the Ising case. For a discussion of the weakly disordered $2 + \epsilon$ state Potts model, see [243].

13.4 Finite-size scaling functions

We now turn to the case of **relevant** perturbations. We write the finite-size scaling variable $z = \vartheta N^{2-x}$, where ϑ measures the distance from the critical point, like e.g. $\vartheta = T - T_c$. The Hamiltonian is

$$H = H_c + \bar{a} z N^{x-2} \int_{-N/2}^{N/2} dv\, \varphi(0, v) = H_c + V \tag{13.37}$$

where $\bar{a}$ is a non-universal constant, related to the metric factors introduced in chapter 3.

For the first order, one just needs the matrix elements of V. For the second order, one has, with $\omega_{i,n} = E_{i,c} - E_{n,c}$ and k denoting the momentum eigenvalue [289]

$$\begin{aligned}
E_n^{(2)}(k) &= -\sum_{i \neq n} \omega_{i,n}^{-1} \langle k, n|V|k, i\rangle \langle k, i|V|k, n\rangle \\
&= -\left(\bar{a} z N^{x-2}\right)^2 \int_0^\infty d\tau \int\!\!\int_{-N/2}^{N/2} dv_1 dv_2 \cdot \\
&\quad \left\{ \sum_i e^{-\omega_{i,n}\tau} \langle k, n|\varphi(0, v_1)|k, i\rangle \langle k, i|\varphi(0, v_2)|k, n\rangle \right. \\
&\quad - \langle k, n|\varphi(0, v_1)|k, n\rangle \langle k, n|\varphi(0, v_2)|k, n\rangle \\
&\quad \left. - \sum_{i<n} \left(e^{-\omega_{i,n}\tau} + e^{\omega_{i,n}\tau}\right) \langle k, n|\varphi(0, v_1)|k, i\rangle \langle k, i|\varphi(0, v_2)|k, n\rangle \right\}
\end{aligned}$$

$$= -\left(\bar{a}zN^{x-2}\right)^2 \int_0^\infty d\tau \int\int_{-N/2}^{N/2} dv_1 dv_2 \cdot$$

$$\Big\{ \langle k,n|\varphi(0,v_1)\varphi(0,v_2)|k,n\rangle - \langle k,n|\varphi(0,v_1)|k,n\rangle\langle k,n|\varphi(0,v_2)|k,n\rangle$$

$$-\sum_{i<n}\left(e^{-\omega_{i,n}\tau}+e^{\omega_{i,n}\tau}\right)\langle k,n|\varphi(0,v_1)|k,i\rangle\langle k,i|\varphi(0,v_2)|k,n\rangle\Big\}$$

$$=: -\left(\bar{a}zN^{x-2}\right)^2 \int_0^\infty d\tau B_{k,n;2}\left(\frac{2\pi}{N}\tau\right) \tag{13.38}$$

using the spectral decomposition. For the scaled gaps one finds

$$\mathcal{E}(\Delta+r,\overline{\Delta}+\bar{r}) = -\left((2\pi)^{x-1}\bar{a}z\right)^2 \int_0^\infty d\tau \left(B_{k,n;2}(\tau) - B_{0,0;2}(\tau)\right) \tag{13.39}$$

and similar expressions can be written down for the higher orders. In general, for the nth order perturbation theory some $(n+2)$-point correlation functions are needed. The n-point correlation functions are known in the infinite plane and must be rewritten into the strip via the standard logarithmic transformation. We shall not give the details of the rather tedious calculation and refer for this to the original work [289].

Ising model: thermal perturbation. The perturbing field $\varphi = \varepsilon$ is the energy density with $x_\varepsilon = 1$. The finite-size scaling variable is $z := (t-1)N$. The results for the gaps are, including the fifth order contributions [289]

$$\begin{aligned}\mathcal{E}(r+1/16,\bar{r}+1/16) &= r+\bar{r}+\frac{1}{8}+\bar{a}z\frac{1}{2}(2\delta_{r,0}-1)(2\delta_{\bar{r},0}-1)\\ &+(\bar{a}z)^2\left(\ln 2+\frac{1-\delta_{r,0}}{2}+\frac{1-\delta_{\bar{r},0}}{2}\right)\\ &+(\bar{a}z)^4\left(\frac{3}{4}\zeta(3)+\frac{1-\delta_{r,0}}{8r^3}+\frac{1-\delta_{\bar{r},0}}{8\bar{r}^3}\right)+\mathcal{O}(z^6)\end{aligned} \tag{13.40}$$

where ζ is the Riemann zeta function and, if $\Delta = 0, 1/2$

$$\begin{aligned}\mathcal{E}(r+\Delta,\bar{r}+\overline{\Delta}) &= \Delta+r+\overline{\Delta}+\bar{r}+(\bar{a}z)^2\left(\mathcal{A}(r,\Delta)+\mathcal{A}(\bar{r},\overline{\Delta})\right)\\ &+(\bar{a}z)^4\left(\mathcal{B}(r,\Delta)+\mathcal{B}(\bar{r},\overline{\Delta})\right)+\mathcal{O}(z^6)\end{aligned} \tag{13.41}$$

where

$$\begin{aligned}\mathcal{A}(r,0) &= 1+\frac{1}{2r-1}\,,\ \mathcal{B}(r,0) = 1+\frac{1}{(2r-1)^3}\ ;\ r=0,2,3\\ \mathcal{A}(r,1/2) &= \frac{1}{2r-1}\,,\ \mathcal{B}(r,1/2) = \frac{1}{(2r-1)^3}\ ;\ r=0,1,2,3\end{aligned} \tag{13.42}$$

Comparing with the Ising finite-size scaling functions eq. (10.46), the non-universal constant $\bar{a}$ is fixed

$$\bar{a} = (-1)^{\bar{Q}}\,\mathrm{sgn}\,(z)\frac{1}{2\pi\eta} \tag{13.43}$$

We see that conformal perturbation theory is capable to find the **exact** expansion of the finite-size scaling functions in terms of the finite-size scaling variable z. The only model-dependent input needed are the values of the conformal weights.

One may now ask whether the series of which the first few terms were derived is convergent and if so, for which values of z it converges. This question has not yet been answered in general, but for the Ising model, the exact solution eq. (10.46) does give an answer. The remnant function $R_{1\frac{1}{2},0}(x)$ involved is analytic at the origin [124]

$$R_{1\frac{1}{2},0}(x) = \Gamma(-1/2)\sum_{n=2}^{\infty}\frac{\zeta(2n-1)}{\Gamma(3/2-n)}\frac{x^n}{n!} \tag{13.44}$$

and has a **finite** radius of convergence $|x| < 1$. Reinterpreting this in terms of the Ising model parameters, one sees that for $\eta = 0$, this representation ceases to converge. This is not surprising, since for $\eta = 0$ the global symmetry and the critical behaviour of the model changes completely.

Ising model: magnetic perturbation. Now consider the case $\varphi = \sigma$ with $x_\sigma = 1/8$. Since the order parameter breaks the global symmetry, only even orders will contribute. The finite-size scaling variable is $\mu := hN^{15/8}$ where h is the magnetic field. The second-order results are [289]

$$\begin{aligned}
\mathcal{E}(r,\bar{r}) &= r+\bar{r}+\left(\frac{\bar{a}\mu}{(2\pi)^{7/8}}\right)^2(1-\gamma_r\gamma_{\bar{r}})\cdot\delta+\mathcal{O}(\mu^4)\ ;\ r,\bar{r}=0,2,3\\
\mathcal{E}(1/2+r,1/2+\bar{r}) &= 1+r+\bar{r}+\left(\frac{\bar{a}\mu}{(2\pi)^{7/8}}\right)^2\cdot\delta+\mathcal{O}(\mu^4)\ ;\ r,\bar{r}=0,1,2,3\\
\mathcal{E}(1/16+r,1/16+\bar{r}) &= \frac{1}{8}+r+\bar{r}+\left(\frac{\bar{a}\mu}{(2\pi)^{7/8}}\right)^2(\delta-g_{r,\bar{r}})+\mathcal{O}(\mu^4)\ ;\ r,\bar{r}=0,1,2
\end{aligned} \tag{13.45}$$

with the constants

$$\gamma_0 = 1\ ,\ \gamma_2 = \frac{19}{124}\ ,\ \gamma_3 = \frac{512}{1457} \tag{13.46}$$

$$\delta = \sum_{m=0}^{\infty}\left(\frac{\Gamma(1/8+m)}{\Gamma(1/8)m!}\right)^2\frac{1}{2m+1/8} = 8.00949725\ldots \tag{13.47}$$

$$\begin{aligned}
g_{00} &= -7.70684920(1)\ ,\ g_{10} = 0.22667203(2)\ ,\ g_{20} = 4.2823790(1)\\
g_{11} &= -0.00666683(2)\ ,\ g_{21} = 0.1508220(2)\ ,\ g_{22} = 0.6177484(6)
\end{aligned} \tag{13.48}$$

and in the last equation the given errors are exact upper and lower bounds.

We now compare this with numerical data in the $(1+1)D$ Ising model. We take $\eta = 1$. In view of the thermal result for the non-universal metric factor $\bar{a}$ eq. (13.43), the choice

$$\bar{a} = (2\pi)^{-1} \tag{13.49}$$

is suggestive. For the comparison with numerical data, some care is needed, however. In principle, there are two different sources for finite-size corrections:

1. The finiteness of the system for $\mu = 0$. These purely finite-size corrections were calculated above.
2. Keeping a finite value of μ on a finite lattice also implies that h is non-zero. This leads to the appearance of **bulk** corrections.

Both correction terms, if present, will break finite-size scaling. A scaling plot is therefore necessary to show that finite-lattice data obtained really yield information on the scaling behaviour. In fact, using the Ising quantum chain with moderately large lattices (up to $N = 15$), finite-size corrections are still apparent. In order to check the conformal predictions, we calculate the ***reduced finite-size scaling functions*** $H_i(\mu)$ for the two lowest gaps

$$\begin{aligned}\xi_\sigma^{-1} &= \frac{\pi}{4}N^{-1} + \frac{\pi^3}{192}N^{-3} + h^{8/15}\frac{1}{2}H_1(\mu)\\ \xi_\varepsilon^{-1} &= 2\pi N^{-1} - \frac{\pi^3}{12}N^{-3} + h^{8/15}\frac{1}{2}H_2(\mu)\end{aligned} \tag{13.50}$$

where the first terms are just the conformal critical point results and the second terms are the leading finite-size corrections for $\mu = 0$. These reduced scaling functions are shown in Figure 32 [179]. The different symbols correspond to different values of h. Clearly the corrected data display a scaling behaviour. The full curves were obtained by using the second order results together with eq. (13.49). There is excellent agreement for $\mu \leq 0.3$. For larger values of μ, deviations appear which are probably due to the higher order corrections.

These results represent an impressive confirmation of the theoretical framework of conformal invariance for the precise quantitative description of critical two-dimensional systems. Extending these results to higher orders may look prohibitive in view of the tedious calculations to be done. However, we shall see in chapter 14 that there is a class of relevant perturbations which preserve the integrability even of the non-critical system. Then the thermodynamic Bethe ansatz can be used to render the computations manageable. However, the requirement of integrability is a very strong one, and is restricted in the case of unitary models to the relevant operators ϕ_{12}, ϕ_{21} and ϕ_{13}. In all other cases, more general techniques as the one presented in this chapter have to be used.

We have seen that conformal perturbation theory only contains contributions analytic in h. It had been suggested [112] that a more general scheme might apply, which argues that because of the operator product expansion in the Ising model $\sigma \cdot \sigma = \mathbf{1} + \varepsilon$, the effective perturbed Hamiltonian should read $H = H_c + h\sigma + Ah^{\psi}\varepsilon$ with $\psi = 8/15$ and A being a free parameter. However, the presence of a free parameter is not in agreement with the conventional scaling theory (see chapters 1 and 3) and also in contradiction to the numerically confirmed universality [99] of the finite-size scaling functions of both the Ising and Potts-3 models when varying t or h. Quantitatively, the idea of [112] would imply for the scaling function of the spin-spin correlation length of the Ising model the form $m_\sigma = h^{8/15}(\pi/(4\mu^{8/15}) + A + \ldots)$. Comparison with Figure 32 shows that A must indeed be very small and an upper bound $A < 3\cdot 10^{-4}$ [179] is obtained. This is consistent with $A = 0$ and reconfirms conventional finite-size scaling.

13.5 Truncation method

We now present a different technique for the approximate calculation of the spectrum of the perturbed Hamiltonian H, following the original work of Yurov and Al. Zamolodchikov [337].

One wants to calculate the spectrum of the perturbed Hamiltonian eq. (13.37). This could be done in an approximate way by selecting a convenient small subspace of the full Hilbert space and diagonalize the interaction matrix V only in that subspace. The matrix elements of V are completely specified in terms of the operator product expansion coefficients $\mathbf{C}_{123}$. What is needed is to find a sensible way to select a truncated space. For simplicity, the perturbing field ϕ is assumed to be scalar.

The conformal fields appearing in a model are classified as follows. Either they are **quasiprimary fields**, that is, its heighest weight state $|\Delta\rangle$, generated by $|\Delta\rangle = \phi(0)|0\rangle$, is annihilated by L_{-1}

$$L_{-1}|\Delta\rangle = 0 \tag{13.51}$$

but it may or may not be annihilated by L_{-2}. Alternatively, they can be written in terms of derivatives of quasiprimary fields. Matrix elements of derivatives can be reduced to matrix elements of quasiprimary fields via

$$\begin{aligned}\langle\Delta_1|L_1^k\phi(0,0)L_{-1}^n|\Delta_3\rangle &= n!k! \\ \cdot\sum_\ell \frac{(\Delta_1+\Delta_3-\Delta)_\ell(\Delta+\Delta_3-\Delta_1)_{n-\ell}(\Delta+\Delta_1-\Delta_3)_{k-\ell}}{\ell!(n-\ell)!(k-\ell)!}&\langle\Delta_1|\phi(0,0)|\Delta_3\rangle\end{aligned} \tag{13.52}$$

where Δ is the conformal weight of the field ϕ and $(a)_n := \Gamma(a+n)/\Gamma(a)$ is the Pochhammer symbol. The matrix elements of the primary fields among themselves are given by eq. (13.9). For the calculation of the matrix elements of V, both left and right conformal algebras as generated by the L_n and the $\bar{L}_n$ must be taken into account. The Hilbert subspace to be considered is taken to include the primary fields appearing in a given sector of the model, if the perturbation respects the global symmetry. Otherwise, all sectors must be considered simultaneously. One then adds the quasiprimary fields and derivatives up to a certain level, which in most applications so far has been the level 5, which leads to matrices of size of about 200 × 200 at most. The matrix elements of V are then calculated and V is diagonalized. A Mathematica^{TM} program to do these computations is available [231].

To illustrate the calculations, consider the example of the Yang-Lee singularity. In chapter 12 it was shown that at the critical point, the model contains as primary fields the identity **1** and a single relevant field φ with $\Delta = -1/5$. The only non-trivial OPE coefficient is

$$\mathbf{C}_{\varphi\varphi\varphi} = i\kappa \ , \ \kappa := \frac{1}{5}\left(\gamma\left(\frac{1}{5}\right)\right)^{3/2}\gamma\left(\frac{2}{5}\right) = 1.91131\ldots \tag{13.53}$$

where $\gamma(x) = \Gamma(x)/\Gamma(1-x)$. Calculating the elements of V, it is enough to consider the field $\varphi(0,0)$ because of translation invariance. The interaction matrix has the form

$$\langle\Delta_1|V|\Delta_3\rangle = -\frac{2\pi}{N}G\, B_{1,3} \tag{13.54}$$

where $G = (2\pi)^{-7/5} h N^{12/5}$ is the dimensionless coupling and corresponds to the finite-size scaling variable $\bar{a}z$ in eq. (13.37), where h is the imaginary magnetic field coupled to φ and N the width of the strip. Since G is kept fixed in these calculations, one is ***automatically inside the finite-size scaling region.*** The symmetric matrix $\hat{B}$ is N-independent. Up to level 4, the quasiprimary states are listed together with their norms in Table 26 [337]. The matrix $\hat{B}$ between these four states is

Table 26. Quasiprimary states up to level four for the Yang-Lee singularity.

State	Norm
φ	1
$(L_{-4} - \frac{625}{624} L_{-1}^4)\varphi$	$\frac{57}{130}$
$\mathbf{1}$	1
$T = L_{-2}\mathbf{1}$	$-\frac{11}{5}$

$$\hat{B} = \begin{pmatrix} \kappa & \kappa/130 & -i & i/5 \\ \kappa/130 & 1083\kappa/8450 & 0 & -57i/130 \\ -i & 0 & 0 & 0 \\ i/5 & -57i/130 & 0 & 0 \end{pmatrix} \tag{13.55}$$

This information is enough to calculate any matrix element of V for any state up to level 5. For illustration, we merely give here the results up to level 3 for scalar states. To shorten notation, we write $\partial = L_{-1}$ and $T = L_{-2}\mathbf{1}$. For the complete table of all non-vanishing matrix elements and eigenvalue calculations, see [337]. The scalar states up to

Table 27. The normalized scalar states and their scaling dimensions for the truncated space up to level 3.

number	State	scaling dimension
1	φ	$-\frac{2}{5}$
2	$\frac{5}{2}\partial\bar{\partial}\varphi$	$\frac{8}{5}$
3	$\frac{25}{12}\partial^2\bar{\partial}^2\varphi$	$\frac{18}{5}$
4	$\mathbf{1}$	0
5	$\frac{5}{11}T\bar{T}$	4

level 3 are given with their scaling dimensions in Table 27 and the symmetric matrix $\hat{B}$

is then, taking into account both the left and the right conformal fields

$$\hat{B} = \begin{pmatrix} \kappa & \frac{\kappa}{10} & \frac{4\kappa}{75} & i & \frac{i}{55} \\ & \frac{4\kappa}{25} & \frac{49\kappa}{750} & \frac{2i}{5} & \frac{72i}{275} \\ & & \frac{289\kappa}{625} & \frac{3i}{25} & \frac{588i}{1375} \\ & & & 0 & 0 \\ & & & & 0 \end{pmatrix} \tag{13.56}$$

where we wrote only the upper part explicitly, since the lower part is given by the **symmetry** of $\hat{B}$.

This technique does provide accurate numerical values for the eigenvalues of H as a function of the finite-size scaling variable G, in particular for the lower ones, provided that the perturbing coupling G does not become too large. It is remarkable that this can be achieved by treating a fairly small matrix. This should be compared to the huge matrices required for the conventional finite-size scaling calculations. In a sense, this technique is a nice and efficient way to implement an "importance sampling" on quantum Hamiltonians, which is normally done by Monte Carlo methods and it is this property which made it very popular in the investigation of a given conformal field theory. On the other hand, the method requires quite some knowledge on the system it is to be applied to, in particular the conformal operator content must be known beforehand. This is not always immediately available for a given statistical lattice system. Further, there is at present no systematic theory on how the truncations affect the energy levels and the truncation normally done at level 5 at present is just empirical. In particular when two levels of the system become very close or cross, then in the context of the truncation method there will be a spurious level repulsion be generated. Finally, the method is up to now mainly applied to study a model for large values of the coupling G, where its disadvantages become more apparent. We postpone until the next chapter a detailed numerical comparison with conventional finite-size scaling results.

Summarising, the truncation method sometimes does provide a useful alternative to conventional numerical techniques.

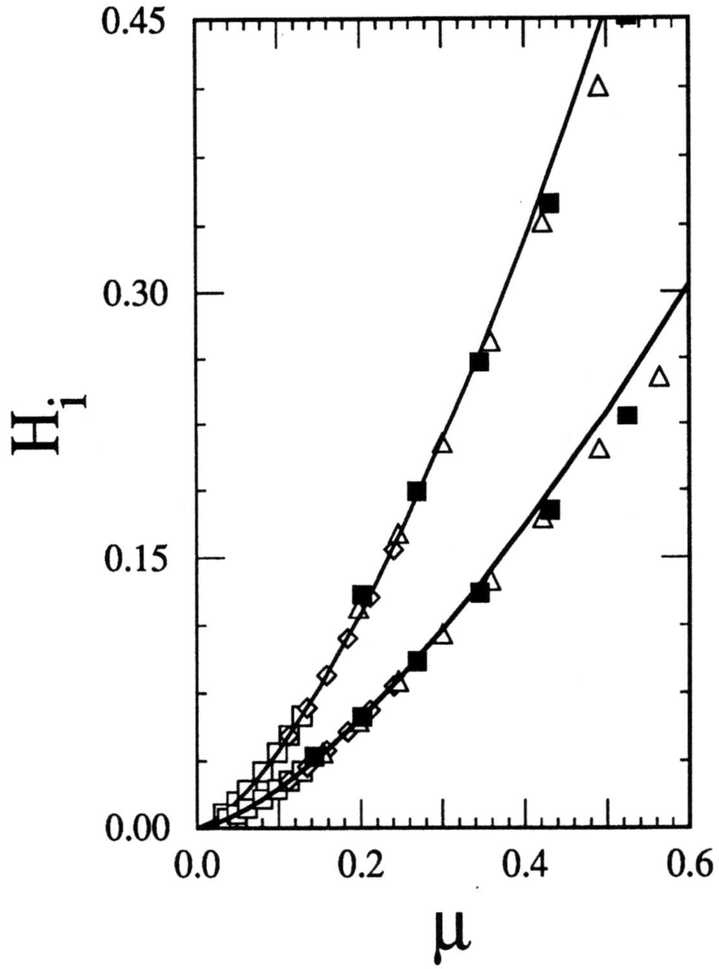

Fig. 32. Reduced finite-size scaling functions H_1 (upper curve) and H_2 (lower curve) for the quantum Ising chain in a magnetic field.

14 The Vicinity of the Critical Point

Up to now, we have concentrated on describing the results of conformal invariance for the universal finite-size scaling functions Y and S_i of the singular free energy density and the correlation lengths

$$\begin{aligned} f^{sin} &= L^{-2} Y(C_1 t L^{1/\nu}, C_2 h L^{(\beta+\gamma)/\nu}) \\ \xi_i^{-1} &= L^{-1} S_i(C_1 t L^{1/\nu}, C_2 h L^{(\beta+\gamma)/\nu}) \end{aligned} \tag{14.1}$$

for the case of **small** values of the finite-size scaling variables $z = tL^{1/\nu}$ and $\mu = hL^{(\beta+\gamma)/\nu}$. We now ask the converse question on their behaviour for z, μ becoming large.

To be specific, take $z \to \infty$, but let $\mu = 0$. Generalizations are obvious. We then have

$$f^{sin} \sim t^{2\nu} C_1^{2\nu} \tilde{Y}(0) = A t^{2-\alpha} \quad , \quad \xi_i^{-1} \sim t^{\nu} C_1^{\nu} \tilde{S}_i(0) = B_i t^{\nu} \tag{14.2}$$

where $2 - \alpha = 2\nu$ and

$$Y(z, 0) = z^{2\nu} \tilde{Y}(1/z) \quad , \quad S(z, 0) = z^{\nu} \tilde{S}_i(1/z) \tag{14.3}$$

Although the *bulk amplitudes* A, B_i are non-universal[13)] one readily identifies the universal amplitude ratios (see eq. (3.32))

$$R_{i,j} := \frac{\tilde{S}_i(0)}{\tilde{S}_j(0)} = \left(\frac{\xi_i}{\xi_j}\right)^{-1} \quad , \quad Q_i := \frac{\tilde{Y}(0)}{\tilde{S}_i^2(0)} = f^{sin} \xi_i^2 \tag{14.4}$$

What are the predictions of conformal invariance for these ?

14.1 The c-theorem

This theorem, due to A. Zamolodchikov [342], shows that, under the assumptions of rotation invariance, positivity and energy-momentum conservation in a two-dimensional theory, there exists a quantity C which is non-decreasing along a renormalization group trajectory, stationary at a RG fixed point and takes as its values at these fixed points the corresponding central charge c. As a corollary, we shall obtain an expression for the Q_i of eq. (14.4), due to Cardy [84].

We begin with energy-momentum conservation,

$$\partial_{\mu} T^{\mu\nu} = 0 \tag{14.5}$$

where $T^{\mu\nu}$ is the energy-momentum tensor and rewrite this in complex coordinates $z, \bar{z}$ for the $2D$ case

$$\partial_{\bar{z}} T + \frac{1}{4} \partial_z \Theta = 0 \quad , \quad \partial_z \bar{T} + \frac{1}{4} \partial_{\bar{z}} \Theta = 0 \tag{14.6}$$

where $\Theta := T_{\mu}^{\mu}$ vanishes at the critical point. At criticality, these relations were seen in chapter 2 to imply analyticity for $T = T(z)$. This does no longer hold for the situation

13) They are independent of the boundary conditions however, at least if these do not change the global symmetry.

under consideration now and $T = T(z, \bar{z})$. Rotation invariance implies for the correlation functions of T and Θ

$$< T(z,\bar{z})T(0,0) > = \frac{F(z\bar{z})}{z^4}$$
$$< \Theta(z,\bar{z})T(0,0) > = < T(z,\bar{z})\Theta(0,0) > = \frac{G(z\bar{z})}{z^3\bar{z}}$$
$$< \Theta(z,\bar{z})\Theta(0,0) > = \frac{H(z\bar{z})}{z^2\bar{z}^2} \tag{14.7}$$

and similarly for $\overline{T}$. Combining the first of these with energy-momentum conservation

$$< \partial_{\bar{z}}T(z,\bar{z})T(0,0) > + \frac{1}{4} < \partial_z\Theta(z,\bar{z})T(0,0) >= 0 \tag{14.8}$$

We adopt the notation $\dot{F} := z\bar{z}F'(z\bar{z})$ where the prime stands for the derivative of $F(x)$ with respect to its argument. One finds

$$\dot{F} + \frac{1}{4}\dot{G} - \frac{3}{4}G = 0 \tag{14.9}$$

and similarly by combining with Θ

$$\dot{G} - G + \frac{1}{4}\dot{H} - \frac{1}{2}H = 0 \tag{14.10}$$

Eliminating G from eqs. (14.9,14.10), and defining

$$C := 2F - G - \frac{3}{8}H \tag{14.11}$$

one finds [342]

$$\dot{C} = -\frac{3}{4}H \leq 0 \tag{14.12}$$

due to positivity. This last condition is for example always satisfied for **unitary** models. C as defined is a non-decreasing function of $R := (z\bar{z})^{1/2}$, for fixed values $\{g\}$ of the couplings (like temperature or magnetic field). The R-dependence is related to the dependence on the $\{g\}$ via the renormalization group (Callan-Symanzik equation)

$$\left(R\frac{\partial}{\partial R} + \sum_i \beta(\{g_i\})\frac{\partial}{\partial g_i}\right) C(R,\{g\}) = 0 \tag{14.13}$$

where $\beta(\{g_i\})$ is the beta-function. Consequently, the quantity $C(\{g\}) := C(1,\{g\})$ is non-decreasing along the RG trajectories. C is stationary if and only if $H = 0$, corresponding to a RG fixed point. At a fixed point, we have seen earlier from conformal invariance that $F = c/2$, $G = H = 0$ and consequently $C = c$ at a RG fixed point. This proves the stated theorem[14)].

[4)] One might contemplate the existence of other quantities with a similar behaviour, but this question goes beyond the scope of these lectures.

In particular, this implies that under the assumptions stated, the RG flow is always going "downhill" and is irreversible. Limit cycles and more complicated behaviour is ruled out. This is not surprising in view of the usual real-space renormalization procedure. Each step of coarse-graining implies a "loss" of information on the ultraviolet behaviour of the theory. One may interpret C as some measure of this. Integrating eq. (14.12), one finds the change in the central charge between two fixed points

$$\Delta c = -\frac{3}{4}\int_0^\infty dR^2\, R^2 < \Theta(R)\Theta(0) > \tag{14.14}$$

We shall see below that

$$\Theta(z,\bar{z}) = -4\pi\lambda(1-\Delta)\phi(z,\bar{z}) \tag{14.15}$$

where ϕ is the perturbing relevant scaling operator with conformal weights (Δ,Δ) and λ the corresponding coupling.

On the other hand, one can define a correlation length $\tilde{\xi}_i$ via the second moment of the scaling operator ϕ_i

$$\tilde{\xi}_i^2 := \frac{\int d^2r\, r^2 < \phi_i(r)\phi_i(0) >_c}{\int d^2r\, < \phi_i(r)\phi_i(0) >_c} \tag{14.16}$$

To simplify the notation, consider the example where $\phi_i = \varepsilon$, the energy density. Then by the fluctuation-dissipation theorem

$$\int d^2r\, < \varepsilon(r)\varepsilon(0) >_c = -\frac{\partial^2 f^{sin}}{\partial t^2} \simeq -(2-\alpha)(1-\alpha)t^2 f^{sin} \tag{14.17}$$

where α is the specific heat critical exponent, which implies that [84]

$$f^{sin}\tilde{\xi}_\varepsilon^2 \simeq -[(2-\alpha)(1-\alpha)]^{-1}t^2 \int d^2r\, r^2 < \varepsilon(r)\varepsilon(0) > \tag{14.18}$$

Since a thermal perturbation leads to an off-critical system with $c = 0$, one has the following "sum rule" expression for c

$$c = 6\pi^2 t^2 (2-x_\varepsilon)^2 \int_0^\infty dR\, R^3 < \varepsilon(R)\varepsilon(0) > \tag{14.19}$$

Combining yields for Q_ε and similarly, considering the case $\phi_i = \sigma$, also for Q_σ of eq. (14.4) [84]

$$Q_\varepsilon = -\frac{c}{12\pi}\frac{2-\alpha}{1-\alpha} \quad , \quad Q_\sigma = -\frac{c}{12\pi}\frac{2-\gamma}{1-\gamma} \tag{14.20}$$

where γ is the susceptibility exponent.

We stress again that the $\tilde{\xi}_i$ used here are only proportional to the correlation lengths ξ_i defined from the exponential decrease of the correlation functions. Their relationship is quite complicated. At the critical point, for example, when $\tilde{\xi}$ is defined by

$$\tilde{\xi}^2\chi^{(2)} = \sum_{u,v} u^2 < \phi(u,v)\phi(0,0) > \tag{14.21}$$

where $\chi^{(2)}$ is the corresponding susceptibility, then $\tilde{\xi} = \tilde{A}L$ where [280]

$$\tilde{A} = \frac{1}{2\pi}\left[\psi'\left(\frac{x}{2}\right) - \frac{\pi^2}{2\sin^2(\pi x/2)}\right]^{1/2} \tag{14.22}$$

and ψ is the digamma function and x the scaling dimension of ϕ. Only in the limit $x \to 0$, one recovers $\tilde{\xi}/\xi \to \sqrt{2}$, which one would also obtain if only the large-distance behaviour of the two-point correlation function had to be considered.

14.2 Conserved currents close to criticality

Having found the Q_i, we now turn to the R_{ij}. Their calculation is done using a theory developed by A. Zamolodchikov [345, 346]. The perturbed Hamiltonian is

$$H = H_c + \lambda \int d^2r\, \phi(r) \tag{14.23}$$

where ϕ is a scalar relevant operator with $x = 2\Delta$. The coupling λ has the scaling dimension $2 - x > 0$. Consider the conformal tower of the identity generated by the left Virasoro generators L_{-n}. Denote the operators in this tower by T_s^α where s refers to the spin and α is a label for further quantum numbers which might be needed. In particular, one has $T_2 = T$. In view of the energy-momentum conservation eq. (14.6), one forms the derivative

$$\partial_{\bar{z}} T_s^\alpha = \lambda R_{s-1}^{(\alpha,1)} + \ldots + \lambda^n R_{s-1}^{(\alpha,n)} + \ldots \tag{14.24}$$

where the $R_{s-1}^{(\alpha,n)}$ are formed from the conformal operators of the model described by H_c. The left side has conformal weights $(s,1)$. It follows that the conformal weights of the $R_{s-1}^{(\alpha,n)}$ are $(s - n(1-\Delta), 1 - n(1-\Delta))$. However, these do become **negative** if n is large enough. This implies that the perturbation series above **must** terminate after a **finite** number of terms, up to irrelevant operators, at least when a unitary or, more generally non-unitary but minimal model, describes the critical Hamiltonian H_c. In fact, for almost all cases just the first term survives

$$\partial_{\bar{z}} T_s^\alpha = \lambda R_{s-1}^{(\alpha,1)} \tag{14.25}$$

one of the few exceptions being the Ising model with a thermal perturbation. The explicit calculation of $R_s^{(\alpha,1)}$ does not pose any problem. As shown by Zamolodchikov, it follows [345] from the Ward identities that there is a set of operators D_n such that

$$\begin{aligned} \partial_{\bar{z}} &= \lambda D_0 \\ [L_n, D_m] &= -\left((1-\Delta)(n+1) + m\right) D_{n+m} \\ D_{-n-1}\mathbf{1} &= \frac{1}{n!} L_{-1}^n \phi(z,\bar{z}) \end{aligned} \tag{14.26}$$

For example

$$\partial_{\bar{z}} T = \lambda D_0 L_{-2}\mathbf{1} = \lambda(\Delta - 1) D_{-2}\mathbf{1} = \lambda(\Delta - 1) L_{-1}\phi \tag{14.27}$$

and since $L_{-1} = \partial_z$ one has indeed eq. (14.15), after taking care of the normalization. In the particular case $\Delta = 1$, corresponding to **marginal** perturbations, all orders in λ will contribute, although the lowest order contribution vanishes.

As a second example, consider the field $T_4 =: T^2 := L_{-2}^2\mathbf{1}$.

$$\begin{aligned}\partial_{\bar{z}} T_4 &= \lambda D_0 L_{-2} L_{-2}\mathbf{1} \\ &= \lambda(\Delta - 1)\,(D_{-2}L_{-2} + L_{-2}D_{-2})\,\mathbf{1} \\ &= \lambda(\Delta-1)\left(2L_{-2}L_{-1} + \frac{\Delta-3}{6}L_{-1}^3\right)\phi \end{aligned} \tag{14.28}$$

but since the right-hand side cannot be rewritten as a derivative, this does not imply the conservation of anything. To make further progress, one can take the perturbing operator to be one of the operators $\phi_{r,s}$ occurring in the Kac table eq. (10.26). As shown in chapter 4, these fields satisfy null vector relations, for example

$$\begin{aligned}\left(L_{-2} - \frac{3}{2(2\Delta+1)}L_{-1}^2\right)\phi_{1,2} &= 0 \\ \left(L_{-3} - \frac{2}{\Delta+2}L_{-1}L_{-2} + \frac{1}{(\Delta+1)(\Delta+2)}L_{-1}^3\right)\phi_{1,3} &= 0 \\ \left(L_{-4} - \frac{4\Delta}{9}L_{-2}^2 - \frac{4\Delta+15}{6\Delta+18}L_{-1}L_{-3} + \frac{2\Delta+3}{3\Delta+9}L_{-1}^2L_{-2} - \frac{1}{4\Delta+12}L_{-1}^4\right)\phi_{2,2} &= 0 \end{aligned} \tag{14.29}$$

and in these equations, $\Delta = \Delta_{r,s}$. For the general expressions for arbitrary r, s, see [217, 203].

If one takes $\phi_{1,3}$ in eq. (14.28), one finds indeed a conservation law $\partial_{\bar{z}} T_4 = \partial_z \Theta_2$ with

$$\Theta_2 = \frac{\lambda(\Delta-1)}{\Delta+2}\left[2\Delta L_{-2} + \frac{(\Delta-2)(\Delta-1)(\Delta+3)}{6(\Delta+1)}L_{-1}^2\right]\phi_{1,3} \tag{14.30}$$

if only the perturbation of the critical Hamiltonian H_c is suitably chosen! To study this more systematically, one now looks for conservation laws of the form

$$\partial_{\bar{z}} T_s = \partial_z \Theta_{s-2} \tag{14.31}$$

where T_s is in the conformal tower of the identity and Θ_s in the conformal tower of the perturbing field $\phi_{r,s}$. The existence of a conservation law is equivalent to that the linear operator $\partial_{\bar{z}}$ acting between these two conformal towers has a non-vanishing kernel, up to derivative fields. In principle, one can calculate its matrix elements in the way indicated, but calculations rapidly become tedious. Very often, however, the first few conserved quantities of this type can be found from a simple *counting criterion* [345], which goes as follows.

Define the numbers $\hat{\Lambda}_s$ and $\hat{\Phi}_s$ by the generating functions

$$\begin{aligned}\sum_{s=0}^{\infty} q^s \hat{\Lambda}_s &= (1-q)q^{c/24}\chi_{1,1}(q) \;+\; q \\ q^{\Delta_{r,s}} \sum_{s=0}^{\infty} q^s \hat{\Phi}_s &= (1-q)q^{c/24}\chi_{r,s}(q)\end{aligned} \tag{14.32}$$

where $\chi_{1,1}(q)$ and $\chi_{r,s}(q)$ are the Virasoro character functions (see eq. (10.28)) of the identity operator $\phi_{1,1}$ and the relevant perturbing operator $\phi_{r,s}$ of conformal weight $\Delta_{r,s}$. If one has for some s that

$$\hat{\Lambda}_{s+1} \geq \hat{\Phi}_s + 1 \tag{14.33}$$

then there does exist at least one conservation law of the above form. This is seen as follows. The $\hat{\Lambda}_s$ and $\hat{\Phi}_s$ defined above are just the dimensions of the spaces of quasiprimary fields constructed in the conformal towers of either the identity or the perturbing field ϕ at the level s. The operator $\partial_{\bar{z}}$ is a linear operator acting between them. Thus, if the condition eq. (14.33) is satisfied, it must have a non-vanishing kernel. This is equivalent to the existence of a conservation law of the type eq. (14.31) [345].

As an example, consider the Ising model perturbed by a magnetic field, corresponding to $\phi_{1,2}$. The characters are taken from Table 12. In Table 28, we illustrate the application of A. Zamolodchikov's counting criterion eq. (14.33). One recognizes that one does indeed have T_s conserved if [345]

$$s = 1, 7, 11, 13, 17, 19 \tag{14.34}$$

One notes that all spins appearing in the conservation laws are odd. This is a consequence

Table 28. Counting criterion for the Ising model perturbed with $\phi_{1,2}$ for odd spin. A $*$ labels a conservation law.

s	1*	3	5	7*	9	11*	13*	15	17*	19*	21
$\hat{\Lambda}_{s+1}$	1	1	1	2	2	3	4	5	7	9	11
$\hat{\Phi}_s$	0	1	1	1	2	2	3	5	6	8	12

of the conservation of charge conjugation in a two-dimensional system. In general, a $2D$ conserved quantity with odd spin has a charge conjugation eigenvalue $C = +1$, while for even spin, it has $C = -1$. Since the Ising model does not have a charge conjugation, all its states are considered to have $C = +1$.

Similarly, the counting criterion may be used to show the existence of conserved quantities T_s with the following spins [345]

$$\begin{aligned} \text{for } \phi_{1,3}: \quad & s = 1, 3, 5, 7 \\ \text{for } \phi_{1,2}, \phi_{2,1}: \quad & s = 1, 5, 7, 11 \;\; ; \;\; m \geq 5 \end{aligned} \tag{14.35}$$

while the cases with $m < 5$ must considered separately. The conservation law for $s = 1$ is always present, since it corresponds to the conservation of energy-momentum. To find any further conserved quantities with $s \neq 1$ is highly nontrivial. We leave it as an exercise to find the first conserved T_s for $m = 3, 4$. It is generally believed that these first few conserved T_s are just the first members in an infinite series of conservation laws. If true, this would be strong evidence for the complete **integrability** of these non-critical systems. We stress that the fact that the central charge $c < 1$ is essential in finding these conservation laws.

14.3 Exact S-matrix approach

We now want to take advantage of the presence of the conserved currents to calculate finite-size quantities like R_{ij} in eq. (14.4). We suppose that we are in the situation where all correlation lengths are finite. The correlation functions of a off-critical theory decay exponentially like those of a relativistic quantum field theory (see section 1.5). Therefore, in the limit of the continuum continuous geometric symmetries are restored, the inverse correlation lengths $(1/\xi_j)$ can be interpreted as the masses m_j of relativistic particles. The ultraviolet behaviour of such massive field theory is given by the critical conformal theory, where the masses become negligible. In the infrared regime, all particles are far away from each other and a physical state is given by a set of *on-shell asymptotic particle momenta* $\vec{p}(j) = (p^0(j), p^1(j))$, with $(p^0(j))^2 - (p^1(j))^2 = m_j^2$. Scattering theory is interested in the mass-shell amplitude, which relates *incoming* asymptotic particles to *outgoing* ones, defined as

$$|\vec{p}_1(j_1), \dots, \vec{p}_n(j_n)\rangle_{\text{in}} = \mathcal{S}(\vec{p}_1(j_1), \dots, \vec{p}_n(j_n); \vec{p}'_1(j'_1), \dots, \vec{p}'_m(j'_m))\, |\vec{p}'_1(j'_1), \dots, \vec{p}'_m(j'_m)\rangle_{\text{out}} \qquad (14.36)$$

The function $\mathcal{S}$ of the momenta is called the S-*matrix*. Our aim is to briefly review the remarkable results of A. Zamolodchikov [345], who first understood the consequences of conserved off-critical quantities on the infrared behaviour of the theory, in the context of perturbed conformal theory. A more detailed discussion can be found in the very abundant literature on the subject.

The existence of conserved quantities (besides the stress-energy tensor) was discussed a long time ago in the context of *integrable quantum field theories* in (1+1) dimensions, see [339]. Though a single new conserved quantity is sufficient to derive the results, we expect in general infinitely many conserved quantities and the consequences can be stated as follows: *only elastic scattering can take place in an integrable quantum field theory. The scattering process only exchanges the momenta and the internal quantum numbers. The scattering matrix $\mathcal{S}$ factorizes into the product of the $n(n-1)/2$ possible two-particle scattering matrices* [198].

Let us restate this in a more convenient language. It is useful to express the particle momenta as function of the *rapidity variable* θ defined by

$$p^0(j) = m_j \cosh\theta\,, \qquad p^1(j) = m_j \sinh\theta \qquad (14.37)$$

A consequence of Lorentz invariance is that the scattering matrix only depends on the rapidity differences $\theta_{ij} := \theta_i - \theta_j$. The *Mandelstam variable* $s(i,j)$ of a pair collision between two particles with masses m_i and m_j is given by

$$s(i,j) = (\vec{p}(i) + \vec{p}(j))^2 = m_i^2 + m_j^2 + 2m_i m_j \cosh\theta_{ij} \qquad (14.38)$$

The physical sheet of the s-plane is mapped into the strip $0 < \text{Im}\theta < \pi$. The two-particle scattering matrix has cuts along $s \geq (m_i + m_j)^2$ and $s \leq (m_i - m_j)^2$, which are required by two-particle unitarity. The mapping from the s-plane to the θ-plane is shown in Figure 33. The two-particle scattering of an integrable theory is a pure elastic phase shift which defines the S-matrix $S(\theta)$

$$|\theta_1(i), \theta_2(j)\rangle_{\text{in}} = S_{ij}(\theta_{12})\, |\theta_1(i), \theta_2(j)\rangle_{\text{out}} \qquad (14.39)$$

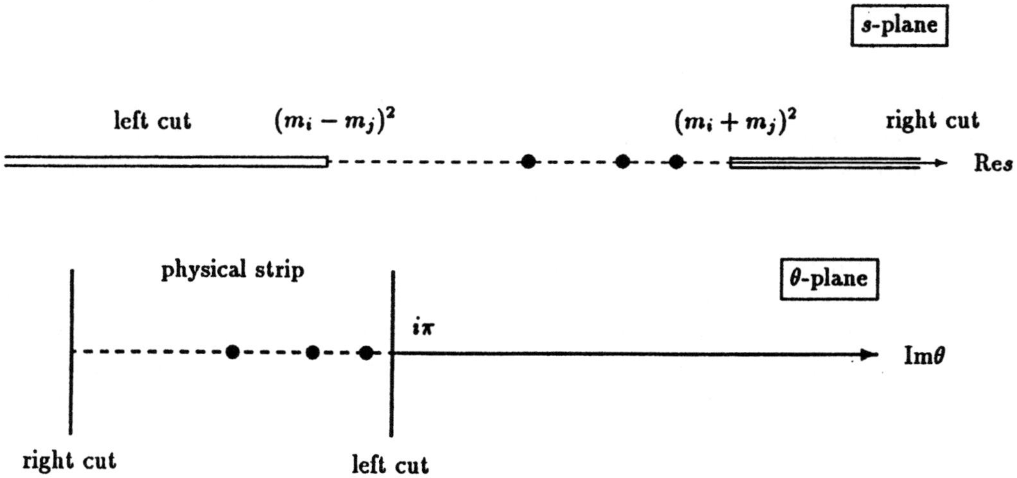

Fig. 33. Mapping the s-plane to the θ-plane, showing the positions of the cuts and the bound states (dots).

This S-matrix reduces to a simple scalar function if the masses are non-degenerate. If the mass-spectrum is degenerate, the elasticity of the scattering process only allows a redistribution of the momenta among the particles with the same masses but different internal quantum numbers. We will assume in the sequel that the mass-spectrum is non-degenerate and that the particles are their own anti-particles but our discussion extends straightforwardly to the general case. Notice that the two-particle scattering matrix, which depends on the Mandelstam variable s, is related to $S(\theta)$ by a relation which takes into account the Jacobian due to the overall energy-momentum δ-function of all the momenta (Lorentz invariance).

$$\mathcal{S}(s(i,j)) = 4m_i m_j \sinh\theta_{12}\, S_{ij}(\theta_{12}) \tag{14.40}$$

The n-particle S-matrix simply decomposes into the product of the possible two-particle processes, as depicted in Figure 34 in a three-particle example, where we have

$$S_{ij\ell}(\theta_1, \theta_2, \theta_3) = S_{ij}(\theta_{12})\, S_{i\ell}(\theta_{13})\, S_{j\ell}(\theta_{23}) \tag{14.41}$$

The *unitarity* of the scattering process can be implemented, by factorization, on the two-particle S-matrix: it is expressed by

$$S_{ij}(\theta)\, S_{ij}(-\theta) = 1 \tag{14.42}$$

The transformation $\theta \to -\theta$ reverses the direction of motion of the particles. Notice that due to parity invariance, we have

$$S_{ij}(\theta) = S_{ji}(\theta) \tag{14.43}$$

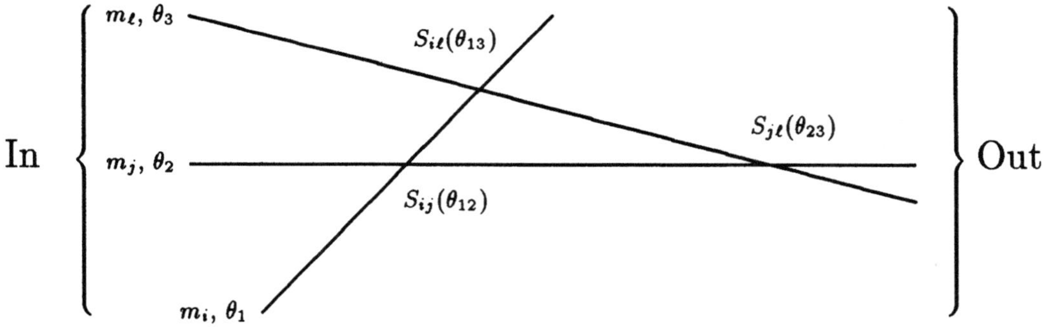

Fig. 34. Three-particle elastic scattering factorization into the product of the three possible two-particle S-matrices. The partic'es move from the left to the right.

We have so far discussed the scattering process in the s-channel. The S-matrix in the t-channel, where $t = 2m_i^2 + 2m_j^2 - s$, expressed in terms of the rapidity variable implies the *crossing condition*

$$S_{ij}(\theta) = S_{ij}(i\pi - \theta) \tag{14.44}$$

It follows from these equations that $S_{ij}(\theta)$ is a $2\pi i$-periodic function of θ.

The set of constraints established so far is very general and they do not tell much about the S-matrices. Information is gained from the *bound-state structure* of the theory. Though particle production is suppressed, the S-matrix is a meromorphic function in θ which may have regular poles in $0 \leq \mathrm{Im}\theta \leq i\pi$, which correspond to physical processes taking place in the intermediate region between the asymptotic regimes and at purely imaginary values of the rapidities. Such poles are shown as dots in Figure 33. They are called *anomalous thresholds*. The particles corresponding to these poles are stable and have also to be considered as asymptotic states. The kinematic constraints on the poles give rise to relations between the different masses of the theory. If $S_{ij}(\theta)$ has a simple pole at $\theta = iu_{ij}^k$, then the mass $(m_k)^2$ is the rest mass of a particle created by m_i and m_j and the total square momentum of the system. Therefore

$$m_k^2 = m_i^2 + m_j^2 + 2m_i m_j \cos(u_{ij}^k) \tag{14.45}$$

Such a bound-state is diagrammatically illustrated in Figure 35a. Notice the identities

$$u_{ij}^k + u_{ik}^j + u_{jk}^i = 2\pi\,, \qquad u_{ij}^k = u_{ji}^k \tag{14.46}$$

The residue of the S-matrix at a physical pole u in the s-channel is positive in a unitary theory. Such a pole is graphically represented in Figure 35b. The second pole at $\bar{u}_{ij}^k := \pi - u_{ij}^k$, due to eq. (14.44), with negative residue, is a physical pole in the t-channel. Such a pole will also appear in any n-particle scattering which involves the particles m_i and m_j. For example in the three particle scattering depicted in Figure 34. A consistency equation for the residue of the pole $\theta_{12} = iu_{ij}^k$ of the S-matrix (eq. (14.41)) is given by the *bootstrap equation* (Figure 36):

$$S_{k\ell}(\theta) = S_{i\ell}(\theta - i\bar{u}_{ik}^j)\, S_{j\ell}(\theta + i\bar{u}_{jk}^i) \tag{14.47}$$

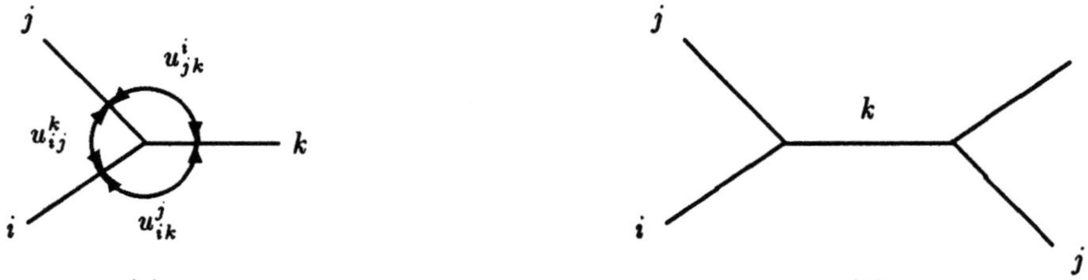

(a) (b)
Fig. 35. Anomalous threshold (a) giving rise to a simple pole in the s-channel (b) of S_{ij}.

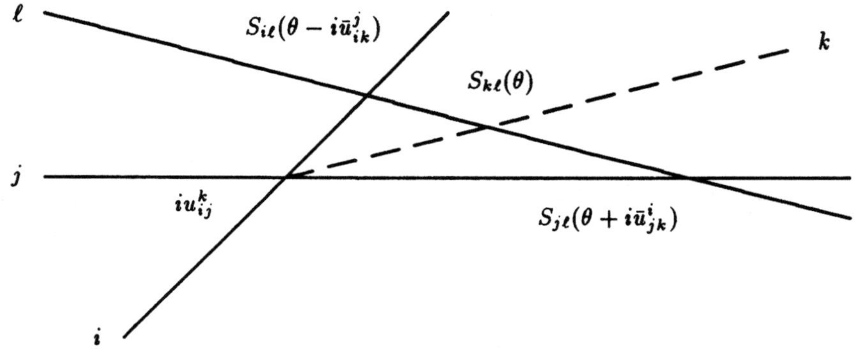

Fig. 36. Bootstrap equation ($\bar{u}^j_{ik} = \pi - u^j_{ik}, \bar{u}^i_{jk} = \pi - u^i_{jk}$).

This beautiful equation makes it possible to compute recursively the *minimal S-matrices*. The knowledge of the bound-state structure is sufficient to determine a generic solution for the S-matrices which minimaly reproduces all poles required by the bound-state structure. They are of course infinitely many solutions to eq. (14.47) because if the knowledge of poles implies eq. (14.47), the converse is not true. The minimal S-matrix is the simplest solution to eq. (14.47) with the proper bound-state structure.

It now remains to connect the information on the spins of the conserved currents to the bound-state structure of the S-matrices. Suppose we have determined some (or all) conserved currents which defines the conserved spin s charges $\mathcal{P}_s$:

$$\mathcal{P}_s = \oint d\bar{z}\, T_{s+1} + \oint dz\, \Theta_{s-1} \tag{14.48}$$

There is in general a complex conjugate set of conserved spin $(-s)$ charges $\overline{\mathcal{P}}_s$ generated by $\overline{T}$. The charges act on the asymptotic states and under a Lorentz transformation $\theta \longrightarrow \theta + v$, they will covariantly transform like

$$\mathcal{P}_s \longrightarrow e^{sv} \mathcal{P}_s \tag{14.49}$$

Notice that $p(j) = p^0(j) + p^1(j) = m_j e^{\theta}$ transforms like a spin one tensor and its quantum operator is obviously identified with $\mathcal{P}_1$, since $T_2 = T$ and $\Theta_0 = \Theta/4$ are the components of the stress-energy tensor. Similarly, a spin -1 tensor is given by $\bar{p}(j) = p^0(j) - p^1(j) = m_j e^{-\theta}$, eigenvalue of the operator $\overline{\mathcal{P}}_1$.

A conserved charge $\mathcal{P}_s$ acts on a one-particle asymptotic state (in or out) as

$$\mathcal{P}_s|\theta(j)\rangle = q(s,j)(m_j)^s e^{s\theta}\,|\theta(j)\rangle \tag{14.50}$$

as a consequence of the Lorentz transformation property (14.49) of $\mathcal{P}_s$. The parameters $q(s,j)$ are yet unknown dimensionless coefficients. It is in general not necessary to determine them exactly. The conserved charges are integrals of local densities. Therefore, they act distributively on a n-particle asymptotic state and we have

$$\mathcal{P}_s|\theta_1(j_1),\ldots,\theta_n(j_n)\rangle = \sum_{r=1}^{n} q(s,j_r)(m_{j_r})^s e^{s\theta_r}\,|\theta_1(j_1),\ldots,\theta_n(j_n)\rangle \tag{14.51}$$

The expectation value of $\mathcal{P}_s$, for a single $s > 1$, taken between different asymptotic states is sufficient to prove the elasticity of the scattering process. While $\mathcal{P}_1$ gives the usual translation invariance law for the sum of the incoming and outgoing momenta, a single further charge $\mathcal{P}_s$ gives a relation between powers of the momenta which can only be solved by forbidding particle production and preserving each individual moment.

We are now in position to connect the conserved charges with the bootstrap equation (14.47) and the bound-state structure. It is possible to identify a bound-state as an asymptotic state defined in the limit where a two-particle state has rapidity variables corresponding to a pole of the S-matrix (see Figure 36)

$$\lim_{\epsilon\to 0}\epsilon\,|\left(\theta - i\bar{u}^j_{ik} + \frac{\epsilon}{2}\right)(i)\,\left(\theta + i\bar{u}^i_{jk} - \frac{\epsilon}{2}\right)(j)\rangle = |\theta(k)\rangle \tag{14.52}$$

Applying $\mathcal{P}_s$ to both sides of this equation gives rise to a linear system of solvable equations for the parameters $q(s,j)$.

$$q(s,i)(m_i)^s e^{-is\bar{u}^j_{ik}} + q(s,j)(m_j)^s e^{+is\bar{u}^i_{jk}} = q(s,k)(m_k)^s \tag{14.53}$$

Some parameters $q(s,i)$ can be zero for a given value of s. However, we can conclude $\mathcal{P}_s = 0$ only if $q(s,i) = 0$ for all values i.

Let us now sketch how the fundamental equation (14.53) is sufficient to determine the ratios

$$R_{ij} = \left(\frac{\xi_i}{\xi_j}\right)^{-1} = \frac{m_i}{m_j} \tag{14.54}$$

and to give a convincing candidate for the S-matrix. We suppose that there are several particles with masses $m_1 < m_2 < \ldots$. A. Zamolodchikov worked out the Ising model in a magnetic field where the existence of conserved charges with spins $s = 1, 7, 11, 13, 19$ (eq. (14.34)) can be established. Zamolodchikov made the reasonable assumption that the lightest particle with mass m_1 is fundamental and generates by "fusion" all other particles. For example the two-particle process of two fundamental particles has a fundamental bound-state. The mass equation (14.45) with all three masses equal implies $u^1_{11} = 2\pi/3$. Equation (14.53) simplifies in this case to

$$e^{-i\frac{\pi}{3}s} + e^{+i\frac{\pi}{3}s} = 1 \tag{14.55}$$

The spectrum of possible spin s reduces to odd numbers, not multiples of three, for instance the set $s = 1, 5, 7, 11, 13, 17, 19, \ldots$. This is compatible with the given set of conserved spins.

Suppose now that the same two-particle scattering has the particle with mass m_2 as a bound-state. We have to solve the equations

$$2m_1^2(1+\cos(u_{11}^2)) = \left[2m_1 \cos\left(\frac{u_{11}^2}{2}\right)\right]^2 = m_2^2$$
$$q(s,1)(m_1)^s \left(e^{-is\bar{u}_{12}^1} + e^{is\bar{u}_{12}^1}\right) = q(s,2)(m_2)^s \tag{14.56}$$

considering $s = 1, 7, 11, 13, 19$ and eq. (14.46). After some simplifications, this system reduces to

$$2\cos\left(s\frac{u_{11}^2}{2}\right) = \left[2\cos\left(\frac{u_{11}^2}{2}\right)\right]^s \left(\frac{q(s,2)}{q(s,1)}\right) \tag{14.57}$$

It is now necessary to find a new physical condition which unequivocally makes the desired spin-spectrum compatible with the all possible spin equations (14.53). The correct assumption consists in chosing that the two-particle process with two masses m_2 has a bound state with mass m_1. This gives rise to similar equations where the indices 1 and 2 are exchanged. We have of course the equations

$$\frac{m_2}{m_1} = 2\cos\left(\frac{u_{11}^2}{2}\right) = \left[2\cos\left(\frac{u_{22}^1}{2}\right)\right]^{-1} \tag{14.58}$$

and

$$2\cos\left(s\frac{u_{22}^1}{2}\right) = \left[2\cos\left(\frac{u_{22}^1}{2}\right)\right]^s \left(\frac{q(s,1)}{q(s,2)}\right) \tag{14.59}$$

We can now solve the problem for u_{11}^2 and we obtain

$$u_{11}^2 = \frac{2\pi}{5}, \qquad u_{22}^1 = \frac{4\pi}{5} \tag{14.60}$$

and the mass-ratio [345, 346]

$$R_{21} = \frac{m_2}{m_1} = 2\cos\left(\frac{\pi}{5}\right) = 1.6180339\ldots \tag{14.61}$$

Besides this remarkable result, A. Zamolodchikov determined the minimal solution to the bootstrap equation and found for example

$$S_{11}(\theta) = \frac{\tanh\left(\frac{\theta}{2}+i\frac{\pi}{3}\right)}{\tanh\left(\frac{\theta}{2}-i\frac{\pi}{3}\right)} \frac{\tanh\left(\frac{\theta}{2}+i\frac{\pi}{5}\right)}{\tanh\left(\frac{\theta}{2}-i\frac{\pi}{5}\right)} \frac{\tanh\left(\frac{\theta}{2}+i\frac{\pi}{30}\right)}{\tanh\left(\frac{\theta}{2}-i\frac{\pi}{30}\right)} \tag{14.62}$$

Each of the three parts of S_{11} gives rise to a physical bound-state. The presence of hyperbolic functions reflects the $2\pi i$-periodicity of the S-matrix. The crossing condition (14.44) doubles the physical poles between 0 and $i\pi$ such that there is one with positive residue in the s-channel and another one with negative residue in the crossed t-channel. The unitarity condition (14.42) implies the existence of one zero in the S-matrix for each physical pole. Due to the negative $-\theta$ in eq. (14.42), this zero is located on the imaginary segment

$-i\pi \leq \mathrm{Im}\theta \leq 0$. These arguments are sufficient to explain the two hyperbolic tangents in each part of the solution (14.62). The first two parts on the r.h.s. of (14.62) correspond to the bound-states with masses m_1 and m_2, already discussed. The last part is a natural consequence of the particular bootstrap equation (14.47)

$$S_{11}(\theta) = S_{11}(\theta - i\pi/3)\, S_{11}(\theta + i\pi/3) \tag{14.63}$$

This enforces the presence of a further bound-state at $\theta = iu_{11}^3 = i\pi/15$ corresponding to a particle with mass

$$\frac{m_3}{m_1} = 2\cos\left(\frac{\pi}{30}\right) = 1.9890437\ldots \tag{14.64}$$

The S-matrix S_{11} is minimal because all poles and zeroes have a specific meaning, necessary for the physical interpretation.

The bootstrap program can be performed step by step, from the lightest particle to the heaviest one, until this one is reached. This can be seen when the bootstrap program is closed, that is when all physical poles of the S-matrices are explained with the knowledge of the existing particles and when all possible bootstrap equations have been shown to be consistent. In this particular case, eight particles are necessary to have a consistent system. This number is in fact not surprising: the conserved spins appear to take the values

$$s = 1, 7, 11, 13, 17, 19, 23, 29 \,\mathrm{mod}(30) \tag{14.65}$$

which are the *Coxeter exponents* of the affine Lie algebra $\hat{E}_8$, and 30 is its *Coxeter number*. Adding one to the Coxeter exponents gives the orders of the invariant polynomials (Casimir invariants) of the Lie algebra E_8. This example illustrates the observed relation between integrable perturbed systems and quantum integrable field theories such as the *affine Toda* field theories [115, 93, 94, 55, 193, 100]. These describe an r-component real scalar field $\Phi^a, a = 1, \ldots r$ with Lagrangian

$$\mathcal{L} = \frac{1}{2}\partial_\mu \Phi^a \partial^\mu \Phi^a - \frac{g}{\beta^2}\sum_{i=1}^{r+1} n_i \left[e^{\beta \vec{\alpha}_i \cdot \vec{\Phi}} - 1\right] \tag{14.66}$$

where the r-dimensional $(r+1)$ vectors $\vec{\alpha}_i$ are the simple roots of some **affine** simple Lie algebra $\hat{\mathbf{g}}$ and r is the rank of the classical Lie algebra $\mathbf{g}$. The integers $\{n_i\}$ verifies the identity $\sum_i n_i \vec{\alpha}_i = 0$, and $\sum_i n_i = h$ is the Coxeter number of $\hat{\mathbf{g}}$. For example, the Ising model in a magnetic field has the same mass-spectrum, the same bound-state structure, and the same conserved spins as the $\hat{E}_8$ affine Toda theory. However, the S-matrices of the Toda models differ from those of the Ising model by so-called CDD-terms which take into account the coupling constant dependence of the Toda model. As $\beta \to 0$, the scattering in the Toda model must become trivial. In this limit, the theory reduces to a set of uncoupled free massive bosons, and the two-particle S-matrix simplifies to

$$S_{ij}(\theta) = \delta_{i,j} \tag{14.67}$$

The subject of Toda theory has been now abundantly reviewed [94, 193] and we will not discuss in more detail this interesting subject.

However, it is useful for later references to give the complete list of mass-ratios for the Ising model in a magnetic field [345, 346].

$$\begin{aligned}
m_2/m_1 &= 2\cos\pi/5 &&= 1.6180339\ldots \\
m_3/m_1 &= 2\cos\pi/30 &&= 1.9890437\ldots \\
m_4/m_1 &= 2m_2/m_1 \cos 7\pi/30 &&= 2.4048671\ldots \\
m_5/m_1 &= 2m_2/m_1 \cos 2\pi/15 &&= 2.9562952\ldots \\
m_6/m_1 &= 2m_2/m_1 \cos \pi/30 &&= 3.2183404\ldots \\
m_7/m_1 &= 4m_2/m_1 \cos\pi/5 \cos 7\pi/30 &&= 3.8911568\ldots \\
m_8/m_1 &= 4m_2/m_1 \cos\pi/5 \cos 2\pi/15 &&= 4.7833861\ldots
\end{aligned} \tag{14.68}$$

One can also show that these mass ratios do not depend on the fact that only the minimal S-matrix was taken.

From a physical point of view, this result is highly unexpected. Previously, we had described the Ising model either by a spin variable or alternatively by a scalar ϕ^4 field theory. From none of these formulations it is obvious that there might be several local conserved quantitites leading to stable, soliton-like excitations. These are completely absent if one considers the Ising model in a vanishing magnetic field and just changes the temperature. It would be desirable to have some intuitive understanding on the origin of these stable states.

This example is quite complicated and to finish with a simpler one, let us recall the Ising model with a thermal perturbation. The energy-density operator, which couples to the temperature, introduces a single correlation length in the system. It is known that this off-critical system is, like the critical one, described by the scalar combinations of Majorana fermions, which are massive in this case. The mass is proportional to the inverse correlation length. Zamolodchikov's counting criterion shows the existence of conserved quantities for all odd spins. In this case, the integrability of the system has been understood a long time ago. The off-critical system is essentially free and therefore the S-matrix is free from any pole structure. However, the underlying fermionic structure of this scalar theory appears through the minus sign of the two-particle S-matrix given by [212, 213, 41]

$$S(\theta) = -1 \tag{14.69}$$

It is easy to verify that this solution is compatible with all constraints discussed in this section.

It is clear that this approach, while being extremely powerful, does rely on quite a few hypotheses and it is not obvious at first sight whether these apply to the study of statistical systems. It seems therefore worthwhile to go back to a lattice formulation and check to what extent these assumptions and their consequences are satisfied for given statistical systems. This will be done in the following.

14.4 Phenomenological consequences

Integrable perturbations. Let us check the predictions for the R_{ij} given in the last section for the Ising model **exactly at** the critical temperature but with a magnetic field

[174]. The quantum Hamiltonian is

$$H = -\frac{1}{\zeta} \sum_{n=1}^{N} \left(\sigma^z(n) + \sigma^x(n)\sigma^x(n+1) + h\sigma^x(n) \right) \tag{14.70}$$

with the conformal normalization $\zeta = 2$ and periodic boundary conditions. It is enough to restrict attention to the ratios $r_i := R_{i+1,1}$. In Figure 37 we display the $r_i, i = 1, \ldots 5$ as functions of the finite-size scaling variable $\mu = hN^{15/8}$. The symbols refer to different values of the magnetic field h. We first note that the data do indeed collapse into a finite-size scaling form. Secondly, for the critical point $\mu = 0$, we see that the mass ratios do reproduce the values expected from critical point conformal invariance

$$r_1 = 8, r_2 = 17, r_3 = 24, r_4 = 32, r_5 = 33, \ldots \tag{14.71}$$

since $R_{ij} = x_i/x_j$ **at** $\mu = 0$. If μ increases, the r_i start to decrease until they pass through some minimum. This finding is at least qualitatively accounted for by conformal perturbation theory. Define the finite-size scaling functions G_i and the reduced scaling functions H_i for the inverse correlation lengths

$$\xi_i^{-1} = h^{1/y} G_i(\mu) \;\; , \;\; G_i(\mu) = 2\pi x_i \mu^{-1/y} + H_i(\mu) \tag{14.72}$$

where $y = 15/8$ for the Ising model with a magnetic field. The first term in G_i is nothing else than the conformal invariance prediction eq. (3.49). For a symmetry-breaking field, the lowest-order term in H_i is of order $\mu^{2-1/y}$. For a magnetic perturbation, one usually has $y \simeq 2$. The minimum is thus produced by the combined effect of the leading singularity and the perturbative corrections. Finally, the r_i tend towards some limit value when $\mu \to \infty$. Already from the figure one recognizes agreement with the prediction eq. (14.68). We see that two mass ratios are in the discrete spectrum, while above $m/m_1 = 2$ there is a continuum of states. This is generated by the multiparticle excitations in which the particles inside the composite state can have a non-vanishing relative momentum. For a more precise check, one has to deal numerically with the **double limit** that $\mu \to \infty$ after the finite-size scaling limit has been taken. In fact, computationally it has proved to be advantageous to replace this by the double limit of taking first $N \to \infty$ with h fixed and letting $h \to 0$ afterwards. The best numerical result presently available is, using quantum chains with up to $N = 21$ sites [182]

$$r_1 = 1.6181(5) \;\; , \;\; r_2 = 1.994(5) \tag{14.73}$$

as obtained from finite-size scaling. For comparison, we give the results found with the truncation method [338]

$$r_1 = 1.61(1) \;\; , \;\; r_2 = 1.98(2) \;\; , \;\; r_3 = 2.43(4) \;\; , \;\; r_4 = 3.03(7) \tag{14.74}$$

This has also been confirmed in a study a of the correlation function in the isotropic $2D$ Ising model [236]. The success of this comparison with eq. (14.68) does give confidence that the theory leading to it is applicable to statistical systems. This is further strengthened by similar results derivable for other systems and perturbations corresponding to $\phi_{1,2}, \phi_{2,1}$ or $\phi_{1,3}$. The S-matrices and in particular their mass ratios are by now

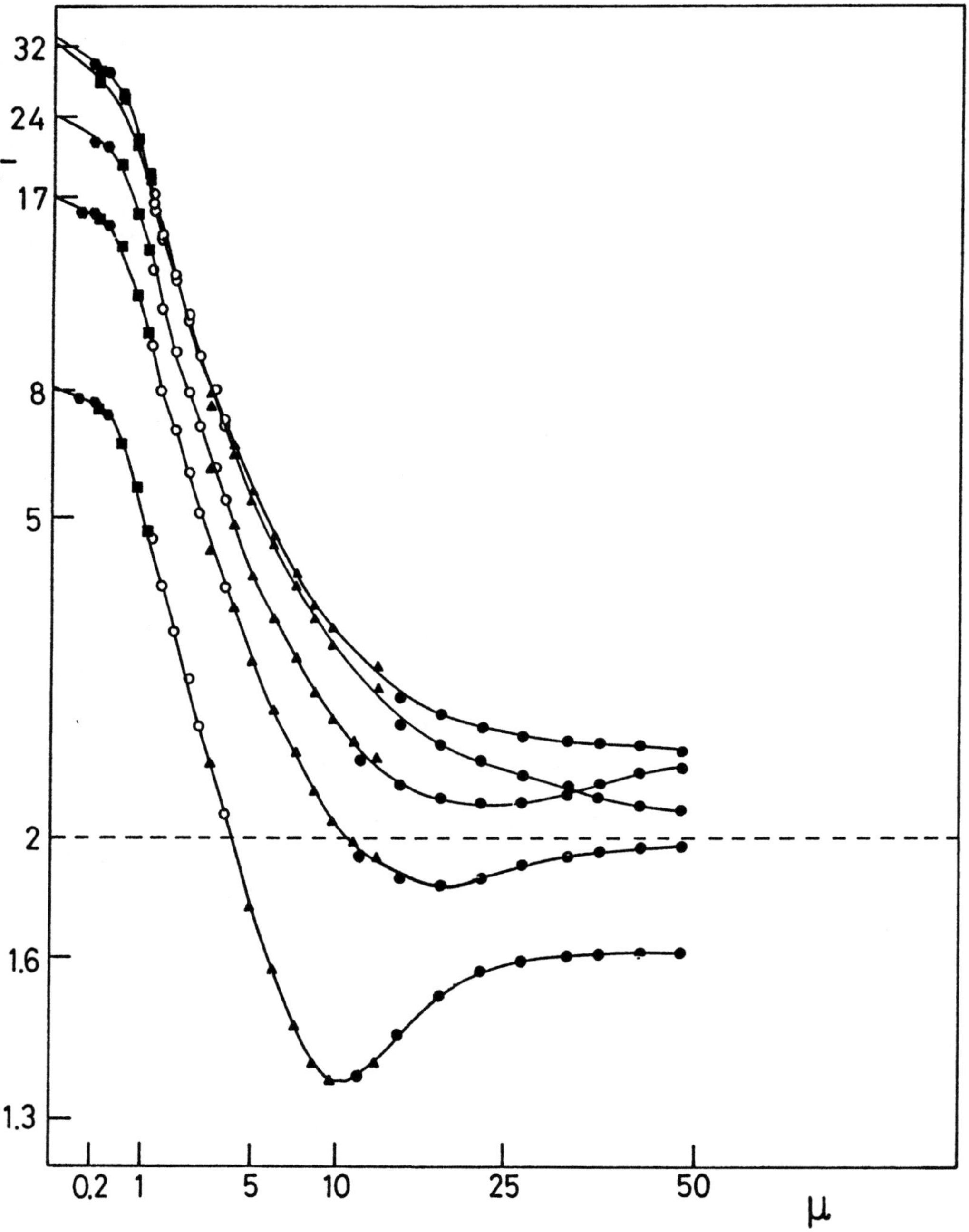

Fig. 37. Mass ratios $r_i = (\xi_{i+1}/\xi_1)^{-1}$ for the Ising model in a magnetic field as a function of μ.

exactly known. Precise numerical checks using either finite-size scaling or the truncation method have been done on a huge variety of both unitary and non-unitary models. The S-matrices for the tricritical Ising model have been calculated [93, 118] and been checked by both finite-size scaling [144] and the truncation method [230]. For S-matrices of the Potts model, see [90]. Studies on non-unitary minimal models include the Yang-Lee singularity [85, 349].

Although these considerations are restricted the models with central charge $c < 1$, there are special cases where the integrability exists even if $c \geq 1$. One example is the Ashkin-Teller model with a thermal perturbation which can be expressed by the integrable $(1+1)D$ sine-Gordon model. Mass ratios computed this way were confirmed numerically [181]. Another example is given by a set of Z_p-invariant systems [340] and the mass ratios have been calculated [16]. Finally, we mention that non-scalar perturbation can be studied as well. This leads for example to a description of the so-called superintegrable chiral Potts model, see [264, 196, 195, 138, 35, 8, 9] and references therein, in terms of a relevant perturbation of the Z_p invariant model of [340] and does reproduce the known critical exponents [88]. It goes beyond the scope of an introduction to attempt to review the entire work currently being done. For a recent review, see e.g. [257].

Non-integrable perturbations. The perturbing fields studied so far all have in common that the existence of several conservation laws involving some T_s was established. However, there are many physical fields, for example the order parameter (with the only exception of the Ising model), which in unitary models is given by $\phi_{2,2}$, for which no conservation law involving T_s with $s \neq 1$ is presently known. These models are commonly referred to as *non-integrable* in the literature.

The first example studied to be in this class was the tricritical Ising model coupled to a magnetic field. The quantum Hamiltonian is [179]

$$H = -\frac{1}{\zeta}\sum_{n=1}^{N} [\, \{t_t\sigma^z(n) - \sigma^x(n)\sigma^x(n+1) + \sigma^x(n)\sigma^x(n+2) + h_t\sigma^x(n)\} + h_s(-1)^n\sigma^x(n)] \tag{14.75}$$

where h_s is the symmetry-breaking staggered field coupled to the order parameter which is described by the primary field $\phi_{2,2}$. The tricritical point (t_t, h_t) was given in chapter 12. The magnetic scaling variable is $\mu = h_s N^{77/40}$. In Figure 38, we show the mass ratios r_i defined analogously to the Ising model case as a function of μ. Comparing with the corresponding plot for the Ising model case (Figure 37), we note that they are quite similar in many but not all aspects. We observe that the data do display finite-size scaling. One reproduces the conformal expectation in the $\mu \to 0$ limit

$$r_1 = 2\frac{2}{3}, r_2 = 11\frac{2}{3}, r_3 = 16, r_4 = 27\frac{2}{3}, \ldots \tag{14.76}$$

However, attempts to find a conservation law with some T_s and $s \neq 1$ have failed so far.[15] Nevertheless, finite-size scaling calculations yield for the lowest two mass ratios in the $\mu \to \infty$ limit [179]

$$r_1 = 1.62(1) \quad , \quad r_2 = 1.98(2) \tag{14.77}$$

[15] It is known that up to $s = 17$, no conservation law exists, but for three exotic values of the central charge which do not correspond to any minimal model: i) $c = -11 \Rightarrow \Delta_{22} = 3/2$ and

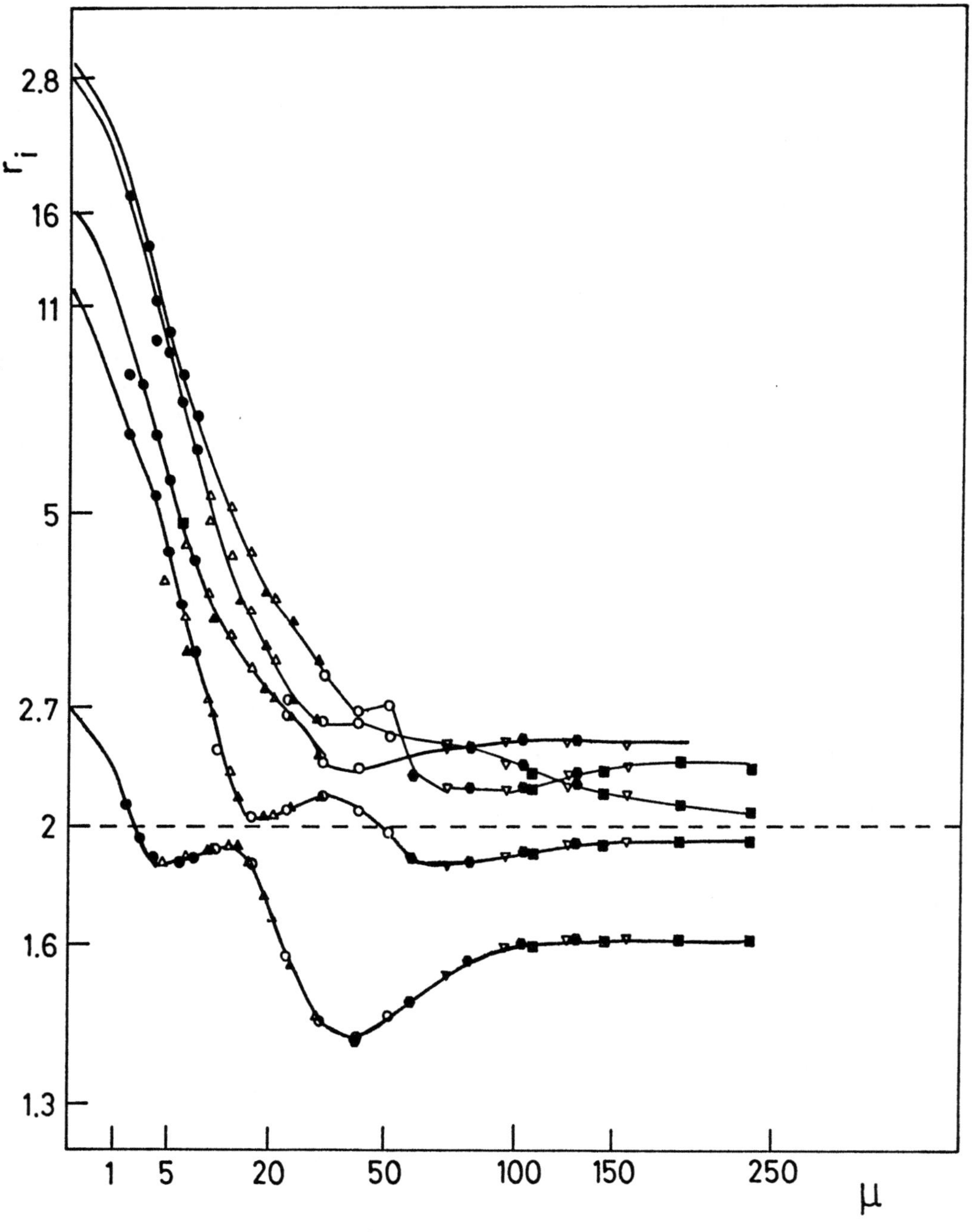

Fig. 38. Mass ratios $r_i = (\xi_{i+1}/\xi_1)^{-1}$ as functions of μ for the tricritical Ising model.

Surprisingly, these are found to be in agreement, within numerical errors, with the Ising model result! This appears to be the first example where two systems being in **different** universality classes yet show the same values for some universal amplitudes, within numerical errors. This could indicate a deeper relation between these two systems. The truncation method has yielded [230]

$$r_1 = 1.6(2) \ , \ r_2 = 1.9(2) \tag{14.78}$$

Martins [250] has extended these calculations, using the truncation method, to the tetracritical ($m = 5$) and pentacritical ($m = 6$) Ising models, perturbed with their order parameter $\phi_{2,2}$. He finds for the lowest mass ratio [250]

$$r_1 = 1.63(2) \text{ if } m = 5 \ , \ r_1 = 1.62(2) \text{ if } m = 6 \tag{14.79}$$

again in agreement with the Ising model result. Finally, the very same mass ratios are also reproduced within the Ashkin-Teller model [180] coupled to a magnetic field, which has central charge $c = 1$. The Hamiltonian is

$$H = -\frac{1}{\zeta}\sum_{n=1}^{N} \left[\{(\Gamma_n + \Gamma_n^3 + \varepsilon\, \Gamma_n^2) + (\sigma_n \sigma_{n+1}^3 + \sigma_n^3 \sigma_{n+1} + \varepsilon\, \sigma_n^2 \sigma_{n+1}^2)\} + h(\sigma_n + \sigma_n^3) \right] \tag{14.80}$$

The order parameter perturbation also breaks a Z_2 symmetry. The finite-size scaling variable is $\mu = hN^{15/8}$. The mass ratios r_i are given as a function of ε in the $\mu \to \infty$ limit. We refer to chapter 12 for the variable definitions and to Figures 30 and 31 for the phase diagram. In contrast to the models considered so far in this subsection, the Ashkin-Teller model does have a charge conjugation and states are labeled by a C-parity. When calculating the mass spectrum in the $\mu \to \infty$ limit, one observes that in the **entire interval** $0 \leq \varepsilon < 1/\sqrt{2}$, the mass ratios in the $C = +1$ sector do again reproduce the Ising model result. This is shown in Table 29 [180], where the $*$ give those $C = +1$ states which are close to the Ising model values. Given that for the Ising model, the integrals of motion all have odd spin and therefore $C = +1$, it is reasonable to find the Ising mass ratios in the $C = +1$ sector only of the Ashkin-Teller model.

This finding is quite surprising. Previously, we had understood the apperance of stable, solitonlike excitations (one can check that they all have momentum zero) as a consequence of the existence of at least one non-trivial integral of motion T_s. For all the models discussed here, however, no integral of motion besides energy-momentum is known. It seems that some important aspect of critical systems perturbed with a relevant, symmetry-breaking scaling operator has yet escaped attention.

Returning to Figure 38, we observe that some of the eigenvalues of H appear to show some kind of level repulsion. Effects of this kind are absent in the Ising model, see Figure 37, but are also seen in the higher multicritical Ising model [250] and the Ashkin-Teller model. If this interpretation of the finite-size scaling functions is correct, this might be taken as a hint for the non-integrability of the $\phi_{2,2}$ perturbation.

---

the conserved spins 1, 7, 9, 11, 13 (twice), 15 (twice), 17 (3 times). ii) $c = 33 \Rightarrow \Delta_{22} = -4$ and the conserved spins 1, 7, 11, 13, 17. iii) $c = -35 \Rightarrow \Delta_{22} = 9/2$ and the conserved spins $1, 7, \ldots$. This follows from the explicit construction of the matrix of $\partial_{\bar{z}}$ in the corresponding subspaces of fixed spin and looking for zeroes of its determinant.

Table 29. Mass ratios $r_i = m_{i+1}/m_1$ for the Ashkin-Teller model in the limit $\mu \to \infty$ for several values of ε.

ε	r_1	r_2	r_3	r_4	r_5	r_6
$-1/\sqrt{2}$	1.934(6)	2.0	-	-	-	-
-0.6	1.652(4)	2.0	-	-	-	-
-0.3	1.221(4)	1.837(6)	1.94(1)	2.0	-	-
0.0	1	1.618*	1.618	1.989*	1.989	2
0.1	1.050(3)	1.618(4)*	1.651(5)	1.99(1)*	<2.0(1)	2.0
0.25	1.125(5)	1.621(5)*	1.71(1)	1.98(2)*	<2.0(1)	2.0
0.5	1.245(5)	1.623(5)*	1.785(10)	1.99(1)*	<2.0(1)	2.0
$1/\sqrt{2}$	1.330(5)	1.615(10)*	1.79(2)	1.91(2)*	>1.95	2.0
0.875	1.370(5)	1.57(1)	1.86(1)	1.85(2)	2.00(5)	2.0
1.0	1.395(5)	1.52(5)	1.75(5)	1.8(1)	?	

We conclude that at least for perturbations which break a global Z_2 symmetry the mass ratios predicted for the integrable Ising model in a magnetic field seem to be quite a general feature[16)]. We tentatively illustrate this diagrammatically in Figure 39. One can show that by perturbing a multicritical Ising model, with a central charge given by m via the Kac formula, by $\phi_{1,3}$ one arrives at the multicritical point given by $m-1$. It has been determined how the primary fields change when m goes to $m-1$ [223, 286]. For a generalization to non-unitary minimal models, see [232]. It remains at present an

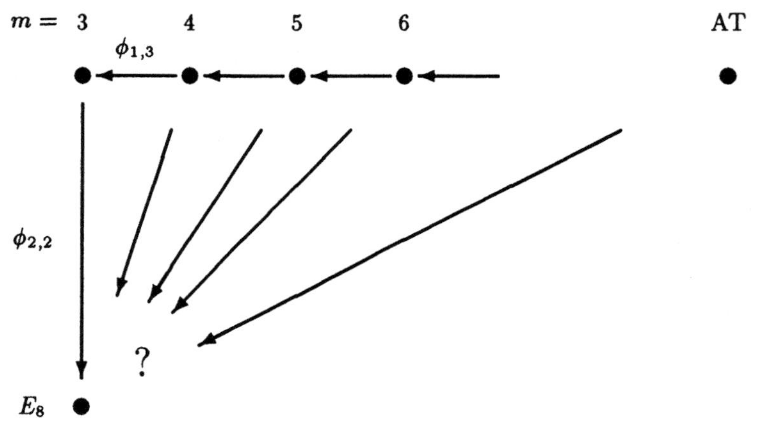

Fig. 39. Relation between Z_2 breaking perturbations $\phi_{2,2}$ in multicritical Ising models with $m = 3, 4, 5, 6$ and the Ashkin-Teller model (AT).

16) Recently, it was shown that perturbing the non-unitary minimal model $\mathcal{M}_{2,11}$ with ϕ_{14}, the same values of the spins of the conserved quantities and the same mass ratios as for the Ising model in a magnetic field are obtained [287].

open question in what sense the apparent relationship between these different universality classes can be understood.

14.5 Thermodynamic Bethe Ansatz

The methods discussed in the last sections were able to confirm the mass-spectrum given by the S-matrix, in the case where the perturbed critical model is integrable and purely massive. It is however not obvious that the minimal S-matrix is the appropriate one to describe the infrared behaviour of the off-critical model, since we have seen that the bootstrap program has infinitely many solutions. A method to test the S-matrix is given by the *Thermodynamic Bethe Ansatz* (TBA) which is able to compute the ultraviolet behaviour of the free energy, therefore giving the value of the conformal anomaly c (see eq. (3.54)) **via** the formula

$$F = f_0 L - \frac{\pi c_{\mathit{eff}}}{6L} + \mathcal{O}(1/L) = f_0 L - \frac{\pi C_{\mathit{eff}}(r)}{6L} \tag{14.81}$$

where f_0 is a non-universal bulk term and $c_{\mathit{eff}} = c - 12x_{\mathit{inf}}$ is the effective central charge. For unitary models, $c_{\mathit{eff}} = c$ (see eq. (12.36) for a the typical Yang-Lee case where $x_{\mathit{inf}} \neq 0$). The Thermodynamic Bethe Ansatz will give us the universal part $C_{\mathit{eff}}(r)$ which depends on the finite-size scaling parameter $r = m_1 L$, L being the size of the system and m_1 the lightest mass of the system. If the perturbation is purely thermal or magnetic, the finite-size scaling variable r corresponds, in the notation of chapter 3, to z^ν or $\mu^{\nu/(\beta+\gamma)}$, respectively. In the limit $r \to 0$, we have $C_{\mathit{eff}}(r) \to c_{\mathit{eff}}$.

Suppose that we choose a specific S-matrix and that we place the particles in a box where they scatter until they reach the thermodynamic equilibrium at temperature T. As we shall see, it is reasonable to assume that these equilibrium states will be described by Bethe wave functions, since all particle are stable. The TBA method traces back to the work of Yang and Yang [334], and regained popularity after the work of Al. Zamolodchikov, who revived it in the framework of perturbed conformal theory [348].

Suppose that we have a set of n well-separated particles located at points r_j on a large torus of size $M \times L$, so that we can assume that they do not interact. This means that for any pair i, j, we have $|r_i - r_j| \gg \xi_1$, where ξ_1 is the largest correlation length of the lightest particle with mass m_1. The particles are essentially free and on-shell and periodic boundary conditions are assumed. There are $n!$ regions in the configuration space were all particles are completely separated. Each region can be described by a n-particle wave function, which depends on the momenta $\vec{p_j}$ of the particles (or equivalently on the rapidities θ_j). Passing from one region to another amounts to cross one or many particles which will therefore scatter. The wave function will consequently be modified by the application of two-particle S-matrix elements on the asymptotic state. Suppose that we start from one region and move one particle, say j, along the cycle of length M. After having completed its motion "around the world", the system will have returned into the initial region, and the wave function, modified by $n-1$ scattering processes, must be the initial one. This gives us the matching condition for $1 \leq j \leq n$:

$$e^{ip_j^1 M} \prod_{k \neq j} S_{jk}(\theta_{jk}) = 1 \tag{14.82}$$

where $\theta_{ik} = \theta_i - \theta_k$. We assume that the mass-spectrum is non-degenerate. In the general case, where degeneracy is considered, the wave function will have internal indices which will appear in the matching equation, turning it into a matrix equation. Notice that the extension L of the torus plays the role of the inverse temperature $T = 1/L$ in the partition function

$$\mathcal{Z} = e^{-LF_0(L)} \tag{14.83}$$

where $F_0(L) = F - f_0 L$. It is useful to introduce explicitly the phase of the S-matrix

$$S_{ij}(\theta) := e^{i\delta_{ij}(\theta)} \tag{14.84}$$

to express eq. (14.82) in the form

$$M m_j \sinh\theta_j + \sum_{k\neq j} \delta_{jk}(\theta_{jk}) = 2\pi J_j \tag{14.85}$$

The integers J_i are to be determined by the equilibrium condition. In the non-interacting case, where $\delta_{ij} \equiv 0$, the last equation reduces to the usual quantification condition for the momenta of particles in a box.

The energy and momentum of the system are given by

$$E = \sum_{j=1}^{n} m_j \cosh\theta_j\,, \qquad P = \sum_{j=1}^{n} m_j \sinh\theta_j \tag{14.86}$$

Now the statistics of identical particles must be taken into account. The wave function must be symmetrized for bosons or antisymmetrized for fermions. On the other side, the unitarity equation (14.42) implies $(S_{ij}(0))^2 = 1$. Two different cases are possible:

$$S_{ij}(0) = -1 \qquad \text{and} \qquad S_{ij}(0) = 1 \tag{14.87}$$

- In the former case, the wave function is symmetric under the exchange of two identical particles with the same rapidities. If the particles are bosons, such states should be excluded. This situation is generally called the *fermionic case*. However, if the identical particles are fermions, they are allowed to have the same rapidities and this situation is called *bosonic case*.
- In the latter case, the situations are reversed, bosons give rise to the bosonic case and fermions to the fermionic cases.

Equation (14.85) forms a complicated system, which must be discussed in the thermodynamic limit where $M \to \infty$ and, for any type of stable asymptotic particle with mass m_j, the number of particles n_j grows proportional to M. The difference between adjacent solutions to eq. (14.85) goes to zero as $\theta_i - \theta_j \sim 1/m_1 M$. It makes sense to introduce in this limit a continuous ***rapidity density*** of particles $\tilde{\rho}_j(\theta)$ defined for a small rapidity interval $\Delta\theta$ around the value θ with $n_j(\theta)$ particles with mass m_j inside as

$$\tilde{\rho}_j(\theta) = \frac{n_j(\theta)}{M\Delta\theta} \tag{14.88}$$

This smooth function of θ appears to be independent of the particular value of $\Delta\theta$ in the interval $1/m_1 M \ll \Delta\theta \ll 1$. The phase sum in eq. (14.85) can be replaced by the integral

$$m_j \sinh(\theta_j) + \sum_k \int_{-\infty}^{+\infty} d\theta' \, \delta_{jk}(\theta_j - \theta')\tilde{\rho}_k(\theta') = 2\pi \frac{J_j}{M} \tag{14.89}$$

The indices do not label anymore each individual particle but only their types, characterized by the masses of the stable particles. The details on each particle is now averaged in the densities. The energy density of the system now reads

$$e(\tilde{\rho}) = \sum_j \int_{-\infty}^{+\infty} d\theta \, m_j \tilde{\rho}_j(\theta) \cosh\theta \tag{14.90}$$

Equation (14.89) is considered for any spectrum of J_j, even for the cases where the solution does not correspond to a physical state. Such solutions are called *holes* while the ones corresponding to physical solutions are called *roots*. A continuous spectrum $J_j(\theta)$ can now be defined, and therefore the thermodynamic limit is given by the *level density* $\rho_j(\theta)$, which satisfies

$$2\pi\rho_j(\theta) = m_j \cosh\theta + \sum_k \int_{-\infty}^{+\infty} d\theta' \, \varphi_{jk}(\theta - \theta')\tilde{\rho}_k(\theta') \tag{14.91}$$

where

$$\rho_j(\theta) = \frac{1}{M}\frac{\partial J_j(\theta)}{\partial\theta} \qquad \text{and} \qquad \varphi_{jk}(\theta) = \frac{\partial \delta_{jk}(\theta)}{\partial\theta} = -i\frac{\partial \ln S_{jk}(\theta)}{\partial\theta} \tag{14.92}$$

The unitarity of the S-matrix implies $\varphi_{jk}(\theta) = \varphi_{jk}(-\theta)$. From the knowledge of explicit examples, we can assume that $\varphi_{ij}(\theta)$ decays exponentially fast at large values of $|\theta|$. Notice that the averaged densities are not sensitive to local redistributions of particles among themselves in any small interval $\Delta\theta$, which contains a number of levels proportional to $M\rho_j(\theta)\Delta\theta$ and a number of rapidities proportional to $M\tilde{\rho}_j(\theta)\Delta\theta$ for the particles with mass m_j. The number of different redistributions in such an interval is given by

$$\begin{aligned} &\frac{[M\rho_j(\theta)\Delta\theta]!}{[M(\rho_j(\theta)-\tilde{\rho}_j(\theta))\Delta\theta]![M\tilde{\rho}_j(\theta)\Delta\theta]!} && \text{fermionic case} \\ &\frac{[M(\rho_j(\theta)+\tilde{\rho}_j(\theta))\Delta\theta]!}{[M\rho_j(\theta)\Delta\theta]![M\tilde{\rho}_j(\theta)\Delta\theta]!} && \text{bosonic case} \end{aligned} \tag{14.93}$$

The logarithms of these numbers of possible states give in the asymptotic limit $M \to \infty$ the *entropy* density $s(\rho, \tilde{\rho})$, given per unit length

$$s(\rho,\tilde{\rho}) = \begin{cases} \sum_j \int_{-\infty}^{+\infty} d\theta \, [\rho_j \ln\rho_j - \tilde{\rho}_j \ln\tilde{\rho}_j - (\rho_j - \tilde{\rho}_j)\ln(\rho - \tilde{\rho}_j)] & \text{fermionic case} \\ \sum_j \int_{-\infty}^{+\infty} d\theta \, [(\rho_j + \tilde{\rho}_j)\ln(\rho + \tilde{\rho}_j) - \rho_j \ln\rho_j - \tilde{\rho}_j \ln\tilde{\rho}_j] & \text{bosonic case} \end{cases} \tag{14.94}$$

The sum in the partition function of the system minimizes the free energy, which can be expressed in terms of the energy density, the temperature $T = 1/L$, and the entropy density by the usual thermodynamic relation

$$f = e(\tilde{\rho}) - Ts(\rho, \tilde{\rho}) \tag{14.95}$$

From eqs. (14.90) and (14.94), we know the functional dependence of the l.h.s. of the last equation in the densities $\tilde{\rho}$ and ρ, which are related to each other by eq. (14.91). Our aim is to minimize the free energy. First of all, we have

$$\frac{\delta e(\tilde{\rho})}{\delta \tilde{\rho}_j(\theta)} = m_j \cosh\theta \tag{14.96}$$

and from eq. (14.91)

$$\frac{\delta \rho_k(\theta')}{\delta \tilde{\rho}_j(\theta)} = \frac{1}{2\pi}\varphi_{kj}(\theta' - \theta) \tag{14.97}$$

The variation of the entropy with respect to $\tilde{\rho}_j$ is therefore

$$\frac{\delta s(\rho,\tilde{\rho})}{\delta \tilde{\rho}_j(\theta)} = \begin{cases} \ln\left(\dfrac{\rho_j(\theta)}{\tilde{\rho}_j(\theta)} - 1\right) - \displaystyle\sum_j \int_{-\infty}^{+\infty} \frac{d\theta'}{2\pi}\varphi_{jk}(\theta-\theta')\ln\left(1 - \frac{\tilde{\rho}_j(\theta)}{\rho_j(\theta)}\right) & \text{fermionic case} \\ \ln\left(\dfrac{\rho_j(\theta)}{\tilde{\rho}_j(\theta)} + 1\right) + \displaystyle\sum_j \int_{-\infty}^{+\infty} \frac{d\theta'}{2\pi}\varphi_{jk}(\theta-\theta')\ln\left(1 + \frac{\tilde{\rho}_j(\theta)}{\rho_j(\theta)}\right) & \text{bosonic case} \end{cases} \tag{14.98}$$

The variation of the free energy with respect to $\tilde{\rho}_j$ only depends on the ratios $\rho_j/\tilde{\rho}_j$ and it is therefore useful to introduce the ***pseudoenergies*** $\epsilon_j^{\pm}$ defined as

$$\epsilon_j^{\pm}(\theta) := \ln\left(\frac{\rho_j(\theta)}{\tilde{\rho}_j(\theta)} \mp 1\right) \tag{14.99}$$

with the + sign taken in the fermionic case and the − sign in the bosonic case. The positivity of the densities implies $\epsilon_j^- \geq 0$ while ϵ_j^+ can take any real value. The minimization equation for the free energy takes the fundamental form

$$\boxed{L m_j \cosh\theta = r\frac{m_j}{m_1}\cosh\theta = \epsilon_j^{\pm}(\theta) \pm \sum_k \int_{-\infty}^{+\infty} \frac{d\theta'}{2\pi}\varphi_{jk}(\theta-\theta')\ln\left(1 \pm e^{-\epsilon_k^{\pm}(\theta')}\right)} \tag{14.100}$$

which can be solved for $\epsilon_j^{\pm}$, which depends on $r = m_1 L$, too. In turn, the minimized free energy can be expressed in terms of the pseudoenergies $\epsilon_j^{\pm}$ only. To show this, we write the entropy in terms of $\epsilon_j^{\pm}$ and $\tilde{\rho}_j$

$$s(\rho,\tilde{\rho}) = \sum_j \int_{-\infty}^{+\infty} d\theta\, \tilde{\rho}_j(\theta)\, \left[\epsilon_j^{\pm} + \left(1 \pm e^{\epsilon_j^{\pm}}\right)\ln\left(1 \pm e^{-\epsilon_j^{\pm}}\right)\right] \tag{14.101}$$

We first replace $L m_j \cosh\theta - \epsilon_j^{\pm}$ in $Lf(L) = Le(\tilde{\rho}) - s(\rho(\tilde{\rho},\epsilon),\tilde{\rho})$ by the last term on the r.h.s. of eq. (14.100). We make use of eq. (14.91) to get rid of the explicit dependence in φ

and of eq. (14.99) to avoid any explicit dependence in ρ. The result beautifully simplifies to

$$Lf(L) = -\frac{\pi C_{eff}(r)}{6L} = \mp \sum_k \int_{-\infty}^{+\infty} \frac{d\theta}{2\pi} \cosh(\theta) m_k \ln\left(1 \pm e^{-\epsilon_k^{\pm}(\theta)}\right) \tag{14.102}$$

and we can finally write

$$C_{eff}(r) = \pm \frac{3}{\pi^2} \sum_k \int_{-\infty}^{+\infty} d\theta \cosh(\theta) \frac{m_k}{m_1} r \ln\left(1 \pm e^{-\epsilon_k^{\pm}(\theta)}\right) \tag{14.103}$$

The computation of the central charge from the knowledge of the S-matrix is obtained as $r \to 0$. In this limit, θ has to be taken very large to have a non-trivial contribution to the l.h.s. of eq. (14.100). One has

$$r\frac{m_j}{m_1} \cosh\theta \sim \frac{r}{2} \frac{m_j}{m_1} e^{\theta} = \frac{m_j}{m_1} \exp(\theta - \ln\frac{2}{r}) \tag{14.104}$$

After some numerical work, one realizes that the pseudoenergies $\epsilon_j(\theta)$ are constant in the range $-\ln(2/r) \ll \theta \ll +\ln(2/r)$, where the l.h.s. of eq. (14.100) can be neglected, and grow exponentially at very large values of $|\theta|$. In the range where they become the constants $\hat{\epsilon}_j^{\pm}$, one has to solve the system of equations

$$\hat{\epsilon}_j^{\pm} = \pm \sum_k N_{jk} \ln\left(1 \pm e^{-\hat{\epsilon}_k^{\pm}}\right) \tag{14.105}$$

where

$$N_{jk} = -\int_{-\infty}^{+\infty} \frac{d\theta}{2\pi} \varphi_{jk}(\theta) = -\frac{1}{2\pi}\left(\delta_{jk}(\infty) - \delta_{jk}(-\infty)\right) \tag{14.106}$$

is the total phase shift of the S-matrix. This region will not contribute to the integral (14.103) for the central charge as $r \to 0$. Only the large velocity region is significant. Due to the parity of the pseudoenergies, we can replace the lower boundary of integration in eq. (14.103) by 0 and multiply the integrand by two. In the large θ limit, we have to solve

$$\frac{m_j}{m_1} \frac{r}{2} e^{\theta} = \tilde{\epsilon}_j^{\pm}(\theta) \pm \sum_k \int_{-\infty}^{+\infty} \frac{d\theta'}{2\pi} \varphi_{jk}(\theta - \theta') \ln\left(1 \pm e^{-\tilde{\epsilon}_k^{\pm}(\theta')}\right) \tag{14.107}$$

where $\tilde{\epsilon}_j^{\pm}(\theta)$ overlaps with the constant solution $\hat{\epsilon}_j^{\pm}$ for $\theta \ll \ln(2/r)$ and grows exponentially as $\theta \to \infty$. In terms of $\tilde{\epsilon}_j^{\pm}(\theta)$, the effective central charge becomes

$$C_{eff}(r) = \pm \frac{3}{\pi^2} \sum_k \int_0^{\infty} d\theta \frac{m_k}{m_1} r e^{\theta} \ln\left(1 \pm e^{-\tilde{\epsilon}_k^{\pm}(\theta)}\right) \tag{14.108}$$

Notice that the derivative of both sides of eq. (14.107) with respect to θ gives a new expression for $(m_j/m_1) r e^{\theta}$.

$$\frac{m_j}{m_1} \frac{r}{2} e^{\theta} = \partial_\theta \tilde{\epsilon}_j^{\pm}(\theta) \pm \sum_k \int_{-\infty}^{+\infty} \frac{d\theta'}{2\pi} \varphi_{jk}(\theta - \theta') \partial_{\theta'} \ln\left(1 \pm e^{-\tilde{\epsilon}_k^{\pm}(\theta')}\right) \tag{14.109}$$

This can be introduced into eq. (14.108). After several integrations by parts, one obtains

$$c_{eff} = \pm\frac{6}{\pi^2}\sum_j\left[\int_{\hat{\epsilon}_j^\pm}^{\infty} dy \ln\left(1\pm e^{-y}\right) + \frac{1}{2}\hat{\epsilon}_j^\pm \ln\left(1\pm e^{-\hat{\epsilon}_j^\pm}\right)\right] \tag{14.110}$$

The effective central charge is often expressed with the ***Roger's dilogarithmic function*** $L_R(x)$ defined as

$$L_R(x) = -\frac{1}{2}\int_0^x dy\left[\frac{\ln y}{1-y} + \frac{\ln(1-y)}{y}\right] \tag{14.111}$$

In the region $[0,1]$, $L_R(x)$ is a continuous growing function of x (with $L_R(0) = 0$) which satisfies the relations

$$\begin{aligned} L_R(x) + L_R(1-x) &= L_R(1) = \frac{\pi^2}{6} \\ L_R\left(\frac{1}{1+x}\right) + L_R\left(-\frac{1}{x}\right) &= \frac{i\pi}{2}\ln\left(1+\frac{1}{x}\right) \end{aligned} \tag{14.112}$$

After a few manipulations of (14.110), one can write

$$c_{eff} = \begin{cases} \dfrac{6}{\pi^2}\displaystyle\sum_j L_R\left(\frac{1}{1+e^{\hat{\epsilon}_j^+}}\right) & \text{fermionic case} \\ \dfrac{6}{\pi^2}\displaystyle\sum_j L_R\left(\frac{1}{e^{\hat{\epsilon}_j^-}}\right) & \text{bosonic case} \end{cases} \tag{14.113}$$

Equation (14.105) must be solved numerically. Several groups have intensively computed the effective central charge for many models [348, 347, 350, 95, 249, 221]. The result is that the minimal matrix is already the proper one to obtain the correct central charge of the critical system. Many systems require the extension of TBA where kink solutions are considered for the asymptotic particles. This gives rise to TBA-equations which slightly differ from the one presented here.

The TBA method can be applied for any integrable massive system with a known S-matrix, minimal or not. Actually, most of them have an ultraviolet behaviour which does not correspond to any physical model of statistical mechanics but make sense in the general framework of quantum field theory. In general, the models have in common that the S-matrices are quite complicated functions of the rapidity, and it is difficult to illustrate TBA with complicated examples. Fortunately, they are two very simple cases where we can verify the TBA procedure without any complicated computation. Consider first the free bosonic theory with $S(\theta) = 1$ (eq. (14.67)). In this case $\varphi(\theta) = 0$ and eq. (14.105) for the bosonic case implies $\hat{\epsilon}^- = 0 \Rightarrow \rho/\tilde{\rho} = 0$. This reflects the fact that infinitely many massless bosons of the conformal ultraviolet limit can have the same momenta. Therefore $c_{eff} = c = (6/\pi^2)L_R(1) = 1$ as expected for the free massless bosonic theory. The second simple example is given by the Ising model with a thermal perturbation and a S-matrix $S(\theta) = -1$ (eq. (14.69)). The scalar bosonic massive particle is of fermionic type with respect to TBA. Again, we have $\varphi(\theta) = 0$, and therefore $\hat{\epsilon}^+ =$

$0 \Rightarrow \rho/\tilde{\rho} = 2$, as a consequence of the presence of a two-component Majorana fermion in the theory. Using the first line of eq. (14.112), we find $c_{eff} = c = (6/\pi^2)L_R(1/2) = 1/2$, which is the correct central charge value for the critical Ising model.

The TBA program is therefore able to pinpoint the correct S-matrix of integrable massive off-critical systems. Actually, the method is able to do more than just verifying the central charge. Higher orders in the scaling parameter r were computed and compared with perturbation theory around the critical point or other numerical methods which can be solved in the vicinity of the critical point. These more advanced applications are beyond the scope of an introduction.

14.6 Asymptotic finite-size scaling functions

In Figures 37 and 38 we have seen that the approach of the scaling functions towards their $\mu \to \infty$ limit seems to be exponential. In fact, the asymptotic behaviour for large values of the finite-size scaling variable can be derived from the S-matrix as well. Formulae of this kind were first derived by Lüscher [G17, 244]. For the special case of a $2D$ strip geometry of finite width L the result can be stated as follows [222]. This result holds generally for arbitrary Lagrangians.

Let $a = 1, \ldots, n$ denote the n stable particles confined to a box of size L in one direction and infinite extent in the other and let $0 < m_1 < m_2 < \ldots < m_n$ be their masses. Then the mass shift $\Delta m_a(L) = m_a(L) - m_a$ is for states with mass $m_a < 2m_1$

$$\Delta m_a(L)/m_1 = M_a^{(1)} + M_a^{(2)} + \mathcal{O}(\exp(-\sigma_a L)) \tag{14.114}$$

$$M_a^{(1)} = -\frac{1}{8m_a^2} \sum \frac{\lambda_{abc}^2}{\mu_{abc}} \exp(-\mu_{abc}\, L) \tag{14.115}$$

where the sum runs over all pairs (b, c) of stable particles such that $|m_b^2 - m_c^2| < m_a^2$ is satisfied,

$$M_a^{(2)} = -\frac{1}{2\pi} \sum_b{}' \int_{-\infty}^{\infty} d\theta \, \exp(-\cosh\theta \, m_b \, L) \, m_b \cosh\theta \left(S_{ab}(\theta + \frac{i\pi}{2}) - 1 \right) \tag{14.116}$$

where $\mu_{abc} = m_b m_c m_a^{-1} \sin u_{bc}^a$, u_{bc}^a is defined by $m_a^2 = m_b^2 + m_c^2 + 2m_b m_c \cos u_{bc}^a$ and

$$\lambda_{abc}^2 = -8i\,(m_b m_c \sin u_{bc}^a)^2 \, \mathrm{Res}_{\theta = iu_{bc}^a} \, S_{bc}(\theta) \tag{14.117}$$

where $S_{ab}(\theta)$ is the two-particle S-matrix describing the scattering process $ab \to ab$ and θ is the rapidity variable. The prime in eq. (14.116) indicates that one should only sum over terms which are larger than the error term $\mathcal{O}(\exp(-\sigma_a L))$, and $\sigma_a \leq 2\mu_{a11}$. For precise estimates of σ_a, we refer to the original work of Klassen and Melzer [222].

The correspondence of this with the finite-size scaling functions $G_i(\mu)$ of the correlation lengths eq. (14.72) is a follows. The system size L is related to the finite-size scaling variable μ via

$$m_1 \, L = \zeta^{-1}\xi_1^{-1} N = \zeta^{-1} G_1(\infty)\mu^{1/y} \tag{14.118}$$

(recall that $y = 15/8$ in the Ising model). The mass shifts are then calculated with $i = a$ for all states below the continuum threshold

$$\delta G_i(\mu) = (G_i(\infty) - G_i(\mu))\,/G_1(\infty) = -\Delta m_a/m_1(\infty) \tag{14.119}$$

This results can be used in a two-fold way. First, one can explicitly test the S-matrix beyond the mass ratios using finite-lattice data. We shall do so below. Secondly, one can also predict the asymptotic finite-size scaling function from the knowledge of the two-particle S-matrix. This gives complementary information to the conformal perturbation theory approach described in the last chapter.

In Table 30 [222], we list the contributions from the simple poles of the S-matrix which contribute in the example of the $2D$ Ising model in a magnetic field, where the **minimal** S-matrix was used. For a comparison with the numerical data, we need the value of

Table 30. Numerical values for μ_{abc} and $\rho_{abc} := \lambda_{abc}^2/(8m_a^2\,\mu_{abc})$ for the $2D$ Ising model.

	abc	μ_{abc}	ρ_{abc}
G_1	111	$0.866025 \cdot m_1$	$1.046154 \cdot 10^2$
	122	$1.538842 \cdot m_1$	$1.463532 \cdot 10^4$
G_2	211	$0.587785 \cdot m_1$	$1.205778 \cdot 10^2$
	212	$0.951057 \cdot m_1$	$9.045125 \cdot 10^3$
	221	$0.951057 \cdot m_1$	$9.045125 \cdot 10^3$
G_3	311	$0.104528 \cdot m_1$	$1.130876 \cdot 10^{-1}$

$G_1(\infty)$. Since finite-size scaling techniques alone require a double extrapolation, its value will normally be known to somewhat lesser precision than the finite-lattice data for $G_i(\mu)$ for μ finite. Further, the comparison with the asymptotic expression $M_a^{(1)}$ requires that the values of μ must be large enough so that the corrections to it are still negligible. If on the other hand μ becomes too large, the numerical errors in the determination of of the $G_i(\mu)$ will become larger than the value of $\delta G_i(\mu)$ itself. The calculation of the $G_i(\mu)$ is improved by knowing the size of the leading finite-size corrections as obtained from conformal perturbation theory, here in the Ising model example

$$\xi_i^{-1} = h^{8/15}\left(G_i(\mu) + N^{-2}R_i(\mu) + \mathcal{O}(N^{-4})\right) \tag{14.120}$$

In Figure 40 [182], we compare the mass shift predicted from the S-matrix with the **minimal** number of poles, using the contribution of $M_a^{(1)}$ only (full curves), with the finite-lattice data. We first note that the mass shifts are surprisingly large. Still, we find an excellent agreement with the asymptotic predictions over several orders of magnitude for G_1 and G_2, while for G_3, due to the smallness of μ_{311}, the available data just reach the beginning of the asymptotic regime. This means that indeed the available evidence is consistent with the **minimal** solution for the S-matrix. This is in agreement with the calculation of the central charge from the S-matrix using the TBA described in the last section and does provide an independent confirmation of the minimal solution to the bootstrap program. However, there are a few more lessons to be drawn from this plot. The data for $\delta G_i(\mu)$ were obtained in using the finite-size estimate $G_1(\infty) = 5.4156(3)$ [182]. Figure 40 shows clearly for μ large enough a **linear** relation between $\delta G_i(\mu)$ and $\mu^{1/y}$ as expected from the asymptotic prediction. This is even independent of any particular S-matrix and can be used by itself as a check on the quality of the data. Further, the slope

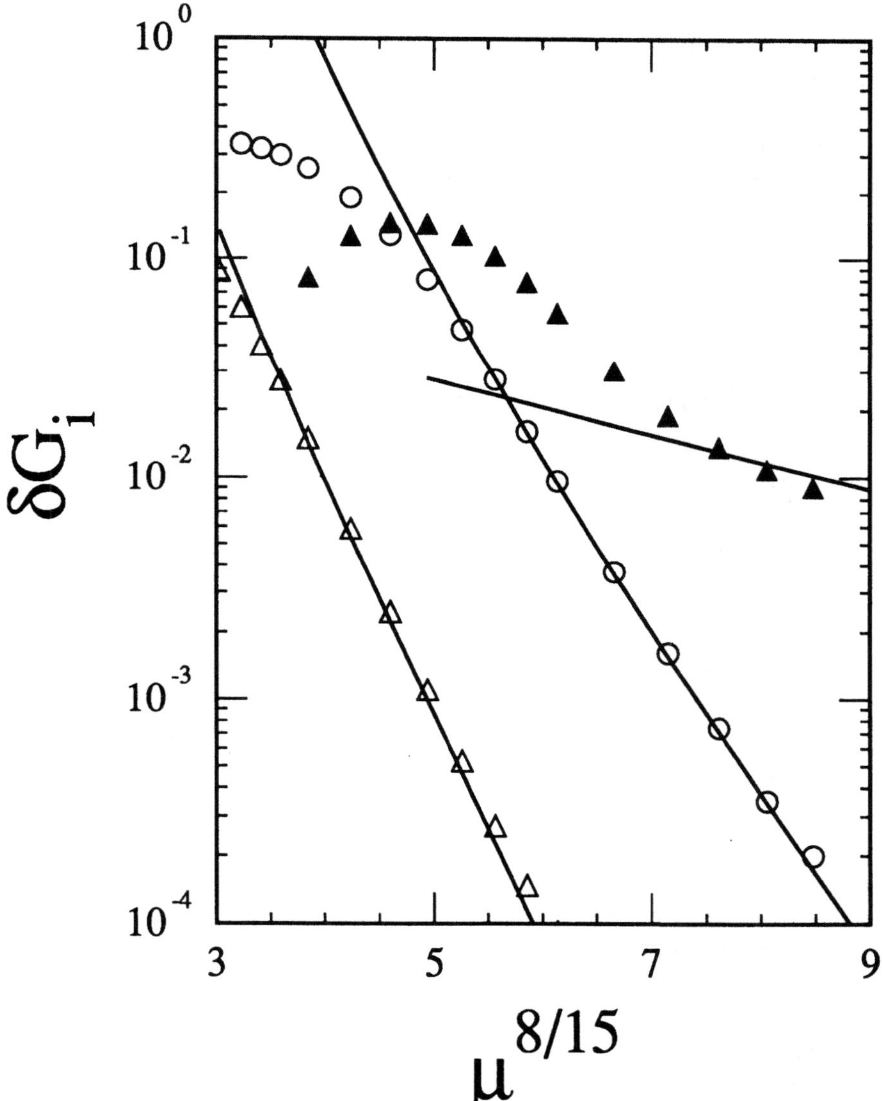

Fig. 40. Mass shift $\delta G_i, i = 1, 2, 3$ in the $(1+1)D$ Ising model with a magnetic field. The lowest curve corresponds to δG_1, the uppermost to δG_3.

only depends on the masses but not on the couplings λ_{abc}, which offers an additional test. However, for very small values of δG_1, the finite-lattice data fall systematically above the asymptotic form. Eventually the data would appear to be independent of μ. This means that the value of $G_1(\infty)$ used was a little overestimated. The order of magnitude of this systematic error can be read off to be $\sim 10^{-4}$. It is important to note that this way one has a means at hand to tell about systematic extrapolation errors and their sizes *a posteriori*.

These remarks are fully confirmed by an extended study which combines the finite-size data with the truncation method. This yields the improved estimate $G_1(\infty) = 5.41550(1)$ [222]. In Table 31 we compare the results obtained from the truncation method (TM) and from the finite-size scaling data (FSS), using the improved value of $G_1(\infty)$, with the asymptotic formula eq. (14.114) (ASY). Comparing these results and recalling that the

Table 31. Mass shifts δG_i for the Ising model in a magnetic field.

	$\delta G_1 \cdot 10^3$			$\delta G_2 \cdot 10^3$		
$m_1 L$	TM	FSS	ASY	TM	FSS	ASY
9.24575	40.56(1)	40.556(6)	57.29			
10.4142	15.02(2)	15.018(3)	17.11			
11.4778	5.85(1)	5.845(3)	6.158			
12.4613	2.44(3)	2.440(3)	2.486	128.97(3)	128.988(6)	208.4
13.3811	1.07(3)	1.084(3)	1.085	80.49(5)	80.53(1)	100.1
14.2486	0.48(4)	0.508(6)	0.5014	47.45(8)	47.54(3)	51.36
15.0722	0.20(6)	0.253(4)	0.2422	27.3(1)	27.5(2)	27.90
15.8582	0.1(1)	0.13(2)	0.1213	15.8(3)	16.08(3)	15.89
16.6114		0.07(4)	0.06268	9.1(5)	9.61(6)	9.422
18.0348				3(1)	3.7(1)	3.646
19.3660					1.6(1)	1.554
20.6216					0.71(9)	0.7113
21.8135					0.33(12)	0.3434

gaps are originally of order unity, the agreement shown is truly impressive. We also see that the truncation method is really pushed beyond its limits in this particular application. This is not yet yet so apparent for the lowest level, but already quite visible for the second one. The finite-size scaling method reproduces the large μ behaviour to at least one order of magnitude better. This success does depend however, on using the improved value of $G_1(\mu)$.

In conclusion, these examples show an unprecedented precision of the theoretical description of the finite-size scaling functions.

15 Surface Critical Phenomena

Up to now we have been always considering a statistical system defined in the infinite two-dimensional plane. However, conformal invariance can also be applied to systems with boundaries present. It is impossible to give on just a few pages a full description on the rich field of surface effects and we will only consider some of the problems for which conformal invariance has proved to be useful. For more background on surface critical phenomena, we refer to the detailed reviews [48, 103]. Wetting and related questions are reviewed in [105].

In this chapter, we shall again follow the same sequence of logical steps done before for the case of bulk critical phenomena. After defining the relevant thermodynamics, we introduce surface critical exponents and discuss scaling. Conformal invariance will then be used to determine completely the two-point correlation function - a problem of the same degree of difficulty as finding the four-point correlation function in the bulk. Finite-size scaling techniques will be described and we shall use them to give the surface operator content of the models discussed before. Finally, defect lines will be studied which interpolate between a free surface and the bulk.

15.1 Systems with a boundary

In general, real experimentally considered systems are always finite and possess a boundary. The presence of boundaries gives rise to new important effects neglected so far. We shall restrict attention to a particularly simple case: a flat surface with free boundary conditions. We are therefore totally neglecting the fluctuation properties of boundaries or interfaces, these being interesting topics in their own right (see e.g. [285] for an introduction and further references). While in itself this is a very crude approximation, this does not affect the universal properties of surface phase transitions. We consider exclusively hypercubic systems within a *slab geometry*, that is, just one direction is of finite length L with free boundary conditions, while the other direction(s) are taken to be infinite in extent.

We further assume that the thermodynamic parameters of the system are such that bulk and surface properties can be well separated. For example, we write for the free energy density

$$f = f_b(T, V, \mathcal{N}) + \frac{2}{L} f_s(T, T_1, V, \mathcal{N}) + \dots \tag{15.1}$$

where f_b is the *bulk free energy density* studied so far and f_s is the *surface free energy density*. While T is the temperature as measured for the bulk system, we have also included a *surface temperature* T_1 to allow for a variation of thermodynamic variables close to the free surface. The factor 2 takes the presence of two free interfaces into account.

Here, we want to limit ourselves to just one rather restricted question, namely about the types of singularities present in f_s and the quantities derived from it close to the bulk critical point. To get an overview over the generic situation, consider the space-dependent variation of the order parameter $m(r)$, where r measures the distance to the free surface, assumed to be at $r = 0$. Taking Landau theory [48] as a guide, there are in general four possible situations, as illustrated in Figure 41 (see [48]). The bulk critical temperature $T_{c,b}$ is the *bulk critical temperature* (we have called this critical temperature up to now).

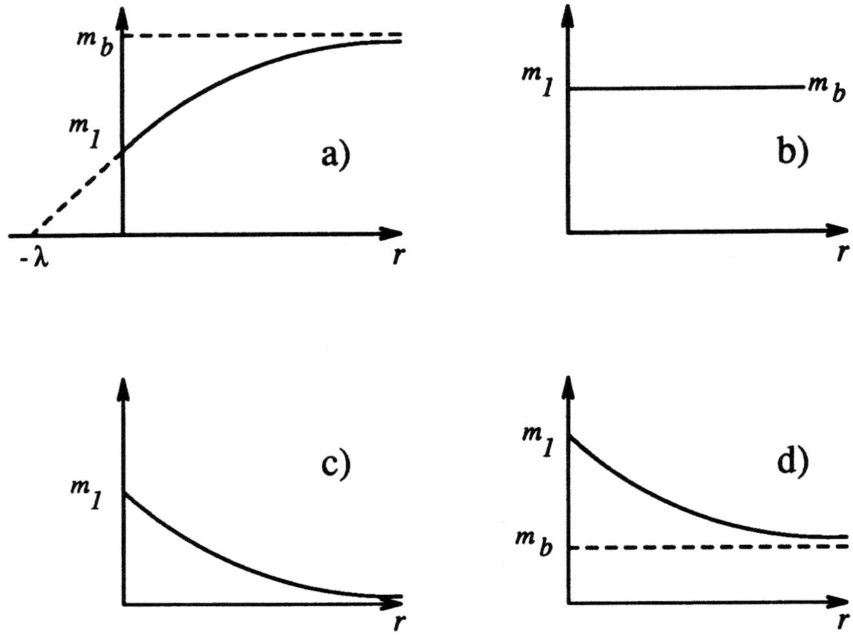

Fig. 41. Schematic local order parameter profiles $m(r)$ in dependence of the temperature and the extrapolation length λ.

1. Let $T < T_{c,b}$ and take a positive *extrapolation length* λ, as shown in Figure 41a. The transition from the disordred state to this state is called the *ordinary transition* O where m_b is the magnetization deep inside the bulk and m_1 the magnetization at the surface.
2. Let $T = T_{c,b}$. The extrapolation length is infinite. The transition from the disordered state to this state is called the *special transition* SP. It is a surface tricritical point. (Figure 41b)
3. Let $T_{c,b} < T < T_{c,s}$. The extrapolation length is negative and $T_{c,s}$ is some surface critical temperature. The transition from the disordered state into this state is called the *surface transition* S. (Figure 41c)
4. Let $T < T_{c,b}$ and take a negative extrapolation length. The transition from the state of Figure 41c to this state is called the *extraordinary transition* E. (Figure 41d)

It becomes apparent that the local behaviour of thermodynamic quantities can be quite distinct from those deep inside the bulk. In particular, we will have to allow for the possibility that the critical exponents close to the surface may be different from their bulk values. This can be already verified in the context of Landau theory [48]. We summarize the generic surface phase diagram in Figure 42. Although the results sketched in Figures 41 and 42 are merely a consequence of mean-field theory, they remain qualitatively correct in three and higher dimensions, at least in the context of the Ising model. For systems like the $2D$ Ising model however, a free surface cannot order by itself at a non-zero temperature (for short-ranged interactions). Therefore for such a system, only the ordinary surface transition exists.

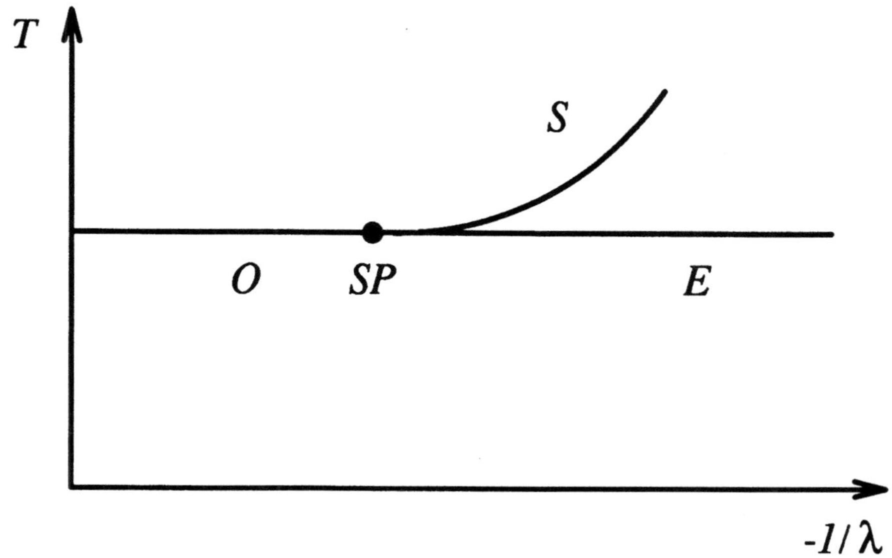

Fig. 42. Schematic surface phase diagram.

We now describe the calculation of the surface thermodynamical quantities, see [48, 103]. This proceeds in quite an analogous manner with those of the bulk as defined in chapter 1. There we had introduced the reduced temperature t and the reduced magnetic field h. Here, we introduce in addition a reduced *surface magnetic field* h_1, which only acts on the boundary spins. For a large class of models, this is the only new relevant scaling field which arises. The new surface quantities can now be defined by taking derivatives of the surface free energy density eq. (15.1). First, we define *excess quantities* by taking derivatives of f_s with respect to bulk scaling fields

$$\begin{aligned} m_s &:= -\frac{\partial f_s}{\partial h} \quad \text{excess magnetization} \\ e_s &:= \frac{\partial f_s}{\partial t} \quad \text{excess internal energy} \\ \chi_s &:= \frac{\partial m_s}{\partial h} \quad \text{excess susceptibility} \\ C_s &:= \frac{\partial e_s}{\partial t} \quad \text{excess specific heat} \end{aligned} \tag{15.2}$$

Second, we define surface or *local quantities* by taking derivatives with respect to surface fields

$$\begin{aligned} m_1 &:= -\frac{\partial f_s}{\partial h_1} \quad \text{surface magnetization} \\ \chi_1 &:= \frac{\partial m_1}{\partial h_1} \quad \text{local susceptibility} \end{aligned} \tag{15.3}$$

Finally, we can define *mixed quantities* such as the *layer susceptibility*

$$\chi_{11} := \frac{\partial m_1}{\partial h} \tag{15.4}$$

In these definitions, it is understood that all those thermodynamic variables not explicitly involved in the derivatives are kept constant.

The distinct nature of these quantities becomes apparent when expressing a space-dependent profile such as $m(r)$ (see Figure 41) in terms of them. For example, for the order parameter one has

$$\begin{aligned} m_b &= m(r = \infty) \\ m_1 &= m(r = 0) \\ m_s &= \int_0^\infty dr\,[m(r) - m_b] \end{aligned} \tag{15.5}$$

Next, we turn to the definition of the critical exponents. Their definition is again quite analogous to the bulk exponents in eq. (1.11). We assume that those of the parameters t, h, h_1 not explicitly mentioned are zero. The surface exponents are then defined by

$$\begin{aligned} C_s &\sim |t|^{-\alpha_s} \\ m_s &\sim (-t)^{\beta_s} \\ m_1 &\sim (-t)^{\beta_1} \\ \chi_s &\sim |t|^{-\gamma_s} \\ \chi_1 &\sim |t|^{-\gamma_1} \\ \chi_{11} &\sim |t|^{-\gamma_{11}} \\ m_s &\sim h^{1/\delta_s} \\ m_1 &\sim h^{1/\delta_1} \\ m_1 &\sim h^{1/\delta_{11}} \\ \frac{\partial^{2\ell} f_s}{\partial h_1^{2\ell}} &\sim |t|^{-2\Delta_1} \frac{\partial^{2\ell-2} f_s}{\partial h_1^{2\ell-2}} \end{aligned} \tag{15.6}$$

For the correlation function, surface critical exponents are defined as follows. Denote $\vec{x}_s = (x_{\parallel}, 0)$ and $\vec{x} = (x_{\parallel}, r)$. Then one has correlations parallel and perpendicular to the free surface at $r = 0$

$$G_{\parallel}(\vec{x}_{\parallel} - \vec{x}'_{\parallel}) := G(\vec{x}_s, \vec{x}'_s) \ , \ G_{\perp}(r, r') := G(\vec{x}, \vec{x}')|_{x_{\parallel} = x'_{\parallel}} \tag{15.7}$$

where $G(x, x')$ is the two point order parameter correlation function defined in eq. (1.6). The surface exponents are then defined exactly **at** the critical point $t = h = h_1 = 0$

$$\begin{aligned} G_{\parallel}(\vec{x}_{\parallel}) &\sim |\vec{x}_{\parallel}|^{-d+2-\eta_{\parallel}} \\ G_{\perp}(r, r') &\sim |r - r'|^{-d+2-\eta_{\perp}} \end{aligned} \tag{15.8}$$

These definitions apply to the ordinary, special and extraordinary transitions. Exponents belonging to the surface transition can be defined by replacing t by $t_s := (T - T_{c,s})/T_{c,s}$, but since the bulk is not critical at a surface transition, critical exponents only exist for the local quantities of eq. (15.3).

As already seen for the bulk critical exponents, there exist many scaling relations, relating different exponents to each other. The renormalization group treatment can be extended to surface critical phenomena and the notions of relevant, marginal and irrelevant scaling fields are introduced as in the bulk, see chapter 1. For notational simplicity, we only consider here the case of the ordinary transition and refer to [48, 103] for the other cases. This is sufficient for us, since later we shall only deal with two-dimensional systems, where there is only an ordinary surface transition. Since for an ordinary transition, there is only a relevant surface magnetic field h_1, we can write the scaling form for the singular part of the surface free energy density

$$f_s^{sin} = |t|^{2-\alpha_s} W_s^{\pm}\left(h|t|^{-\beta-\gamma}, h_1|t|^{-\Delta_1}\right) \tag{15.9}$$

For the bulk system, hyperscaling was seen to be equivalent to the dimensional argument that $f_b^{sin} \sim \xi_b^{-d}$, valid below the upper critical dimension $d < d^*$, since one assumes that there is only one relevant length scale. For the free surface, the analogous statement is $f_s^{sin} \sim \xi_b^{-(d-1)}$ which implies

$$\alpha_s = \alpha + \nu \tag{15.10}$$

Since α_s is related to bulk exponents, it follows from scaling eq. (15.9) that there is only one new independent surface critical exponent. Knowing one of them, for example $\eta_{\parallel}$, all the others can be readily found by scaling relations. These read, see [48, 103]

$$\begin{array}{ccc} \beta_s = \beta - \nu & \gamma_s = \gamma - \nu & \delta_s = (\beta+\gamma)/\beta_s \\ \beta_1 = (\nu/2)(d-2+\eta_{\parallel}) & \gamma_1 = \nu(2-\eta_{\perp}) & \gamma_{11} = \nu(1-\eta_{\parallel}) \\ \Delta_1 = (\nu/2)(d-\eta_{\parallel}) & \delta_1 = (\beta+\gamma)/\beta_1 & \delta_{11} = \Delta_1/\beta_1 \\ & \eta_{\perp} = (\eta+\eta_{\parallel})/2 & \end{array} \tag{15.11}$$

There exists a plethora of ϵ-expansion results for the critical exponents. We merely mention as an example the result for the exponent $\eta_{\parallel}$ of the $O(n)$ vector model at the ordinary transition in $d = 4 - \epsilon$ dimensions [288, 102, 103]

$$\eta_{\parallel} = 2 - \frac{n+2}{n+8}\epsilon - \frac{(n+2)(17n+76)}{2(n+8)^3}\epsilon^2 + O(\epsilon^3) \tag{15.12}$$

15.2 Conformal invariance close to a free surface

Repeating the program followed in the bulk, we now ask for the consequences of local, space-dependent scaling close to a free surface, thereby following the original work of Cardy [76]. As we did in chapter 2, we first examine the projective conformal transformations. Consider a scaling operator $\phi(\vec{r})$ and its two-point correlation function

$$G(\vec{r}_1, \vec{r}_2) = \langle \phi(\vec{r}_1)\phi(\vec{r}_2) \rangle \tag{15.13}$$

The coordinate space is restricted to a semi-infinite geometry, namely the half-plane $\vec{r} = (x, y)$ with $y \geq 0$.

We want to find the general form of the two-point function G from the requirement that it transforms covariantly under those conformal transformations which keep the free surface $y = 0$ invariant. Due to translations parallel to the surface, we have $G = G(x_1 - x_2, y_1, y_2)$. Writing $u = x_1 - x_2$, scale invariance requires that

$$u\frac{\partial G}{\partial u} + y_1\frac{\partial G}{\partial y_1} + y_2\frac{\partial G}{\partial y_2} + 2xG = 0 \tag{15.14}$$

Finally, recall that the special conformal transformations were obtained as an inversion, a translation and another inversion. The infinitesimal transformation is

$$\begin{aligned} x' &= x + \epsilon(x^2 - y^2) \\ y' &= y + 2\epsilon xy \end{aligned} \tag{15.15}$$

The Jacobian of this is, to leading order, $1 + 4\epsilon xy$. From the covariance of G, and expanding to first order in ϵ, one finds

$$(x_1^2 - x_2^2 - y_1^2 + y_2^2)\frac{\partial G}{\partial u} + 2x_1y_1\frac{\partial G}{\partial y_1} + 2x_2y_2\frac{\partial G}{\partial y_2} + 2x(x_1 + x_2)G = 0 \tag{15.16}$$

Consequently, we have two conditions for the function G, which depends on three variables. Therefore, the projective conformal group will leave a scaling function Φ undetermined. The general form of G can be found from the method of characteristics and turns out to be [76]

$$G(\vec{r}_1, \vec{r}_2) = (y_1y_2)^{-x}\,\Phi\left(\frac{(x_1 - x_2)^2 + y_1^2 + y_2^2}{y_1y_2}\right) \tag{15.17}$$

Although this result was only derived for a two-dimensional system, the derivation readily generalizes to higher dimensions as well.

Although $\Phi(v)$ is not yet determined, its form can be used to related different limiting cases. Comparing with the scaling predictions of the last section, one expects that $G \sim |x_1 - x_2|^{-2x_s}$ if $|x_1 - x_2| \to \infty$ with y_1 and y_2 fixed, where the surface exponent x_s is

$$x_s := \frac{1}{2}\left(d - 2 + \eta_{\|}\right) \tag{15.18}$$

This implies that $\Phi(v) \sim v^{-x_s}$ as $v \to \infty$. However, this limit is also reached by $y_2 \to 0$ with y_1 and $x_1 - x_2$ fixed, describing a correlation of a density close to the surface with one in the bulk. Writing $y_1 = R\cos\vartheta, x_1 - x_2 = R\sin\vartheta$, one has

$$G \sim R^{-x-x_s}(\cos\vartheta)^{x_s - x} \tag{15.19}$$

so that the form of G is completely determined one the exponent x_s is known. Note, however, that the projective conformal transformation do not determine the value of x_s.

To find both x_s and Φ in two dimensions, we have to impose covariance under the full conformal group. In doing so, we have to restrict to systems which are two-dimensional in the bulk. This implies that all what follows only applies to the ordinary transition. In order to keep the free surface geometry invariant, we have to restrict to those conformal transformations $z \to w(z) = z + \epsilon(z)$ where $\epsilon(z)$ is real analytic ($\overline{\epsilon(z)} = \epsilon(\bar{z})$). As we have seen in chapter 2, the central relation from which the whole theory was built was the conformal Ward identity eq. (2.40). While this remains valid if the contour C is restricted to lie in the upper half-plane, the subsequent separation of $T(z)$ and $\overline{T}(\bar{z})$ of the bulk system no longer applies.

Rather, one can extend the domain of definition of $T(z)$ into the lower half-plane via

$$T(z) = \overline{T}(z) \;; \quad \text{if } \mathrm{Im} z < 0 \tag{15.20}$$

and relabel $\bar{z}_i = z'_i$. Using this, the Ward identity eq. (2.40) becomes [76]

$$\frac{1}{2\pi i}\int_C dz\hat{\epsilon}(z) < T(z)\phi_1(z_1, z'_1)\cdots > + \frac{1}{2\pi i}\int_{\overline{C}} dz\hat{\epsilon}(z) < T(\bar{z})\phi_1(z_1, z'_1)\cdots >$$
$$= \sum_i \left[\hat{\epsilon}'(z_i)\Delta_i + \hat{\epsilon}(z_i)\partial_{z_i} + \hat{\epsilon}'(z'_i)\Delta_i + \hat{\epsilon}(z'_i)\partial_{z'_i}\right] < \phi(z_1, z'_1)\cdots > \tag{15.21}$$

where $\overline{C}$ is a contour lying in the lower half-plane which is oriented oppositely to the contour C and encloses the points z'_i. Now consider the physical meaning of the defintion eq. (15.20). It states that $T = \overline{T}$ when $\mathrm{Im} z = 0$. Reexpressed in cartesian coordinates, this implies that $T_{xy} = 0$ across the boundary, which means that there is **no** flux of energy across the surface. Obviously, this is the correct boundary condition to use.

As done in chapter 2, the l.h.s of eq. (15.21) can be rewritten as an integral along a contour enclosing all the z_i and z'_i. The r.h.s. may be written the same way using Cauchy's theorem. The result is, for **all** values of $\mathrm{Im} z$

$$< T(z)\phi_1(z_1, z'_1)\cdots > = \left\{\sum_i \left[\frac{\Delta_i}{(z-z_i)^2} + \frac{1}{z-z_i}\frac{\partial}{\partial z_i}\right]\right. \tag{15.22}$$
$$\left. + \sum_i \left[\frac{\Delta_i}{(z-z'_i)^2} + \frac{1}{z-z'_i}\frac{\partial}{\partial z'_i}\right]\right\} < \phi_1(z_1, z'_1)\cdots >$$

Let us compare this to the corresponding result for the bulk case, eq. (2.41). In particular, the short-distance expansion and operator product algebra only use local properties of the theory and should therefore be independent of the presence of the boundary. So the analysis presented in chapters 4 and 5 can be taken over and the correlation functions can be calculated along the same lines. So we arrive at the important conclusion that the correlation function $< \phi(z_1, z'_1)\cdots\phi(z_n, z'_n) >$ in the semi-infinite geometry, regarded as a function of $z_1, \ldots, z_n, \bar{z}_1, \ldots, \bar{z}_n$, satisfies the same differential equation as the bulk correlation function $< \phi(z_1, \bar{z}_1)\cdots\phi(z_{2n}, \bar{z}_{2n}) >$ taken as a function of $z_1, \ldots, z_{2n}$. This means that besides the points $z_1, \ldots, z_n$ at which the conformal fields are taken, their *mirror images* obtained by reflecting them at the free boundary lying on the real axis must also be taken into account.

For example, the two-point function will have the form

$$G_s = \left[(z_1-\bar{z}_1)(z_2-\bar{z}_2)/\,(z_1-z_2)(\bar{z}_1-\bar{z}_2)(z_1-\bar{z}_2)(\bar{z}_1-z_2)\right]^{2\Delta} F_s(v) \tag{15.23}$$

where $v := (z_1-z_2)(\bar{z}_1-\bar{z}_2)/(z_1-\bar{z}_1)(z_2-\bar{z}_2)$ and $F_s(v)$ satisfies the same ordinary holomorhphic differential equation as the corresponding bulk four-point correlation function. Below, we shall give an example how the free surface boundary conditions are taken into account. This will select the appropriate solution of the holomorphic differential equation.

15.3 Finite-size scaling with free boundary conditions

In analogy to what we have seen in chapter 3, finite-size effects can also be introduced and studied for the case of free boundary conditions. The main distinction with respect to periodic boundary conditions is the presence of a non-universal surface free energy term for the free energy density

$$f = A + \frac{2}{L}B + \frac{1}{L^d}U + \dots \tag{15.24}$$

where A and B are non-universal amplitudes. The amplitude A is identical to the corresponding one for periodic boundary conditions, while B is due to the presence of the free surface. The leading universal contribution is described by U, which comes from the singular part of the free energy density. The considerations of universality for the **singular** free energy density can be repeated almost literally from chapter 3. So we can write for free boundary conditions

$$\begin{aligned} f_{\mathrm{fr}}^{sin}(t,h,h_1) &= L^{-d} Y_{\mathrm{fr}}(C_1 t L^{1/\nu}, C_2 h L^{(\beta+\gamma)/\nu}, C_s h_1 L^{\Delta_1/\nu}) \\ \xi_{\mathrm{i,fr}}^{-1}(t,h,h_1) &= L^{-d} S_{\mathrm{i,fr}}(C_1 t L^{1/\nu}, C_2 h L^{(\beta+\gamma)/\nu}, C_s h_1 L^{\Delta_1/\nu}) \end{aligned} \tag{15.25}$$

for both the free energy density and the correlation lengths as measured **parallel** to the free surface, where C_1, C_2, C_s are non-universal metric factors and $Y_{\mathrm{fr}}(x_1, x_2, x_3)$ and $S_{\mathrm{i,fr}}(x_1, x_2, x_3)$ are universal finite-size scaling functions.

Here, we want to consider how finite-size effects can be calculated from conformal invariance. We need a conformal transformation mapping the upper half-plane onto the strip geometry. This is achieved by the following transformation

$$w = \frac{L}{\pi} \ln z \tag{15.26}$$

quite similar to the one considered in Figure 17 which implies that on the infinitely long strip of width L one has indeed free boundary conditions. Now, since the covariance of the correlation functions and the partition function is unchanged, the calculation of chapter 3 can be almost literally repeated. We find the analogues of the universal finite-size scaling amplitudes[17)] of eqs. (3.49) and (3.54) [74, 50]

$$\boxed{S_{\mathrm{i,fr}}(0,0,0) = \pi x_{i,s} \; , \; Y_{\mathrm{fr}}(0,0,0) = -\pi c/24} \tag{15.27}$$

where $x_{i,s}$ is the surface scaling dimension of the scaling operator ϕ_i and c is the central charge. We can also rephrase this into a statement for the critical point spectrum of the quantum Hamiltonian H in the strip geometry with finite width $L = Na$ and **free** boundary conditions

$$H = \frac{\pi}{N} L_0 - \frac{\pi c}{24} \frac{1}{N} + \text{ const.} \tag{15.28}$$

where L_0 is a generator of the Virasoro algebra. We stress that due to the presence of the free boundary conditions, the spectrum is given here by the irreducible representations of a **single** Virasoro algebra.

17) Recently, a universal amplitude-exponent relation was found for the transverse correlation length of an interface confined in a strip geometry at a second-order wetting transition [265].

These results allow to calculate surface critical exponents directly from finite-size scaling data, where the relationship between $x_s := x_{\sigma,s}$ and the exponents defined earlier is given by eq. (15.18). It is also possible to verify explicitly the values of the central charge calculated from systems with periodic boundary conditions. Finally eq. (15.27) gives a further means to fix the correct conformal normalization of a given quantum Hamiltonian with free boundary conditions. This relies on the result that for the energy density, the surface critical exponent is

$$x_{\varepsilon,s} = d \tag{15.29}$$

This has been proven to all orders in ε-expansion [104] for the $O(n)$ vector model. We shall also see below that it is satisfied for all two-dimensional spin systems we consider.

15.4 Surface operator content

By now we have everything at hand to be able to give the surface operator content for the models we have considered before. So we are not going to repeat the model definitions, but refer the reader for this and the notations used to the previous chapters.

Ising model. Since we have already discussed the Ising model correlation functions in great detail, we shall use this model again here to illustrate the calculation of surface correlation functions. We shall discuss as an example here the spin-spin correlation function [76]. The order parameter σ is a primary operator of level 2. The operator product algebra reads $\sigma \cdot \sigma = \mathbf{1} + \varepsilon$. Recall the form of the two-point correlation function eq. (15.23). The solution for F_s was already given in chapter 6, it is a linear combination of the elementary functions $\sqrt{\sqrt{1-v} \pm 1}$, where, using the notation of the previous section,

$$v = -\left[(x_1 - x_2)^2 + (y_1 - y_2)^2\right] / 4 y_1 y_2 \tag{15.30}$$

so that the physical region corresponds to $v < 0$. The correct solution is determined by the boundary conditions.

As $v \rightarrow 0$, we expect that $G_s \sim |z_1 - z_2|^{-1/4}$, which is satisfied by both solutions. On the other hand, for $v \rightarrow -\infty$ and using that $F_s(v) \sim v^{-\alpha}$, we have from (15.23) that

$$G_s \sim |x_1 - x_2|^{-1/2-2\alpha} \tag{15.31}$$

where the allowed values of $\alpha = \pm 1/4$. We know that the order parameter vanishes at the surface for an ordinary transition. This is the boundary condition which selects the physical solution. So only $\alpha = +1/4$ is acceptable. In terms of the critical exponents, this implies

$$\eta_{\parallel} = 1 \tag{15.32}$$

and the solution for F_s satisfying this boundary condition is

$$F_s(v) = \sqrt{\sqrt{1-v} + 1} - \sqrt{\sqrt{1-v} - 1} \tag{15.33}$$

Collecting all this, the final result for the two-point spin-spin correlation function is [76]

$$G_s(x_1 - x_2, y_1, y_2) = (y_1 y_2)^{-1/8} \Psi(\rho) \qquad (15.34)$$
$$\rho = \left[(x_1 - x_2)^2 + (y_1 + y_2)^2\right] / \left[(x_1 - x_2)^2 + (y_1 - y_2)^2\right]$$
$$\Psi(\rho) = \sqrt{\rho^{1/4} - \rho^{-1/4}}$$

in agreement with earlier exact Ising model calculations of McCoy and Wu [252].

We now consider the quantum Ising chain with free boundary conditions. As for the periodic ones, this provides a further check on the conformal techniques and adds another point of view. We first have to recall how the Ising quantum chain can be diagonalized for free boundary conditions. This is very simple: almost all the calculations of chapter 10 can be taken over. The quantum Hamiltonian is, see eq. (10.1)

$$H = -\frac{1}{2\eta}\left[\sum_{n=1}^{N} t\sigma^z(n) + \sum_{n=1}^{N-1}\left(\frac{1+\eta}{2}\sigma^x(n)\sigma^x(n+1) + \frac{1-\eta}{2}\sigma^y(n)\sigma^y(n+1)\right)\right] \qquad (15.35)$$

Here the coupling constant η should not be confused with a critical exponent. Both the Jordan-Wigner transformation and the diagonalization of the resulting quadratic form are independent of the boundary conditions. The diagonalized Hamiltonian is still given by eq. (10.12) where the eigenvalues Λ_k are obtained from eq. (10.18). The only new feature arises in the form of the matrix $\widehat{\mathcal{M}}$ to be diagonalized, which for free boundary conditions is

$$\widehat{\mathcal{M}} = \begin{pmatrix} t^2 + \frac{(1+\eta)^2}{2} & -t & \frac{1-\eta^2}{4} & & & & \\ -t & t^2 + \frac{1+\eta^2}{2} & -t & \ddots & & & \\ \frac{1-\eta^2}{4} & -t & \ddots & \ddots & \ddots & & \\ & \ddots & \ddots & \ddots & \ddots & \ddots & \\ & & \ddots & \ddots & \ddots & \ddots & \frac{1-\eta^2}{4} \\ & & & \ddots & \ddots & t^2 + \frac{1+\eta^2}{2} & -t \\ & & & & \frac{1-\eta^2}{4} & -t & t^2 + \frac{(1-\eta)^2}{2} \end{pmatrix} \qquad (15.36)$$

The solution of this eigenvalue problem can be done is close analogy to solving the one-dimensional Schrödinger equation with a step-like potential. In fact, the technique readily generalizes to other boundary conditions or even interactions which are no longer fully translation invariant. It is enough to illustrate the idea for the special case $\eta = 1$, to which we restrict from now on. For the eigenfunctions, we make the ansatz, with $r = 1, 2, \ldots, N$

$$(\vec{\Phi}_k)_r = Ae^{ikr} + Be^{-ikr} \qquad (15.37)$$

and the constants A and B must be conveniently chosen. For $2 \leq r \leq N-1$, this ansatz leads again to the explicit form of the lattice dispersion relation, see eq. (10.20)

$$\Lambda_k^2 = (t - \cos k)^2 + \sin^2 k \qquad (15.38)$$

The quantization condition for the k is obtained by considering the eigenvalue equations for $r = 1$ and $r = N$. This leads to the following linear system of equations

$$\begin{pmatrix} 2te^{ik}\cos k - te^{2ik} & 2te^{-ik}\cos k - te^{-2ik} \\ (1+2t\cos k)e^{ikN} - te^{ik(N-1)} & (1+2t\cos k)e^{-ikN} - te^{-ik(N-1)} \end{pmatrix} \begin{pmatrix} A \\ B \end{pmatrix} = 0 \quad (15.39)$$

and it follows that the determinant of this has to vanish. This condition determines the discretization of the k. Calculating the determinant yields [52, 64]

$$\sin kN = t \sin\left(k(N+1)\right) \quad (15.40)$$

For general values of t, this equation cannot be solved explicitly for k. At the critical point $t = 1$, however, the solution is

$$k = \frac{2m+1}{2N+1}\pi \ , \quad m = 0, 1, \ldots, N-1 \quad (15.41)$$

Then the ground state energy and the first two correlation lengths are [64]

$$\begin{aligned} E_0^{(F)} &= -\frac{1}{2}\sum_{m=0}^{N-1} 2\sin\left(\frac{2m+1}{2(2N+1)}\pi\right) \\ &\simeq -\frac{2}{\pi}N + \frac{1}{2} - \frac{1}{\pi} - \frac{\pi}{48}\frac{1}{N} + \mathcal{O}(N^{-2}) \end{aligned} \quad (15.42)$$

$$\xi_{\sigma,s}^{-1} = 2\sin\frac{\pi}{2(2N+1)} \simeq \frac{\pi}{2N} + \mathcal{O}(N^{-2}) \quad (15.43)$$

$$\begin{aligned} \xi_{\varepsilon,s}^{-1} &= 2\sin\frac{\pi}{2(2N+1)} + 2\sin\frac{3\pi}{2(2N+1)} \\ &\simeq \frac{2\pi}{N} + \mathcal{O}(N^{-2}) \end{aligned} \quad (15.44)$$

Let us discuss these results. For the ground state energy, the leading term in the number of sites N is the non-universal bulk free energy density. It is the same as obtained for $\eta = 1$ for both periodic and antiperiodic boundary conditions. The next term is the surface free energy density, which is non-universal as well. The term of order $1/N$ is universal and can be used to extract the central charge. Comparing with eq. (15.27), we read off

$$c = 1/2 \quad (15.45)$$

as expected from the periodic case. Similarly, the surface critical exponents of the order parameter and the energy density are

$$x_{\sigma,s} = 1/2 \ , \quad x_{\varepsilon,s} = 2 \quad (15.46)$$

The first of these reproduces the result $\eta_{\parallel} = 1$ via eq. (15.18), in agreement with earlier numerical results [113] and the second one reconfirms beyond perturbation theory the result eq. (15.29). In a similar way, the higher energy gaps can be calculated. Comparing with the Virasoro characters of Table 12, we find the complete operator content for the Ising model with free boundary conditions [141]

$$H_0^{(F)} = (0) \ , \ H_1^{(F)} = \left(\frac{1}{2}\right) \tag{15.47}$$

in terms of the Virasoro characters for $c = 1/2$. From this, we can understand why $x_{\varepsilon,s} = 2$ always. The corresponding state should have the same quantum numbers of charge, parity and C-parity as the ground state. The first excited state in this sector is $L_{-2}\mathbf{1}$, the stress-energy tensor, which has $x = 2$.

We digress briefly towards yet further boundary conditions. We consider the cases of fixed boundary conditions, where the boundary spins are either both "up" (++) or else "up" and "down" (+−), and mixed (M) boundary conditions, which are free boundary conditions on one side and fixed on the other. Fixing a boundary spin can be said to come from an infinite surface magnetic field. It is an easy exercise to show, using eq. (9.12), that the quantum chain with free boundary condition is dual to the chains with fixed boundary conditions. The operator content can be worked out following the same lines as above and is found to be [81, 61]

$$H^{(++)} = (0) \ , \ H^{(+-)} = \left(\frac{1}{2}\right) \ , \ H^{(M)} = \left(\frac{1}{16}\right) \tag{15.48}$$

The finite-size scaling functions for all these boundary conditions can be worked out fairly easily [65].

We remark that all these operator contents can also be found from modular invariance considerations [81]. This completes the conformal description of this system.

Three-states Potts model. The quantum Hamiltonian of the three states Potts model was given in eq. (12.1). Free boundary conditions imply $\Gamma_{N+1} = 0$. The charge sectors are $H_Q^{(F)}$, with $Q = 0, 1, 2$. These can be further decomposed by classifying the states according to their parity (see chapter 9). The model is therefore decomposed into the sectors $H_{0,\pm}^{(F)}$ and $H_{1,\pm}^{(F)} = H_{2,\pm}^{(F)}$, where $\pm$ labels the parity. The central charge can be obtained by the numerical techniques described in chapter 9 and one finds that $c = 4/5$, as expected. The surface operator content reads [141, 81]

$$H_{0,+}^{(F)} = (0) \ , \ H_{0,-}^{(F)} = (3) \ , \ H_1^{(F)} = H_{1,+}^{(F)} + H_{1,-}^{(F)} = \left(\frac{2}{3}\right) \tag{15.49}$$

in terms of $c = 4/5$ Virasoro characters. In particular we have for the leading surface exponents

$$x_{\sigma,s} = \frac{2}{3} \ , \ x_{\varepsilon,s} = 2 \tag{15.50}$$

and thus $\eta_{\|} = 4/3$, as was also confirmed numerically [113].

We also mention the results for some further boundary conditions. For fixed boundary conditions, the boundary spins can be the same, like (00), or different, like (01). Mixed boundary conditions (M) correspond to free boundary conditions on one side and fixed on the other. For the cases (00) and (M), the system has a global Z_2 symmetry and the spectrum can be decomposed into charge sectors $Q = 0, 1$. The operator content is then [81]

$$H_0^{(00)} = (0)\ ,\quad H_1^{(00)} = (3)\ ,\quad H^{(01)} = \begin{pmatrix}\frac{2}{3}\end{pmatrix}$$

$$H_0^{(M)} = \begin{pmatrix}\frac{1}{8}\end{pmatrix}\ ,\quad H_1^{(M)} = \begin{pmatrix}\frac{13}{8}\end{pmatrix} \tag{15.51}$$

The remaining non-toroidal boundary conditions were already discussed in chapter 12.

Temperley-Lieb algebra and relation with the XXZ chain. The Temperley-Lieb algebra introduced in chapter 8 can be used to yield an interesting relationship of the p-state Potts model with the XXZ quantum chain [12]. Consider the (unnormalized) quantum Hamiltonian of the p-state Potts model with free boundary conditions

$$H_{\text{Potts}} = -\sum_{n=1}^{N}\sum_{k=1}^{p-1}\Gamma_n^k - \sum_{n=1}^{N-1}\sum_{k=1}^{p-1}\sigma_n^k\sigma_{n+1}^{p-k} = 2N - 1 - \sqrt{p}\sum_{\ell=1}^{2N-1} e_\ell \tag{15.52}$$

where the operators e_ℓ are defined by [44]

$$e_{2\ell-1} = \frac{1}{\sqrt{p}}\sum_{k=0}^{p-1}\Gamma_\ell^k\ ;\quad \ell = 1, 2, \ldots, N$$

$$e_{2\ell} = \frac{1}{\sqrt{p}}\sum_{k=0}^{p-1}\sigma_\ell^k\sigma_{\ell+1}^{p-k}\ ;\quad \ell = 1, 2, \ldots, N-1 \tag{15.53}$$

These satisfy the Temperley-Lieb algebra eq. (8.52). An alternative realization of (8.52) in terms of Pauli matrices is

$$e_\ell = \frac{1}{2}\left(\sigma_\ell^x\sigma_{\ell+1}^x + \sigma_\ell^y\sigma_{\ell+1}^y\right) + \frac{1}{2}\cosh\theta\left(1 - \sigma_\ell^z\sigma_{\ell+1}^z\right) + \frac{1}{2}\sinh\theta\left(\sigma_{\ell+1}^z - \sigma_\ell^z\right) \tag{15.54}$$

where

$$\cosh\theta = \frac{1}{2}\sqrt{p} \tag{15.55}$$

It then follows [12]

$$H_{\text{Potts}} = (2N-1)\left(1 - \frac{p}{4}\right) + \sqrt{p}\,H_{\text{XXZ}} \tag{15.56}$$

where

$$H_{\text{XXZ}} = -\frac{1}{2}\sum_{n=1}^{2N-1}\left(\sigma_n^x\sigma_{n+1}^x + \sigma_n^y\sigma_{n+1}^y - \frac{1}{2}\cosh\theta\,\sigma_n^z\sigma_{n+1}^z\right) + \frac{1}{2}\sinh\theta\left(\sigma_1^z - \sigma_{2N}^z\right) \tag{15.57}$$

describes the XXZ quantum chain with a surface field. Equation (15.56) is an operator equivalence, both H_{Potts} and H_{XXZ} are, up to trivial constants, just the sum of Temperley-Lieb operators. Consequently, the possible eigenvalues of H_{Potts} and H_{XXZ} are determined by the algebra eq. (8.52). However, the degeneracy of a given eigenvalue is not determined from the algebra but rather from the representation chosen for the e_ℓ. The XXZ chain does contain eigenvalues which do not appear in the spectrum of the

Potts quantum chain. For example, the XXZ eigenstates with $\sigma_\ell^z|0\rangle = \pm|0\rangle$ uniformly for all ℓ have no analogue in the Potts model. The same correspondence can also be obtained by comparing the isotropic partition functions, see [G11].

The systematic construction of quantum systems for Temperley-Lieb algebras has been studied in detail [248] but goes beyond the scope of this introduction.

Since the XXZ chain can be diagonalized exactly using the Bethe ansatz, many of the numerical results in the models described have been confirmed analytically by mapping them onto a XXZ chain. As an example, consider the correct normalization of a quantum Hamiltonian such that conformal invariance can be applied. The correct normalization of the XXZ chain is known analytically. Then the correct normalization of H_{Potts} follows from eq. (15.57) and is given by [12]

$$\zeta = \frac{\pi\sqrt{p(4-p)}}{4\arccos(\sqrt{p}/2)} \tag{15.58}$$

For $p = 3$, we recover the normalization already applied in eq. (12.1).

Similar results apply for periodic or toroidal boundary conditions as described in detail in the literature [12].

Tricritical Ising model. The surface exponents of the tricritical Ising model were obtained from the Blume-Capel quantum chain eq. (12.6) with free boundary conditions $S_{N+1}^{x,z} = 0$ [24]. The central charge is again found to be $c = 7/10$, in agreement with the bulk value. The surface operator content is [24]

$$H_0^{(F)} = (0) \ , \ H_1^{(F)} = \left(\frac{3}{2}\right) \tag{15.59}$$

from which we read off the surface exponents

$$x_{\sigma,s} = \frac{3}{2} \ , \ x_{\varepsilon,s} = 2 \tag{15.60}$$

and therefore $\eta_{\|} = 3$.

As in the Ising model, one can introduce fixed and mixed boundary conditions. The operator content is [154]

$$H^{(++)} = (0) \ , \ H^{(+-)} = \left(\frac{3}{2}\right) \ , \ H^{(M)} = \left(\frac{7}{16}\right) \tag{15.61}$$

Yang-Lee edge singularity. The Yang-Lee singularity of the Ising model was described by the quantum chain eq. (12.31) with $\sigma_{N+1}^{x,z} = 0$. The surface operator content is simply [145]

$$H^{(F)} = (0) \tag{15.62}$$

and thus $x_{\sigma,s} = \eta_{\|}/2 = 2$.

Ashkin-Teller model. The quantum Hamiltonian of the Ashkin-Teller model was given in eq. (12.45). We have described how its spectrum can be decomposed into the sectors of the model as given by the irreducible representation of the dihedral group D_4, see eqs. (12.51). Free boundary conditions imply $\sigma_{N+1} = 0$. We have the sectors $H_{Q,\pm}^{(F)}$, where $\pm$ labels the C-parity. The surface operator content is then [20]

$$H_{0,+}^{(F)} = \sum_{m=0}^{\infty} (4m^2) + \sum_{m=1}^{\infty} (4m^2/g)$$

$$H_{0,-}^{(F)} = \sum_{m=0}^{\infty} ((2m+1)^2/g)$$

$$H_{2,+}^{(F)} = \sum_{m=0}^{\infty} ((2m+1)^2) + \sum_{m=1}^{\infty} (4m^2/g)$$

$$H_{2,-}^{(F)} = \sum_{m=0}^{\infty} ((2m+1)^2/g)$$

$$H_1^{(F)} = H_3^{(F)} = \sum_{m=0}^{\infty} \left(\frac{(2m+1)^2}{4g}\right) \tag{15.63}$$

using the $c = 1$ Virasoro characters eq. (12.57) and the coupling g depends on the model parameters as given in eq. (12.46). The surface exponents can be read off

$$x_{\sigma,s} = \frac{1}{4g} \quad , \quad x_{\varepsilon,s} = 2 \tag{15.64}$$

and thus $\eta_{\|} = 1/2g$.

This result contains quite a few interesting special cases. These can be selected by specifying g or alternatively, via (12.46), the four-spin coupling ε. For $\varepsilon = 0$, the model decouples into two Ising models. This corresponds to $g = 1/2$ and we recover $x_{\sigma,s} = 1/2$. The four-state Potts model corresponds to $g = 1/4$ while the Kosterlitz-Thouless transition has $g = 1$. We thus have

$$\eta_{\|} = \begin{cases} 2 & \text{four-states Potts model} \\ 1/2 & \text{XY model} \end{cases} \tag{15.65}$$

Finally, we mention that for peculiar values of g the spectrum of the model can be written in terms of irreducible representations of a larger algebra. We only mention here the case of superconformal invariance, which is realized for $g = 1/6, 2/3, 3/2, 6$. For example, for $g = 2/3$, one has [20]

$$H_{0,+}^{(F)} = [0] \quad , \quad H_{2,+}^{(F)} = [1] \quad , \quad H_{0,-}^{(F)} = H_{2,-}^{(F)} = [3/2] \quad , \quad H_1^{(F)} = H_3^{(F)} = [3/8]^R \tag{15.66}$$

where the $[\Delta]$ are superconformal characters which decompose in terms of $c = 1$ Virasoro characters as follows

$$[0] = \{0\} + \sum_{m=0}^{\infty} (6m^2), \; [1] = \{1\} + \sum_{m=1}^{\infty} (6m^2), \tag{15.67}$$

$$[3/2] = \sum_{m=0}^{\infty} \left(\frac{3}{2}(2m+1)^2\right), \; [3/8]^R = \sum_{m=0}^{\infty} \left(\frac{3}{8}(2m+1)^2\right)$$

where the notation of eq. (12.56) was used.

XXZ quantum chain. At last, we discuss the XXZ quantum chain. For the bulk, we had already seen that the unitary minimal conformal spectra could be obtained by a suitable projection. We now give the analogue for free, fixed and mixed boundary conditions [14, 154]. We repeat the quantum Hamiltonian

$$H = -\frac{\gamma}{2\pi \sin\gamma} \sum_{n=1}^{N-1} [\sigma^x(n)\sigma^x(n+1) + \sigma^y(n)\sigma^y(n+1) - \cos\gamma\sigma^z(n)\sigma^z(n+1)$$
$$+v\sigma^z(1) + v'\sigma^z(N)] \tag{15.68}$$

where v, v' are the surface external fields and $v = v' = 0$ corresponds to free boundary conditions. These are conveniently written [14, 154]

$$v = i \sin\gamma \coth\alpha \quad , \quad v' = i\sin\gamma \coth\alpha' \tag{15.69}$$

where α and α' are complex parameters. The spectrum is real for the following choices

$$\begin{aligned} \alpha &= \frac{i\pi}{2}(a-\psi) \ , \ \alpha' = -\frac{i\pi}{2}(a+\psi) \\ \alpha &= \frac{\pi}{2}(b - i\psi) \ , \ \alpha' = -\frac{\pi}{2}(b + i\psi) \end{aligned} \tag{15.70}$$

where a, b, ψ are real. Free boundary conditions are recovered for $a = b = 0, \psi = 1$.

Denote by $E_{Q,i}(N)$ the ith level in the charge sector Q, where the $U(1)$ charge operator is given by eq. (12.73), obtained for N sites. Let $E_0^{(F)}(N)$ be the ground state energy for the Hamiltonian with free boundary conditions. Consider the scaled gaps

$$\mathcal{E}_{Q,i} := \frac{N}{\pi}\left(E_{Q,i}(N) - E_0^{(F)}(N)\right) \tag{15.71}$$

and define the partition function as the $N \to \infty$ limit of the quantities [15, 166]

$$\mathcal{Z}_Q(q) := \sum_i q^{\mathcal{E}_{Q,i}(N)} = P(q) q^{(Q+\phi)^2/(4g)} \tag{15.72}$$

for a, b finite and $\phi = 2g(1-\psi)$, while $P(q)$ was defined in eq. (10.29). For free boundary conditions, we have again the surface exponent $x_{\sigma,s} = 1/(4g)$. We also note that for a fixed value of Q, the operator content is given by a single irreducible representation of a so-called *shifted* $U(1)$ Kac-Moody algebra [21], where the shift ϕ can be written in terms of the model parameters. We shall discuss in the next section shifted Kac-Moody algebras in connection with defect lines.

We now give the projectionchain to the surface operator content of the unitary minimal models. The procedure is quite analogous to the one of chapter 12. Take $E_{0,0}(N)$ as the ground state energy and define [154]

$$\mathcal{F}_{R,i}(N) := \frac{N}{\pi}\left(E_{R,i}(N) - E_{0,0}(N)\right) \tag{15.73}$$

As before, one gets the partition functions in the $N \to \infty$ limit

$$\mathcal{Y}_R(q) := \sum_i q^{\mathcal{F}_{R,i}(N)} = \sum_{k=-\infty}^{\infty} q^{[(nk+R+\phi)^2-\phi^2]/(4g)} P(q) \tag{15.74}$$

where $R = 0, 1, \ldots, n-1$ for N even and $2R = 1, 3, \ldots, 2n-1$ for N odd. Choose

$$g = \frac{n^2}{4m(m+1)} \; , \; \phi = \frac{n}{2m(m+1)} \tag{15.75}$$

and the partition functions $\mathcal{Y}_R$ can now be expressed in terms of the unitary Virasoro characters $\chi_{r,s}(q)$. This defines the desired projection rules.

As we did for the bulk system in chapter 12, we now give the projection rules for the p-state Potts model and the tricritical p-states Potts model. For the p-state Potts model (which had been called the $\mathbb{Z}_R$ system before), one has (see eq. (12.81)) [154]

$$\begin{aligned} \chi_{1,2u+1} &= \mathcal{Y}_{-u} - \mathcal{Y}_{u+1} \quad \text{free, fixed} \\ \chi_{1,2u} &= \mathcal{Y}_{-u+1/2} - \mathcal{Y}_{u+1/2} \quad \text{mixed} \end{aligned} \tag{15.76}$$

For the tricritical p-state Potts model, one has (see eq. (12.81)) [154]

$$\begin{aligned} \chi_{2u+1,1} &= \mathcal{Y}_{u} - \mathcal{Y}_{-u-1} \quad \text{free, fixed} \\ \chi_{2u,1} &= \mathcal{Y}_{u-1/2} - \mathcal{Y}_{-u-1/2} \quad \text{mixed} \end{aligned} \tag{15.77}$$

These results reproduce all the explicit examples given above. It is now a straightforward matter to read off the surface critical exponents of any desired unitary minimal model. Having found this, the conformal techniques outlined in the context of the Ising model can be used to calculate explicitly the multipoint correlation functions.

15.5 Defect lines

Having treated the case of a free surface in some detail, we can now ask what might happen if two critical statistical systems are weakly coupled together on their free surfaces. The line where the two systems are joined is referred to as a *defect line*. The *defect strength* κ measures the coupling between the two systems. For $\kappa = 0$, we are back to a free surface while for $\kappa = 1$, the defect line has disappeared completely. We can define local critical exponents close to the defect line in analogy to the surface exponents. We now ask for the values of these *defect line exponents*.

Here, we want to consider the effect of a single defect line. Since the defect strength couples the two systems via the interactions between neighboring spins, we obtain for the renormalization group eigenvalue y_κ, if the infinitely long defect line is treated as a perturbation of a two-dimensional system [63]

$$y_\kappa = 1 - x_\varepsilon \tag{15.78}$$

where x_ε is the scaling dimension of the energy density. Consequently, if $x_\varepsilon > 1$, the defect perturbation is irrelevant and even close to the defect line, we shall find the critical exponents of the unperturbed system. Conversely, if $x_\varepsilon < 1$, the defect perturbation will be relevant and close to the defect line the system will be described by a new fixed point and exponents.

Let us consider the role of a defect line for a bulk system as well as for a free surface. For a free surface, we have just seen above that always $x_{\varepsilon,s} = 2$. It follows that a defect line is irrelevant close to a free surface. On the other hand, the bulk exponent $x_{\varepsilon,b} < 1$ for most of the systems considered in earlier chapters. For these, the defect line is relevant. In these cases, we expect the local exponents close to the defect line to be the **same** as the ones close to a free surface [63]. A notable exception to this is the $2D$ Ising model, where $x_{\varepsilon,b} = 1$. In that case, the defect perturbation is marginal and the defect exponents are expected to be functions of the defect strength κ. In fact, the defect line exponent $\beta(\kappa)$ of the local order parameter was calculated exactly for the $2D$ Ising model with one infinite defect line [29, 254] and found to depend continuously on κ.

Turban [316] suggested that conformal invariance might be useful to study this problem. Consider the Ising model with a single infinitely long defect line in the two-dimensional plane. By the familiar logarithmic transformation $w = N/2\pi \ln z$, this system can be mapped into the strip geometry, where the system has now **two** defect lines, which are $N/2$ sites apart from each other. The calculation is parallel to the one done for free boundary conditions before, if the scaling operators still transform covariantly even in the presence of the defect lines. If this is true, then one finds again the amplitude-exponent relation between the correlation length measured parallel to the defect line and the defect exponents [316]

$$\xi_i^{-1} = L^{-1} 2\pi x_i(\kappa) \tag{15.79}$$

This relation can (and has been) explicitly checked by comparing the exactly known critical exponents as calculated in the plane [211] with the results following [157, 171, 201] from the conformal relation eq. (15.79). Here, we shall prove the conformal invariance of the $2D$ Ising model with a defect line in another way [172], which has the advantage that the calculations to be done are shorter. Furthermore, this approach sheds further light on the algebraic aspects of conformal invariance.

We consider the Ising quantum chain with a single defect line, described by the quantum Hamiltonian

$$H = -\frac{1}{2}\left[\sum_{n=1}^{N} t\sigma^z(n) + \sum_{n=1}^{N-1} \sigma^x(n)\sigma^x(n+1)\right] - \frac{\kappa}{2}\sigma^x(N)\sigma^x(1) \tag{15.80}$$

In the infinite plane, this corresponds to a system with a semi-infinite defect line. The system interpolates between periodic boundary conditions (for $\kappa = 1$) and antiperiodic boundary conditions (for $\kappa = -1$) and we expect to recover in these two cases the known operator content eq. (10.44). The off-critical behaviour of this system has received considerable attention as well and quite a complex structure for the correlation length was found [68, 351, 28, 69].

The diagonalization of H [273, 2, 175] is essentially the same as for the free boundary case treated above. The matrix to be diagonalized is

$$\widehat{\mathcal{M}} = \begin{pmatrix} t^2+1 & -t & & & & & \mathcal{Q}\kappa t \\ -t & t^2+1 & -t & & & & \\ & -t & \ddots & \ddots & & & \\ & & \ddots & \ddots & \ddots & & \\ & & & \ddots & \ddots & \ddots & \\ & & & & \ddots & t^2+1 & -t \\ \mathcal{Q}\kappa t & & & & & -t & t^2+\kappa^2 \end{pmatrix} \tag{15.81}$$

where $\mathcal{Q} = 1 - 2Q$. Repeating the ansatz eq. (15.37), we reobtain the lattice dispersion relation eq. (15.38) and the following condition for the discretization of the lattice momenta k

$$2\mathcal{Q}\kappa t \sin k + \kappa^2 t \sin(N-1)k + (\kappa^2 - 1)\sin Nk + \sin(N+1)k = 0 \tag{15.82}$$

To discuss the critical point behaviour, we put $t = 1$. Since we are merely interested in the large N limit, we make the ansatz $k = \alpha/N$ and find, to leading order in $1/N$

$$\cos\alpha = -\frac{2\kappa\mathcal{Q}}{1+\kappa^2} \tag{15.83}$$

Introducing this into the dispersion relation (15.38), we obtain the critical point Hamiltonian [172, 175], where we have subtracted the non-universal bulk and surface terms, of order $\mathcal{O}(N)$ and $\mathcal{O}(1)$, respectively

$$H = \frac{2\pi}{N}\sum_{r=0}^{\infty}\left[\left(r+\frac{1}{2}-\mathcal{D}(Q,\kappa)\right)n_r^{(-)} + \left(r+\frac{1}{2}+\mathcal{D}(Q,\kappa)\right)n_r^{(+)}\right] - \frac{\pi c_{eff}}{6N} \tag{15.84}$$

where $n_r^{(\pm)}$ are fermionic number operators and

$$\mathcal{D}(0,\kappa) = D(\kappa) \ , \ \mathcal{D}(1,\kappa) = \frac{1}{2} - D(\kappa) \tag{15.85}$$

with

$$D(\kappa) = \left|\frac{1}{\pi}\arctan\frac{1}{\kappa} - \frac{1}{4}\right| \tag{15.86}$$

Some care is necessary to select the correct branch of the inverse tangent, for example $\arctan(-1/\kappa) = \pi - \arctan(1/\kappa)$. Finally, we have [172]

$$c_{eff} = \frac{1}{2} - 6D(\kappa)^2 \tag{15.87}$$

In Figures 43 and 44 [172] we display the defect exponents as obtained from the Hamiltonian spectrum using the conformal invariance result eq. (15.79). While for $\kappa = 1$ and $\kappa = -1$, respectively, the known operator content as derived from the representations of the pair of $c = 1/2$ Virasoro algebras is recovered, these levels split into distinct multiplets for all other values of κ. Nevertheless, the spectrum is seen to show still a

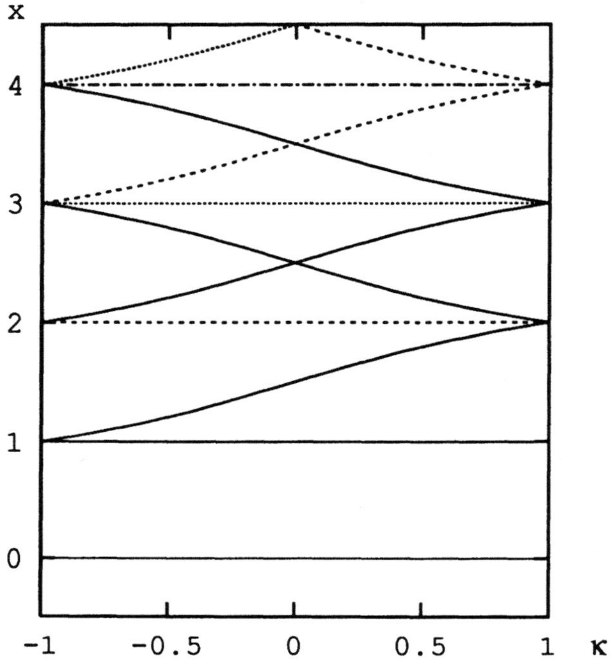

Fig. 43. Scaled spectrum $x(\kappa)$ of the $Q = 0$ sector of the Ising quantum Hamiltonian with a single defect line as a function of the defect strength κ. The degeneracies are as follows: 1 (full lines), 2 (dashed lines), 3 (dotted lines) and 5 (dash-dotted lines).

structure compatible with conformal towers. The degeneracies of the excited secondary states are those of a system with central charge $c \geq 1$, including the excited states of the vacuum. The explanation for these surprising observations follows from the explicit construction of the Virasoro generators in terms of the fermionic oscillators.

In analogy to what we have done in chapter 10, we write the Virasoro generators in terms of the normal-ordered fermionic oscillators $a_{-k} = a_k^+$ [172]

$$L_n(\mathcal{D}) = \frac{1}{2} \sum_{k \in Z+1/2} (k - \mathcal{D}) : a_{n-k+\mathcal{D}} a_{k-\mathcal{D}} : \tag{15.88}$$

In the limit $\mathcal{D} \to +0$, we recover the $c = 1/2$ Virasoro generators L_n, while for $\mathcal{D} \to -0$, we get back to the $\bar{L}_n$, and similarly for $\mathcal{D} \to \pm 1/2$. Form now on, we always take $\mathcal{D} \neq 0, 1/2$. Then the hermiticity condition is no longer valid but is replaced by

$$L_n^+(\mathcal{D}) = L_{-n}(-\mathcal{D}) \tag{15.89}$$

The commutation relations of the $L_n(\mathcal{D})$ can now be worked out following the lines of

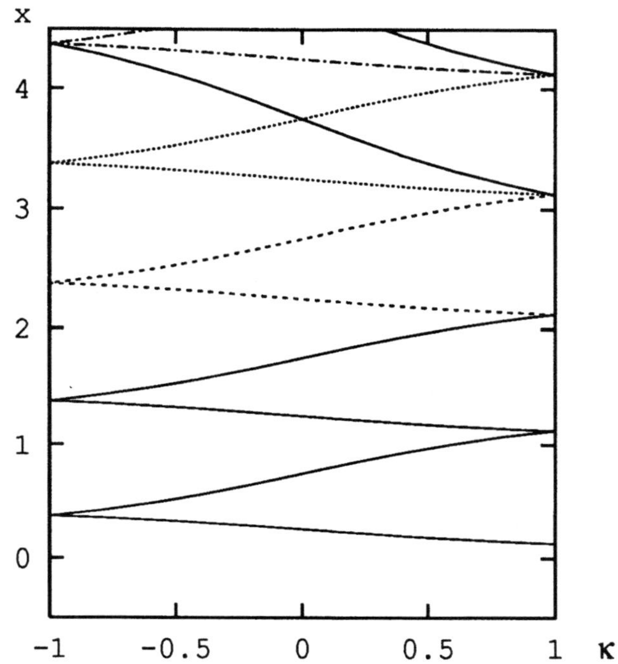

Fig. 44. Scaled spectrum $x(\kappa)$ of the $Q = 1$ sector of the Ising quantum Hamiltonian with a single defect line. The degeneracies are as for $Q = 0$.

chapter 10, with the result [172]

$$[L_n(\mathcal{D}), L_m(\mathcal{D})] = \frac{n-m}{2} L_{n+m}(\mathcal{D}) - \frac{1}{24}(n^3 - n)\delta_{n+m,0} + \frac{n}{4}\left(\mathcal{D}^2 - \frac{1}{4}\right)\delta_{n+m,0} \quad (15.90)$$

To recast this into the conventional form of the Virasoro algebra, we define

$$\tilde{L}_n(\mathcal{D}) := 2L_n(\mathcal{D}) + \frac{1}{2}\left(\mathcal{D}^2 - \frac{1}{4}\right)\delta_{n,0} \quad (15.91)$$

which satisfy a Virasoro algebra with a **negative** central charge $c = -2$, thereby illustrating the general theorem from chapter 4 that for negative central charges hermiticity is not possible [133]. Equation (15.90) is supplemented by

$$[L_n(\mathcal{D}), L_m(-\mathcal{D})] = \quad (15.92)$$
$$\frac{n-m}{2} L_n(-\mathcal{D}) - \frac{n^2}{2} W_{n+m}(-\mathcal{D}) + \frac{1}{4}\left(\frac{n^3}{3} + 2n^2\mathcal{D} + n\left(\mathcal{D}^2 - \frac{1}{12}\right)\right)\delta_{n+m,0}$$

where we have introduced a further set of generators $W_n(\mathcal{D})$

$$W_n(\mathcal{D}) = \frac{1}{2} \sum_{k \in Z+1/2} :a_{n-k+\mathcal{D}} a_{p-\mathcal{D}}: - \frac{\mathcal{D}}{2}\delta_{n,0} = -W_n(-\mathcal{D})$$
$$W_n^+(\mathcal{D}) = W_{-n}(\mathcal{D}) \tag{15.93}$$

and, using the same techniques as before, we can verify that the algebra closes

$$[W_n(\mathcal{D}), W_m(\mathcal{D})] = \frac{n}{4}\delta_{n+m,0}$$
$$[L_n(\pm\mathcal{D}), W_m(\mathcal{D})] = -\frac{m}{2}W_{n+m}(\mathcal{D}) \pm \frac{n^2}{2}\delta_{n+m,0} \tag{15.94}$$

We thus see that the presence of the defect lines does change the structure of the conformal algebra. We do no longer have a simple commuting pair of hermitian generators and have furthermore picked up the additional generators W_n, which form a $U(1)$ Kac-Moody algebra eqs. (7.9) and (7.13) among themselves. However, our quantum Hamiltonian is hermitian. If the model is conformal invariant, we must be capable to express H in terms of the generators of some Virasoro algebra. To do so, we define new hermitian generators [172]

$$K_n(\mathcal{D}) = L_n(\mathcal{D}) + L_n(-\mathcal{D}) + \frac{1}{2}\mathcal{D}^2\delta_{n,0} \tag{15.95}$$

Comparing with the spectrum of H eq. (15.84), we can relate it to $K_0(\mathcal{D})$

$$H = \frac{2\pi}{N}\left(K_0(\mathcal{D}) - \frac{1}{24}\right) \tag{15.96}$$

and the hermitian conformal algebra becomes finally

$$[K_n(\mathcal{D}), K_m(\mathcal{D})] = (n-m)K_{n+m}(\mathcal{D}) + \frac{1}{12}\left(n^3 - n\right)\delta_{n+m,0}$$
$$[K_n(\mathcal{D}), W_m(\mathcal{D})] = -mW_{n+m}(\mathcal{D})$$
$$[W_n(\mathcal{D}), W_m(\mathcal{D})] = \frac{n}{4}\delta_{n+m,0} \tag{15.97}$$

and we read off the central charge $c = 1$. This also explains the degeneracies observed in Figures 43 and 44. The infinitely many primary operators of the Virasoro algebra generated by the K_n are labelled by the eigenvalues of the generator W_0. The presence of the Kac-Moody generators leads to non-trivial relations between the scaling dimensions of the order and the disorder operators [171], the first examples being observed in [60]. We can now conclude that the $2D$ Ising model with a half-infinite defect line and its corresponding quantum chain are conformal invariant at the critical point. As previously seen for free boundary conditions, it follows fron the construction eq. (15.95) that the conformal transformations permitted include those which are real analytic. This justifies the previous proposal [316] to use the conformal mapping from the two-dimensional plane onto the strip geometry and the operator content can be obtained from the Hamiltonian spectrum using eq. (15.79) in the by now familiar way. We also see in Figures 43 and 44 that for $\kappa = 0$, corresponding to free boundary conditions, several of the $U(1)$ primary fields recombine into conformal towers of the $c = 1/2$ Virasoro algebra. This can be understood from the relation [172]

$$K_n(1/4) = \frac{1}{2}L_{2n}(0) + \frac{1}{32}\delta_{n,0} \tag{15.98}$$

since the $L_n(0)$ satisfy a $c = 1/2$ Virasoro algebra. This relation can be used to define branching rules when going from a $c = 1$ to a $c = 1/2$ Virasoro algebra, see [22].

Before writing down the operator content, we present an alternative derivation [23] of the conformal algebra (15.97), which is close in spirit to the Coulomb gas approach of chapter 7. At the same time, this clarifies the origin of the shift terms introduced in the definition of the K_n and the W_n. We go back to $\kappa = 1$ for a moment. As we have seen in chapter 10, the Ising quantum Hamiltonian can be constructed in terms of fermionic oscillators. On the other hand, we have seen in chapter 7 that the stress-energy tensor of any minimal model could be obtained as a bilinear expression in terms of a conserved $U(1)$ Kac-Moody current, see eq. (7.11). The corresponding $U(1)$ Kac-Moody algebra is

$$\begin{aligned} [L_n, L_m] &= (n-m)L_{n+m} + \frac{1}{12}\left(n^3 - n\right)\delta_{n+m,0} \\ [L_n, J_m] &= -mJ_{n+m} \\ [J_n, J_m] &= \frac{n}{4}\delta_{n+m,0} \end{aligned} \tag{15.99}$$

where J_n are the components of the current J. We remark that the L_n as written here are the **sum** of the two Virasoro algebras usually encountered before, which also accounts for $c = 1$ in the Ising model context. The components L_n of the stress-energy tensor are, as a consequence of eq. (7.11), bilinear in the J_n and read in normal-orderd form

$$L_n = \frac{1}{2}\sum_{r\in Z} : J_{n-r}J_r : \tag{15.100}$$

and the J_n have for the Ising model the fermionic representation [23]

$$J_n = i \sum_{k\in Z+1/2} : a_{n-k}b_k : \tag{15.101}$$

where $a_k^+ = a_{-k}$, $b_k^+ = b_{-k}$, $\{a_k, a_\ell\} = \{b_k, b_\ell\} = \delta_{k+\ell,0}$ and $\{a_k, b_\ell\} = 0$. We can interpret the a_k to correspond to the excitations with k small while the b_k correspond to those with k of order N, see chapter 10. At first sight, it might seem that the L_n should be of forth order in the fermionic operators. These terms cancel out, however, as a consequence of the identity [23]

$$\sum_{m'}\sum_{k,\ell} : a_{m-m'-k}a_{m'-\ell}a_k a_\ell := 0 \tag{15.102}$$

and the remaining terms are

$$L_n = \frac{1}{2}\sum_{k\in Z+1/2} k : a_{n-k}a_k + b_{n-k}b_k : \tag{15.103}$$

Now, the commutation relations of the L_n and J_n are left invariant under the automorphism [23]

$$J_n \to W_n(\mathcal{D}) := J_n + \mathcal{D}\,\delta_{n,0}$$
$$L_n \to K_n(\mathcal{D}) := L_n + \mathcal{D}J_n + \frac{1}{2}\mathcal{D}^2\delta_{n,0} \tag{15.104}$$

which introduces the ***shifted*** $U(1)$ ***Kac-Moody algebra*** [21]. The operator content is then given as follows. Starting from periodic boundary conditions in the spin variables, it follows from the Jordan-Wigner transformation that $\mathcal{D}(Q,\kappa)$ does depend on the charge sector Q. On the other hand, $\mathcal{D}$ only depends on the combination $Q+\tilde{Q}$, where $\tilde{Q}$ was used in the chapters 9 and 10 to specify the toroidal boundary conditions. The operator content for the Ising quantum chain with a single defect line then is [174, 23]

$$H^{(\kappa)}_{Q+\tilde{Q}} = \sum_{m=-\infty}^{\infty} \left(\frac{1}{2}\left(m + \mathcal{D}(Q+\tilde{Q},\kappa)\right)^2\right) \tag{15.105}$$

in terms of characters of the $U(1)$ Kac-Moody algebra eq. (15.97). These are of the form

$$\chi_{\Delta,\mathcal{D}}(r,q) = \mathrm{tr}\left(q^{K_0(\mathcal{D})} r^{W_0(\mathcal{D})}\right) = P(q)q^{\mathcal{D}^2/2} r^{\mathcal{D}} \tag{15.106}$$

in contradistinction to the Virasoro characters for $c = 1$, which are given by eq. (12.57). It follows that $\Delta = \mathcal{D}^2/2$. One has the following decomposition of the $U(1)$ Kac-Moody characters (evaluated for $r = 1$) [21]

$$\left(\frac{1}{4}m^2\right) = \sum_{k=0}^{\infty}\left(\frac{1}{4}(m+2k)^2\right)_{Vir} \tag{15.107}$$

in terms of Virasoro characters (Vir). Carrying out the sum, the stated form (15.106) is obtained.

We close the discussion with a brief comment on the maximal spectrum generating algebra. First, the algebra can be enlarged by adding the fermionic operators a_k and b_k as generators. This then yields a supersymmetric extension of the $U(1)$ Kac Moody algebra, whose structure is quite similar to the $N = 2$ superconformal algebra. If $\kappa = 1$, the quantum Hamiltonian provides an oscillator representation $< 0 >^{1/2}$ of $\widehat{U(1)}$ with the decomposition

$$< 0 >^{1/2} = \sum_{m=-\infty}^{\infty}\left(\frac{1}{2}m^2\right) \quad , \quad \kappa = 1 \tag{15.108}$$

in terms of the $U(1)$ Kac Moody characters. Extended symmetry algebras of this type have been studied in detail in the literature. For arbitrary κ, one has a shifted representation [23]

$$H^{(\kappa)} = < \mathcal{D} >^{1/2} = \sum_{m=-\infty}^{\infty}\left(\frac{1}{2}(m+\mathcal{D})^2\right) \tag{15.109}$$

giving the complete spectrum in a **single** irreducible representation.

In a similar way, the problem of n defect lines in the $2D$ Ising model can be studied and solved. This was done for line defects [175], extended regions of changed couplings with [40, 302] or without [190] an interface introduced. Defects with twists have been studied [155]. One might investigate the influence of defects on surface critical phenomena [39]. Random boundary fields are treated in [87]. Further generalizations include space-dependent couplings varying from the surface or the defect line towards the interior of the system, see [187, 49, 66, 199, 31] or else consider radially symmetric defects [32, 317]. These systems display some kind of non-universal criticality. Another possibility, but leaving the domain of applicability of conformal invariance, is to let n grow faster than the lattice size [128], eventually ariving at random systems [253, 315, 328]. Quasiperiodic [107, 38, 184] and even some aperiodic [89, 108, 239] alterations do not change the universality class but the spectra of the quantum Hamiltonians display multifractal [162, 192, 151] features. One can also generalize to other geometric forms of the surfaces [318, 274, 98], using more general conformal transformations than the ubiquitous logarithmic transformations. One particular case of this is the calculation of *corner exponents* [73]. While for free surfaces, these are at least in two dimensions simply related to the surface exponents [76], this no longer necessarily holds for more complicated cases, see [175]. Correlation functions have been calculated as well, see [225, 226, 30, 3]. A conformal field theory approach for calculating the correlation function has been started [67]. This list of further developments is certainly not complete. We refer to the literature for further information.

Defect lines were also studied numerically in the three-states Potts [300] model and the Yang-Lee singularity [145]. The results indicate that the defect exponents are equal to the free surface exponents in agreement with the renormalization group expectations [63]. In higher dimensions, defects were studied as well in the $O(n)$ model either in the $n \to \infty$ limit [63] or via ε-expansion techniques [54].

16 Outlook: Beyond the Conformal Group

The techniques of conformal invariance presented in this book are restricted to systems which are, at least in the critical regime, isotropic and static. We want to finish with an outlook towards possibilities how these restrictions might be overcome.

Consider the scaling form of a two-point correlation function at criticality, see [G2, 161]

$$G\left(br, b^{\theta}t\right) = b^{-2x} G\left(r,t\right) \tag{16.1}$$

where r and t denote "spatial" and "temporal" coordinates, x is the scaling dimension and θ is the *anisotropy exponent*. We are back to isotropic scaling for $\theta = 1$, but if $\theta \neq 1$, the system is strongly anisotropic. This anisotropy cannot be made to disappear by merely rescaling spatial and temporal coupling constants. A system which satisfies (16.1) with $\theta \neq 1$ is called a ***strongly anisotropic critical system***.

Systems of this kind are quite common. Examples include the time-dependent behaviour either at a static critical point [161, G2] or else when the system is quenched from a disordered state to a final one through a first-order transition [47, 246]. In the first case, $\theta = z$ is referred to as *dynamical critical exponent*. We only consider here exclusively systems **without** any macroscopic conservation laws. Another example is provided by a reaction-diffusion process of a molecule species A with diffusive motion and a reaction $A + A \to \emptyset$, where the reaction rates display strongly anisotropic scaling, see [240, 96]. There exist also strongly anisotropic **static** systems, we mention directed percolation [219] or systems being at a Lifshitz point [194, 303] as examples.

Equation (16.1) can be cast into the form

$$G(r,t) = t^{-2x/\theta} \Phi(u) \tag{16.2}$$

defining the scaling function $\Phi(u)$ where

$$u = r^{\theta}/t \tag{16.3}$$

is the scaling variable. While global scale invariance leaves Φ undetermined, we propose to study local, space-time dependent rescaling factors $b = b(r,t)$, generalising (16.1).

Here, we restrict attention to the case $\theta = 2$. For notational simplicity, we merely consider the case $d = 1$ of one spatial dimension, but the generalisation to arbitrary d is immediate. For $\theta = 2$, a convenient group of local scale transformations is the *Schrödinger group* defined by [259, 160]

$$r \to r' = \frac{r + vt + a}{\gamma t + \delta} \quad , \quad t \to t' = \frac{\alpha t + \beta}{\gamma t + \delta} \quad ; \quad \alpha\delta - \beta\gamma = 1 \tag{16.4}$$

where $\alpha, \beta, \gamma, \delta, v, a$ are parameters. This is the maximal group [259] which transforms solutions of the free Schrödinger equation

$$\left(i\frac{\partial}{\partial t} + \frac{1}{2m}\frac{\partial^2}{\partial r^2} \right) \psi = 0 \tag{16.5}$$

into other solutions of (16.5), viz. $(r,t) \mapsto g(r,t)$, $\psi \to T_g \psi$

$$\left(T_g \psi\right)(r,t) = f_g\left(g^{-1}(r,t)\right) \psi\left(g^{-1}(r,t)\right) \tag{16.6}$$

and the companion function f_g is known explicitly. For diffusive processes, write $m^{-1} = 2iD$, where D is the diffusion constant. The Schrödinger group can be obtained as the non-relativistic limit of the conformal group [34] and it is known that non-relativistic free field theory is Schrödinger invariant [160].

The Lie algebra of (16.4) is spanned by the set $\{X_{-1}, X_0, X_1, Y_{-1/2}, Y_{1/2}, M_0\}$ of generators which read [185]

$$\begin{aligned} X_n &= -t^{n+1}\partial_t - \frac{n+1}{2}t^n r\partial_r - \frac{n(n+1)}{4}\mathcal{M}t^{n-1}r^2 \\ Y_m &= -t^{m+1/2}\partial_r - \left(m+\frac{1}{2}\right)\mathcal{M}t^{m-1/2}r \\ M_n &= -t^n\mathcal{M} \end{aligned} \tag{16.7}$$

where the terms $\sim \mathcal{M}$ in X_n and Y_m come from the companion function and $n(m)$ is an integer (half-integer). The commutation relations are

$$\begin{aligned} [X_n, X_m] &= (n-m)X_{n+m} \\ [X_n, Y_m] &= \left(\frac{n}{2}-m\right)Y_{n+m} \\ [X_n, M_m] &= -m\,M_{n+m} \\ [Y_n, Y_m] &= (n-m)M_{n+m} \\ [Y_n, M_m] &= [M_n, M_m] = 0 \end{aligned} \tag{16.8}$$

which is the analogue of the loop algebra eq. (2.13) found in chapter 2. While M_0 is sometimes referred to as central charge of the finite-dimensional subalgebra (16.8), it is clear from the infinite algebra that it is just a component of a $U(1)$ current. From the Jacobi identities it follows that in (16.8) only the commutator $[X_n, X_m]$ admits a central extension, which must be of the familiar Virasoro form [185].

Taking over the conformal terminology of [37], the primary operators $\phi(r,t)$ of (16.8) with scaling dimension x transform as follows

$$\begin{aligned} [X_n, \phi(r,t)] &= \left(t^{n+1}\partial_t + \frac{n+1}{2}t^n r\partial_r + \frac{n(n+1)}{4}\mathcal{M}t^{n-1}r^2 + (n+1)\frac{x}{2}t^n\right)\phi(r,t) \\ [Y_m, \phi(r,t)] &= \left(t^{m+1/2}\partial_r + \left(m+\frac{1}{2}\right)\mathcal{M}t^{m-1/2}r\right)\phi(r,t) \end{aligned} \tag{16.9}$$

where $n(m)$ is integer (half-integer). Those operators, which satisfy (16.9) only for $n = 0, \pm 1$ and $m = \pm 1/2$, are quasi-primary. Following the lines of chapter 2, the two- and three-point functions of quasi-primary operators can be found. The result is [185]

$$< \phi_a(r,t)\phi_b(0,0) >= \delta_{a,b}\,\Phi_0\,t^{-x_a}\exp\left(-\frac{\mathcal{M}}{2}\frac{r^2}{t}\right) \tag{16.10}$$

$$\begin{aligned} < \phi_a(r,u)\phi_b(s,v)\phi_c(0,0) > &= u^{-(x_a+x_c-x_b)/2}v^{-(x_b+x_c-x_a)/2}(u-v)^{-(x_a+x_b-x_c)/2} \\ &\cdot \exp\left[-\frac{\mathcal{M}}{2}\left(\frac{r^2}{u}+\frac{s^2}{v}\right)\right]\Psi\left(\frac{(u-v)uv}{(rv-su)^2}\right) \end{aligned} \tag{16.11}$$

where Ψ is an arbitrary function and Φ_0 a normalization constant. This result is quite similar to the corresponding result eqs. (2.28,2.30) of conformal invariance, but we stress the important role played by the (non-universal) constant $\mathcal{M}$.

Finally, we mention an implication for the stress-energy tensor T_{ij} (we use $i = 0$ for the time coordinate). From the Ward identity (see chapter 2), scale invarance implies[18)]

$$2T_{00} + T_{11} = 0 \tag{16.12}$$

and is confirmed in free field theory[19)] [160], while from Galilei invariance it follows that

$$T_{01} = 0 \tag{16.13}$$

Then Schrödinger invariance is implied [185] and it follows that at criticality and for local interactions, Schrödinger invariance is as natural for a Galilei-invariant theory as is conformal invariance for a Lorentz-invariant system.

We now give an example of a statistical system which reproduces eq. (16.10). Consider the time-dependent $1D$ Ising model with the Hamiltonian $\mathcal{H} = -J\sum_{i=1}^{N} s_i s_{i+1}$. We are interested in the time-dependent probability distribution $P(s_1, \ldots, s_N; t)$ of the spin configuration $\{s\} = (s_1, \ldots, s_N)$. It is convenient to describe the evolution of $P(\{s\}; t)$ in terms of a *master equation* describing the rates by which either $\{s\}$ changes into another configuration $\{s'\}$ by single spin flips or $\{s'\}$ changes in $\{s\}$

$$\frac{d}{dt}P(s_1, \ldots, s_N; t) = \tag{16.14}$$

$$-\left(\sum_i w_i(s_i)\right) P(s_1, \ldots, s_N; t) + \sum_i w_i(-s_i) P(s_1, \ldots, -s_i, \ldots, s_N; t) \tag{16.15}$$

The transition rates $w_i(s_i)$ have to satisfy two conditions. First, the normalization condition

$$\sum_{\{s\}} P(\{s\}; t) = 1 \tag{16.16}$$

has to be kept at all times. Second, the equilibrium distribution $P_{eq} = e^{-\mathcal{H}/T}$ has to be a stationary solution of eq. (16.14). This is implemented by the *detailed balance* condition

$$\frac{w_i(s_i)}{w_i(-s_i)} = \exp - \left(\mathcal{H}(-s_i) - \mathcal{H}(s_i)\right)/T \tag{16.17}$$

Time-dependent average values are obtained from

$$< X > (t) = \sum_{\{s\}} X(\{s\})\, P(\{s\}; t) \tag{16.18}$$

These short remarks of course cannot substitute a systematic introduction to time-dependent critical behaviour, see for example [158, 161, G2].

---

18) Anomalies may modify this relation in the case of interacting non-relativistic theories [43].

19) Note that (16.12) refers to Euclidean theories, and not to a Minkowskian metric.

Glauber [148] made the choice

$$w_i(s_i) = \frac{\alpha}{2}\left(1 - \frac{\gamma}{2}s_i\left(s_{i-1} + s_{i+1}\right)\right) \tag{16.19}$$

where α is the transition rate and $\gamma = \tanh(2J/T)$, which satisfies the above conditions. The $1D$ Ising model with the dynamics (16.19) is completely integrable. For example, if the system is in thermal equilibrium at temperature T, one has [148]

$$G(i-j, t_1 - t_2) = < s_i(t_1)s_j(t_2) >_T = e^{-\alpha t} \sum_{\ell=-\infty}^{\infty} \eta^{|r-\ell|} I_\ell(\alpha\gamma t) \tag{16.20}$$

where $r = i - j$, $t = t_1 - t_2$, $\eta = \tanh J/T$ and I_ℓ is a modified Bessel function. The translation invariance in both time and space is already apparent. To analyse this further, recall the asymptotic expansion, as $x \to \infty$ [305]

$$I_\ell(x) \simeq (2\pi x)^{-1/2} \exp\left(x - \frac{\ell^2}{2x}\right)\left(1 + \mathcal{O}(x^{-1})\right) \tag{16.21}$$

and we have in the scaling limit $r \to \infty$, $t \to \infty$ with $u = r^2/t$ fixed

$$\begin{aligned} G(r,t) &\simeq e^{-(1-\gamma)\alpha t}(2\pi\gamma\alpha t)^{-1/2}\left\{\exp\left(-\frac{r^2}{2\gamma\alpha t}\right) + \sum_{\ell\neq 0} \eta^{|\ell|} \exp\left(-\frac{(r+\ell)^2}{2\gamma\alpha t}\right)\right\} \\ &\sim (2\pi\alpha t)^{-1/2} \exp\left(-\frac{1}{2\alpha}\frac{r^2}{t}\right) \end{aligned} \tag{16.22}$$

(up to a constant), where in the last line the limit $\gamma \to 1$, corresponding to $T \to 0$, the critical point of the $1D$ Ising model, was also performed. We recover indeed eq. (16.10).

Very little is known for arbitrary θ. An attempt [79] was made to generalise (16.1) by taking $b = b(r)$ to be a **conformal** transformation in $d = 2$ space dimensions. This already requires the system to be at a static critical point. The two-dimensional infinite space is then mapped onto the strip (see chapter 3) and it is then argued [79] that, since the strip has a finite width, the two-point function on the strip could be obtained from van Hove theory. This leads to [79] $\Phi(u) = e^{-u}$ independent of θ, where a non-universal constant was absorbed into u. Such a result would be attractive because of its great simplicity. On the other hand, heuristic arguments and evidence from the **static**, but strongly anisotropic systems of $(1+1)D$ directed percolation ($\theta \simeq 1.58$) and of Lisfshitz points in the d-dimensional spherical model ($\theta = 1/2, 2, 3$) rather suggest that, at least for $u \to \infty$, see [185]

$$\Phi(u) \sim \exp\left(-\text{const.}\ u^{1/(\theta-1)}\right) \tag{16.23}$$

independent of d. It might appear plausible that the assumptions underlying [79] are too restricted, but until now $\Phi(u)$ was only calculated in systems with $d \neq 2$. Which of the two proposed forms for $\Phi(u)$ is generally realized, if any, remains to be seen.

We stop the discussion with a remark relating the time-dependent systems studied to quantum chains. As an example, consider the $1D$ Ising model with Glauber dynamics eq. (16.19). The probability $P(s_1, \ldots, s_N; t)$ is taken to be an element of the vector $\mathcal{P}(t)$

of 2^N elements. The master equation takes the form $\partial_t \mathcal{P} = H\mathcal{P}$. For the special choice (16.19), H is nothing else than the Ising model quantum chain eq. (10.1) with the model parameters [122, 304]

$$t = \tanh(2J/T) \ , \ \eta = 1/\cosh(2J/T) \tag{16.24}$$

(not to be confused with the different use of these variables in this chapter). In other words, as T is varied, the model moves on the **disorder line** $t^2 + \eta^2 = 1$. From the spectrum of H Glauber's solution [148] can be completely recovered [122]. In general, the time-dependent $1D$ Ising model with a more general dynamics than eq. (16.19) corresponds to the XYZ chain in a magnetic field [206]. Quantum Hamiltonians also appear for reaction-diffusion problems. We mention but the simplest example here. For the one-species reaction $A + A \rightarrow \emptyset$, the quantum Hamiltonian appearing in the master equation is

$$H = -D \sum_{i=1}^{N} \left\{ \sigma_n^+ \sigma_{n+1}^- + \sigma_n^- \sigma_{n+1}^+ - 2(\varepsilon + 1)\sigma_n^+ \sigma_n^- + 2\delta\, \sigma_n^- \sigma_{n+1}^- \right\} \tag{16.25}$$

where $\sigma^\pm = (\sigma^x \pm i\sigma^y)/2$, D is the diffusion constant, δ measures the reaction rate and $\varepsilon \geq 0$ measures the left-right asymmetry of the diffusion process. From the exact solution [245, 114] quantities like the mean concentration $< c_A > (t) \sim t^{-1/2} e^{-4D\varepsilon t}$ and correlators $< c_{A;i} c_{A;i+1} > (t) \sim t^{-3/2} e^{-4D\varepsilon t}$ are obtained. The Hamiltonian (16.25) has the structure $H = H_0 + \delta\, H_1$. The global symmetry of H is Z_2, while the one of H_0 is $U(1)$. However, the perturbation H_1 is such that the spectrum of H is independent of δ. This feature is generic and a lot of information on the system can be obtained from integrable Hamiltonians as H_0.

Similar Hamiltonians can be obtained for much more general dynamical or chemical systems. Their symmetries point towards links with integrable systems and a lot of work on these problems remains to be and is currently being done.

References

[ ] The references are arranged as follows. First we give a list of general introductions to the fields of critical phenomena and conformal invariance. The works cited are roughly ordered in increasing technical complexity. The original papers used in the main text then follow in alphabetical order.

General overviews and introductory texts

[G1] M.E. Fisher in F.J.W. Hahne (Ed), *Critical Phenomena*, Springer Lecture Notes in Physics, Vol. 186 (Berlin 1983), p. 1

[G2] S.K. Ma, *Modern Theory of Critical Phenomena*, Benjamin (Reading 1976)

[G3] G.A. Baker Jr., *Quantitative Theory of Critical Phenomena*, Academic Press (New York 1990)

[G4] J. Yeomans, *Statistical Mechanics of Phase Transitions*, Oxford University Press (Oxford 1992)

[G5] C. Domb and M.S. Green (Eds.), *Phase Transitions and Critical Phenomena*, Vols. 1-6, Academic Press (New York 1972-1976)

[G6] C. Domb and J. Lebowitz (Eds.), *Phase Transitions and Critical Phenomena*, Vols. 7-15, Academic Press (New York 1983-1992)

[G7] D.J. Amit, *Field Theory, the Renormalization Group and Critical Phenomena*, 2nd ed., World Scientific (Singapore 1984)

[G8] V. Privman (Ed.), *Finite Size Scaling and Numerical Simulation of Statistical Systems*, World Scientific (Singapore 1990)

[G9] J.L. Cardy in [G6], Vol. 11, p. 55

[G10] B.M. McCoy and T.T. Wu, *The Two-dimensional Ising Model*, Harvard University Press (Cambridge 1973)

[G11] R.J. Baxter, *Exactly Solved Models in Statistical Mechanics*, Academic Press (London 1982)

[G12] C. Itzykson and J.M. Drouffe, *Statistical Field Theory*, Vols. 1,2, Cambridge University Press (Cambridge 1988)

[G13] E. Brézin and J. Zinn-Justin (Eds), *Fields, Strings and Critical Phenomena*, Les Houches XLIX, North Holland (Amsterdam 1990)

[G14] J.L. Cardy in [G13], p. 169

[G15] P. Ginsparg in [G13], p. 1

[G16] I. Affleck in [G13], p. 563

[G17] M. Lüscher in [G13], p. 451

[G18] J. Zinn-Justin, *Quantum Field Theory and Critical Phenomena*, Clarendon Press (Oxford 1989)

[G19] A.M. Polyakov, *Gauge Fields and Strings*, Harwood (Chur 1987)

[G20] J. Fröhlich, G. 't Hooft, A. Jaffe, G. Mack, P.K. Mitter and R. Stora (Eds), *New Symmetry Principles in Quantum Field Theory*, NATO-ASI, Vol. B 295, Plenum Press (New York 1992)

Reprint volumes

[R1] J.L. Cardy (Ed), *Finite-size Scaling*, North Holland (Amsterdam 1988)

[R2] C. Itzykson, H. Saleur and J.-B. Zuber (Eds), *Conformal Invariance and Applications to Statistical Mechanics*, World Scientific (Singapore 1988)

Experimental work

[E1] I. Affleck, in K. Dietz and V. Rittenberg (Eds) *Infinite-dimensional Lie Algebras and Conformal Invariance in Condensed Matter and Particle Physics*, World Scientific (Singapore 1987), p. 1

[E2] M.F. Collins, *Magnetic Critical Scattering*, Oxford University Press (Oxford 1989)

[E3] W. Dürr, D. Kerkmann and D. Pescia, Int. J. Mod. Phys. **B4**, 401 (1990)

[E4] T.L. Einstein, in K. Dietz and V. Rittenberg (Eds) *Infinite-dimensional Lie Algebras and Conformal Invariance in Condensed Matter and Particle Physics*, World Scientific (Singapore 1987), p. 17

Research papers

1. M. Abramowitz and I.A. Stegun (Eds), *Handbook of Mathematical Functions*, Dover (New York 1965)
2. D.B. Abraham, L.F. Ko and N.M. Švrakić, Phys. Rev. Lett. **61**, 2393 (1988); J. Stat. Phys. **56**, 563 (1989)
3. D.B. Abraham, N.M. Švrakić and P. Upton, HLRZ preprint 69/91
4. I. Affleck, Phys. Rev. Lett. **56**, 746 (1984)
5. I. Affleck, Phys. Rev. Lett. **56**, 2763 (1986)
6. I. Affleck and A.W.W. Ludwig, Phys. Rev. Lett. **67**, 191, 3160 (1991); Nucl. Phys. **B360**, 641 (1991)
7. L.V. Ahlfors, *Complex Analysis*, 3rd edition, McGraw-Hill (London 1979)
8. G. Albertini, B.M. McCoy and J.H.H. Perk, Phys. Lett. **135A**, 159 (1989); **139A**, 204 (1989); Adv. Stud. Pure Math. **19**, 1 (1989)
9. G. Albertini and B.M. McCoy, Nucl Phys. **B350**, 745 (1991)
10. F.C. Alcaraz, J.R. Drugowich de Felicio, R. Köberle and J.F. Stilck, Phys. Rev. **B32**, 7469 (1985)
11. F.C. Alcaraz and H.J. Herrmann, J. Phys. **A20**, 5735 (1987)
12. F.C. Alcaraz, M.N. Barber and M.T. Batchelor, Ann. of Phys. **182**, 280 (1988)
13. F.C. Alcaraz, M. Baake, U. Grimm and V. Rittenberg, J. Phys. **A21**, L117 (1988)
14. F.C. Alcaraz, U. Grimm and V. Rittenberg, Nucl. Phys. **B316**, 735 (1989)
15. F.C. Alcaraz, M. Baake, U. Grimm and V. Rittenberg, J. Phys. **A22**, L5 (1989)
16. F.C. Alcaraz, J. Phys. **A23**, L1105 (1990)
17. C.R. Allton and C.J. Hamer, J. Phys. **A21**, 2417 (1988)
18. L. Alvarez-Gaumé and P. Zaugg, Ann. of Phys. **215**, 171 (1992)
19. G.E. Andrews, R.J. Baxter and P.J. Forrester, J. Stat. Phys. **35**, 193 (1984)
20. M. Baake, G.v. Gehlen and V. Rittenberg, J. Phys. **A20**, L479, L487 (1987)
21. M. Baake, P. Christe and V. Rittenberg, Nucl. Phys. **B300**[FS22] 637 (1988)
22. M. Baake, J. Math. Phys. **29**, 1753 (1989)
23. M. Baake, P. Chaselon and M. Schlottmann, Nucl. Phys. **B314**, 625 (1989)
24. D.B. Balbão and J.R. Drugowich de Felicio, J. Phys. **A20**, L207 (1987)
25. Yu. A. Bashilov and S.V. Pokrovsky, Comm. Math. Phys. **113**, 115 (1987)
26. M.N. Barber and C.J. Hamer, J. Austr. Math. Soc. **B23**, 229 (1982)
27. M.N. Barber in [G6], Vol. 8, p. 145
28. M.N. Barber and M.E. Cates, Phys. Rev. **B36**, 2024 (1987)
29. R.Z. Bariev, Sov. Phys. JETP **50**, 613 (1979)
30. R.Z. Bariev, J. Phys. **A22**, L397 (1989)
31. R.Z. Bariev and I. Peschel, Phys. Lett. **153A**, 166 (1991)
32. R.Z. Bariev and I. Peschel, J. Phys. **A24**, L87 (1991)
33. E. Barouch and B.M. McCoy, Phys. Rev. **A3**, 786 (1971)
34. A.O. Barut, Helv. Phys. Acta **46**, 496 (1973)

35. R.J. Baxter, Phys. Lett. **133A**, 185 (1988); J. Stat. Phys. **57**, 1 (1989)
36. P.D. Beale, J. Phys. **A17**, L335 (1984)
37. A.A. Belavin, A.M. Polyakov and A.B. Zamolodchikov, Nucl. Phys. **B241**, 333 (1984); J. Stat. Phys. **34**, 763 (1984)
38. V.G. Benza, Europhys. Lett. **8**, 321 (1989)
39. B. Berche and L. Turban, J. Stat. Phys. **56**, 589 (1989)
40. B. Berche and L. Turban, J. Phys. **A24**, 245 (1990)
41. B. Berg, M. Karowski and P. Weisz, Phys. Rev. **D19**, 2477 (1979)
42. W.A. Berger, H.G. Müller, K.G. Kreuzer and R.M. Dreizler, J. Phys. **A10**, 1089 (1977)
43. O. Bergman, Phys. Rev. **D**, to be published
44. E. Berkcan, Nucl. Phys. **B215**[FS7], 68 (1983)
45. T.H. Berlin and M. Kac, Phys. Rev. **86**, 821 (1952)
46. M.A. Bershadsky, V.G. Knizhnik and M.G. Teitelman, Phys. Lett. **151B**, 31 (1985)
47. K. Binder and D. Stauffer, Phys. Rev. Lett. **33**, 1006 (1974)
48. K. Binder, in [G6], Vol. 8, p. 1
49. H.W. Blöte and H.J. Hilhorst, Phys. Rev. Lett. **51**, 2015 (1983)
50. H.W. Blöte, J.L. Cardy and M.P. Nightingale, Phys. Rev. Lett. **56**, 742 (1984)
51. H.W. Blöte and M.P.M. den Nijs, Phys. Rev. **B37**, 1766 (1988)
52. N. Boccara and G. Sarma, J. Physique Lett. **35**, L95 (1974)
53. P. Bouwknegt, Jim McCarthy and K. Pilch, Prog. Theor. Phys. Suppl. **102**, 67 (1990)
54. D. Boyanovsky and J.L. Cardy, Phys. Rev. **B26**, 154 (1982)
55. H.W. Braden, E. Corrigan, P.E. Dorey, and R. Sasaki, Phys. Lett. **227B**, 441 (1989); Nucl. Phys. **B338**, 689 (1990); **B356**, 469 (1991)
56. J.G. Brankov and V.B. Priezzhev, J. Phys. **A25**, 4297 (1992)
57. E. Brézin, J. Physique **43**, 15 (1982); Ann. New York Acad. Sciences **410**, 339 (1983)
58. E. Brézin and J. Zinn-Justin, Nucl. Phys. **B257**[FS17], 867 (1985)
59. R.C. Brower, M.A. Furman and K. Subbarao, Phys. Rev. **D15**, 1756 (1977)
60. A.C. Brown, Phys. Rev. **B25**, 331 (1982)
61. A.I. Bugrij and V.N. Shadura, Phys. Lett. **150A**, 171 (1990)
62. R. Bulirsch and J. Stoer, Num. Math. **6**, 413 (1964)
63. T.W. Burkhardt and E. Eisenriegler, Phys. Rev. **B24**, 1236 (1981); **B25**, 3283 (1982)
64. T.W. Burkhardt and I. Guim, J. Phys. **A18**, L33 (1985)
65. T.W. Burkhardt and I. Guim, Phys. Rev. **B35**, 1799 (1987)
66. T.W. Burkhardt and F. Igloi, J. Phys. **A23**, L633 (1990)
67. T.W. Burkhardt and J.-Y. Choi, Nucl. Phys. **B376**, 447 (1992)
68. G.G. Cabrera and R. Jullien, Phys. Rev. Lett. **57**, 393 (1986); Phys. Rev. **B35**, 7062 (1987)
69. G.G. Cabrera, Int. J. Mod. Phys. **B4**, 1671 (1990)
70. A. Cappelli, C. Itzykson and J.B. Zuber, Nucl. Phys. **B280**, 445 (1987); Comm. Math. Phys. **113**, 1 (1987)
71. A. Cappelli, Phys. Lett. **185B**, 82 (1987)
72. J.L. Cardy and R.L. Sugar, J. Phys. **A13**, L423 (1980)
73. J.L. Cardy, J. Phys. **A16**, 3617 (1983)
74. J.L. Cardy, J. Phys. **A17**, L385 (1984)
75. J.L. Cardy, J. Phys. **A17**, L961 (1984)
76. J.L. Cardy, Nucl. Phys. **B240**[FS12], 514 (1984)
77. J.L. Cardy, Phys. Rev. Lett. **54**, 1354 (1985)
78. J.L. Cardy, J. Phys. **A18**, L757 (1985)
79. J.L. Cardy, J. Phys. **A18**, 2771 (1985)
80. J.L. Cardy, Nucl. Phys. **B270**[FS16], 186 (1986)

81. J.L. Cardy, Nucl. Phys. **B275**[FS17], 200 (1986)
82. J.L. Cardy, J. Phys. **A20**, L891 (1987)
83. J.L. Cardy and I. Peschel, Nucl. Phys. **B300**[FS22], 377 (1988)
84. J.L. Cardy, Phys. Rev. Lett. **60**, 2709 (1988)
85. J.L. Cardy and G. Mussardo, Phys. Lett. **225B**, 275 (1989)
86. J.L. Cardy, Nucl. Phys. **B366**, 403 (1991)
87. J.L. Cardy, J. Phys. **A24**, L1315 (1991)
88. J.L. Cardy, Santa Barbara preprint UCSBTH-92-37
89. H.A. Cecatto, Z. Phys. **B75**, 253 (1989)
90. L. Chim and A.B. Zamolodchikov, Rutgers preprint RU-91-24
91. P. Christe and R. Flume, Phys. Lett. **188B**, 219 (1987)
92. P. Christe and F. Ravanini, Phys. Lett. **217B**, 252 (1989)
93. P. Christe and G. Mussardo, Nucl. Phys. **B330** 465 (1990)
94. P. Christe and G. Mussardo, Int. J. Mod. Phys. **A5** 4581 (1990)
95. P. Christe and M.J. Martins, Mod. Phys. Lett. **A5** 2189 (1990)
96. S. Cornell, B. Chopard, and M. Droz, Phys. Rev. **A44**, 4826 (1991); Physica **188A**, 322 (1992)
97. M. de Crombrugghe and V. Rittenberg, Ann. of Phys. **151**, 99 (1983)
98. B. Davies and I. Peschel, Ann. der Physik **2** to appear
99. J.M. Debierre and L.Turban, J. Phys. **A20**, 1819 (1987)
100. G.W. Delius, M.T. Grisaru, and D. Zanon, Phys. Lett. **277B**, 414 (1992); Nucl. Phys. **B382**, 365 (1992)
101. B. Derrida and L. de Seze, J. Physique **43**, 475 (1982)
102. H.W. Diehl and S. Dietrich, Phys. Lett. **80A**, 408 (1980)
103. H.W. Diehl, in [G6], Vol. 10, p. 75
104. S. Dietrich and H.W. Diehl, Z. Phys. **B43**, 315 (1981)
105. S. Dietrich, in [G6], Vol. 12, p. 1
106. E. Domany, M. Schick, J.S. Walker and R.B. Griffiths, Phys. Rev. **B18**, 2209 (1978)
107. M.M. Doria and I. Satija, Phys. Rev. Lett. **60**, 444 (1988)
108. M.M. Doria, F. Nori and I. Satija, Phys. Rev. **B39**, 6802 (1988)
109. Vl. S. Dotsenko and V.A. Fateev, Nucl. Phys. **B240**, 312 (1984)
110. Vl. S. Dotsenko and V.A. Fateev, Nucl. Phys. **B251**, 691 (1985)
111. Vl. S. Dotsenko and V.A. Fateev, Phys. Lett. **154B**, 291 (1985)
112. Vl. S. Dotsenko, Int. J. Mod. Phys. **B4**, 1039 (1990)
113. M. Droz, A. Malaspinas and A.L. Stella, J. Phys. **C18**, L245 (1985)
114. M. Droz, M. Henkel and V. Rittenberg, to be published
115. T. Eguchi and S.-K. Yang, Phys. Lett. **235B**, 282 (1990)
116. H. Eichenherr, Phys. Lett. **151B**, 26 (1985)
117. e.g. S. Elitzur, R.B. Pearson and J. Shigemitsu, Phys. Rev. **D19**, 3698 (1979)
118. V.A. Fateev and A.B. Zamolodchikov, Int. J. Mod. Phys. **A5**, 1025 (1990)
119. B.L. Feigin and D.B. Fuks, Funct. Anal. and Appl. **16**, 114 (1982); **17**, 241 (1983)
120. G. Felder, Nucl. Phys. **B317**, 215 (1989), erratum **B324**, 548 (1989)
121. G. Felder, J. Fröhlich, and G. Keller, Comm. Math. Phys. **124**, 417 (1989); **130**, 1 (1990)
122. B.U. Felderhof, Rep. on Math. Phys. **1**, 215 (1971)
123. M.E. Fisher and M.N. Barber, Phys. Rev. Lett. **28**, 1516 (1972)
124. M.E. Fisher and M.N. Barber, Arch. Rat. Mech. Anal. **47**, 205 (1972)
125. M.E. Fisher, Phys. Rev. Lett. **40**, 1610 (1978)
126. see M.E. Fisher in [G1], Appendix D
127. M.E. Fisher in *Collective Properties of Physical Systems*, Proc. 24th Nobel Symposium, Academic Press (New York 1973), p. 16

128. L. Frachebourg and M. Henkel, J. Phys. **A24**, 5121 (1991)
129. E. Fradkin and L. Susskind, Phys. Rev. **D17**, 2637 (1978)
130. P. di Francesco, H. Saleur and J.B. Zuber, J. Stat. Phys. **49**, 57 (1987)
131. P. di Francesco, H. Saleur and J.B. Zuber, Nucl. Phys. **B290**, 527 (1987)
132. P. di Francesco, Int. J. Mod. Phys. **A7**, 407 (1992)
133. D. Friedan, Z. Qiu and S. Shenker, Phys. Rev. Lett. **52**, 1575 (1984); Comm. Math. Phys. **107**, 535 (1986)
134. D. Friedan, Z. Qiu and S. Shenker, Phys. Lett. **151B**, 37 (1985)
135. J. Fröhlich and F. Gabbiani, Rev. Math. Phys. **2**, 251 (1990)
136. Y. Gefen, Y. Imry and D. Mukamel, Phys. Rev. **B23**, 6099 (1981)
137. G.v. Gehlen, C. Hoeger and V. Rittenberg, J. Phys. **A17**, L469 (1984); **A18**, 187 (1985) (err.)
138. G.v. Gehlen, V. Rittenberg and H. Ruegg, J. Phys. **A18**, 107 (1985)
139. G.v. Gehlen and V. Rittenberg, Nucl. Phys. **B257**[FS14], 351 (1985)
140. G.v. Gehlen and V. Rittenberg, J. Phys. **A19**, L625 (1986)
141. G.v. Gehlen and V. Rittenberg, J. Phys. **A19**, L631 (1986)
142. G.v. Gehlen and V. Rittenberg, J. Phys. **A20**, 227 (1987)
143. G.v. Gehlen, V. Rittenberg and T. Vescan, J. Phys. **A20**, 2577 (1987)
144. G.v. Gehlen, Nucl. Phys. **B330**, 741 (1990)
145. G.v. Gehlen, J. Phys. **A24**, 5371 (1991)
146. B. Germain-Bonne, Rev. Fran. Autom. Inform. Rech. Operat. **7**(R-1), 84 (1973)
147. P. Ginsparg, Nucl. Phys. **B295**, 153 (1988)
148. R. Glauber, J. Math. Phys. **4**, 294 (1963)
149. P. Goddard and D. Olive, Int. J. Mod. Phys. **A1**, 303 (1983)
150. P. Goddard, A. Kent and D. Olive, Comm. Math. Phys. **103**, 105 (1986)
151. C. Godrèche and J.M. Luck, J. Phys. **A23**, 3769 (1990)
152. S.G. Gorishny, S.A. Larin and F.V. Tkachov, Phys. Lett. **101A**, 120 (1984)
153. I.S. Gradshteyn and I.M. Ryzhik, *Table of Integrals, Series, and Products*, 4th edition, Academic Press (New-York 1965)
154. U. Grimm and V. Rittenberg, Int. J. Mod. Phys. **B4**, 969 (1990)
155. U. Grimm, Nucl. Phys. **B340**, 633 (1990)
156. J.C. Le Guillou and J. Zinn-Justin, J. Physique **50**, 1365 (1989)
157. L.G. Guimarães and J.R. Drugowich de Felicio, J. Phys. **A19**, L341 (1986)
158. J.D. Gunton and M. Droz, *Introduction to the Theory of Metastable and Unstable States*, Springer Lecture Notes in Physics, Vol. 183, Springer (Heidelberg 1983)
159. A.J. Guttmann in [G6], Vol. 13, p. 1
160. C.R. Hagen, Phys. Rev. **D5**, 377 (1972)
161. B.I. Halperin and P.C. Hohenberg, Rev. Mod. Phys. **49**, 435 (1977)
162. T.C. Halsey, M.H. Jensen, L.P. Kadanoff, I.Procaccia and B.I. Shraiman, Phys. Rev. **A33**, 1141 (1986)
163. C.J. Hamer and M.N. Barber, J. Phys. **A14**, 2009 (1981)
164. C.J. Hamer, J. Phys. **A15**, L675 (1982)
165. C.J. Hamer, J. Phys. **A16**, 3085 (1983)
166. C.J. Hamer and M.T. Batchelor, J. Phys. **A21**, L173 (1988)
167. E.R. Hansen, *A Table of Series and Products*, Prentice Hall (Englewood Cliffs, 1975)
168. M. Henkel, J. Phys. **A20**, 995 (1987)
169. M. Henkel, J. Phys. **A20**, 3569 (1987)
170. M. Henkel, J. Phys. **A20**, L769 (1987)
171. M. Henkel and A. Patkós, J. Phys. **A20**, 2199 (1987)
172. M. Henkel and A. Patkós, Nucl. Phys. **B285**[FS19], 29 (1987)

173. M. Henkel and G. Schütz, J. Phys. **A21**, 2617 (1988)
174. M. Henkel and H. Saleur, J. Phys. **A22**, L513 (1989)
175. M. Henkel, A. Patkós and M. Schlottmann, Nucl. Phys. **B314**, 609 (1989)
176. M. Henkel and V. Privman, Phys. Rev. Lett. **65**, 1777 (1990)
177. M. Henkel in [G8], ch. VIII, p. 353
178. M. Henkel and H.J. Herrmann, J. Phys. **A23**, 3719 (1990)
179. M. Henkel, J. Phys. **A23**, 4369 (1990); Phys. Lett. **247B**, 567 (1990)
180. M. Henkel and A.W.W. Ludwig, Phys. Lett. **249B**, 463 (1990)
181. M. Henkel and H. Saleur, J. Phys. **A23**, 791 (1990)
182. M. Henkel, J. Phys. **A24**, L133 (1991)
183. M. Henkel and R.A. Weston, J. Phys. **A25**, L207 (1992)
184. M. Henkel and A. Patkós, J. Phys. **A25**, 5223 (1992)
185. M. Henkel, Int. J. Mod. Phys. **C3**, 1011 (1992)
186. H.J. Herrmann, Phys. Lett. **100A**, 256 (1984)
187. H.J. Hilhorst and J.M.J. van Leeuwen, Phys. Rev. Lett. **47**, 1188 (1981)
188. E. Hille, *Lectures on Ordinary Differential Equations*, Addison-Wesley (New-York 1969)
189. E. Hille, *Ordinary Differential Equations in the Complex Domain*, Wiley (New-York 1976)
190. H. Hinrichsen, Nucl. Phys. **B336**, 377 (1990)
191. H. Hinrichsen and V. Rittenberg, CERN preprint TH-6453-92
192. H. Hiramoto and M. Kohmoto, Int. J. Mod. Phys. **B6**, 281 (1992)
193. T.J. Hollowood and P. Mansfield, Phys. Lett. **226B**, 73 (1989)
194. R.M. Hornreich, M. Luban and S. Strikman, Phys. Rev. Lett. **35**, 1678 (1975); Phys. Lett. **55A**, 269 (1975)
195. S. Howes, L.P. Kadanoff and M. den Nijs, Nucl. Phys. **B215**[FS7], 169 (1983)
196. D.A. Huse, Phys. Rev. **B24**, 5180 (1981)
197. D.A. Huse, Phys. Rev. **B30**, 3908 (1984)
198. D. Iagolnitzer, Phys. Rev. **D18**, 1275 (1978)
199. F. Igloi, B. Berche and L. Turban, Phys. Rev. Lett. **65**, 1773 (1990)
200. E. Ising, Z. Phys. **31**, 253 (1925)
201. A.C. Irving, G. Odor and A. Patkós, J. Phys. **A22**, 4665 (1989)
202. C. Itzykson, H. Saleur and J.-B. Zuber, Europhys. Lett. **2**, 91 (1986)
203. C. Itzykson, Int. J. Mod. Phys. **B6**, 1969 (1992)
204. W. Janke and K. Nather, HLRZ preprint 37/92
205. A. Jennings, *Matrix Computation for Engineers and Scientists*, Wiley (New York 1977)
206. J.D. Johnson and B.M. McCoy, Phys. Rev. **A6**, 1613 (1973)
207. V.G. Kac, Springer Lecture Notes in Physics, Vol. 94, Springer (Berlin 1979), p. 441
208. V.G. Kac *Infinite Dimensional Lie Algebras*, 2nd ed. Cambridge University Press (Cambridge 1985)
209. L.P. Kadanoff in [G5], Vol. 5A, p. 1
210. L.P. Kadanoff and A.C. Brown, Ann. of Phys. **121**, 318 (1979)
211. L.P. Kadanoff, Phys. Rev. **B24**, 5382 (1981)
212. M. Karowski and P. Weisz, Nucl. Phys. **B139**, 445 (1978)
213. M. Karowski, Phys. Rep. **49C**, 229 (1979)
214. D. Kastor, Nucl. Phys. **B280**[FS18], 304 (1987)
215. S. Katsura, Phys. Rev. **127**, 1508 (1962)
216. A. Kato, Mod. Phys. Lett. **A2**, 585 (1987)
217. A. Kent, Phys. Lett. **273B**, 56 (1991)
218. W. Kerner, J. Comp. Phys. **85**, 1 (1989)
219. W. Kinzel, in G. Deutscher, R. Zallen and J. Adler (Eds) *Percolation Structures and Processes*, Adam Hilger (Bristol 1983), p. 425

220. Y.Kitazawa, N. Ishibashi, A. Kato, K. Kobayashi, Y. Matsuo and S. Odake, Nucl. Phys. **B306**, 425 (1988)
221. T. Klassen and E. Melzer, Nucl. Phys. **B338**, 485 (1990); **B350**, 635 (1991)
222. T. Klassen and E. Melzer, Nucl. Phys. **B362**, 329 (1991)
223. T. Klassen and E. Melzer, Nucl. Phys. **B370**, 511 (1992)
224. C.M. Knobler and R.L. Scott, in [G6], Vol. 9, p. 164
225. L.-F. Ko, H. Au-Yang and J.H.H. Perk, Phys. Rev. Lett. **54**, 1091 (1985)
226. L.-F. Ko, Phys. Lett. **131A**, 285 (1988)
227. J.B. Kogut, Rev. Mod. Phys. **51**, 659 (1979)
228. M. Kohmoto, M. den Nijs and L.P. Kadanoff, Phys. Rev. **B24**, 5229 (1981)
229. J.M. Kosterlitz and D.J. Thouless, J. Phys. **C6**, 118 (1973)
230. M. Lässig, G. Mussardo and J.L. Cardy, Nucl. Phys. **B348**, 591 (1991)
231. M. Lässig and G. Mussardo, Comp. Phys. Comm. **66**, 71 (1991)
232. M. Lässig, Phys. Lett. **278B**, 439 (1992)
233. D.P. Landau, Phys. Rev. Lett. **28**, 449 (1972)
234. D.P. Landau and R.H. Swendsen, Phys. Rev. Lett. **46**, 1437 (1981)
235. K. Lang and W. Rühl, Nucl. Phys. **B377**, 371 (1992); Z. Phys. **C50**, 285 (1991)
236. P.G.M. Lauwers and V. Rittenberg, Phys. Lett. **233B**, 197 (1989)
237. I.D. Lawrie and S. Sarbach in [G6], Vol. 9, p. 1
238. E. Lieb, T. Schultz and D. Mattis, Ann. of Phys. **16**, 407 (1961)
239. Z. Lin and R. Tao, Phys. Lett. **150A**, 11 (1990)
240. K. Lindenberg, W.-S. Sheu and R. Kopelman, J. Stat. Phys. **65**, 1269 (1991)
241. J.M. Luck, J. Phys. **A15**, L169 (1982)
242. J.M. Luck, Phys. Rev. **B31**, 3069 (1985)
243. A.W.W. Ludwig and J.L. Cardy, Nucl. Phys. **B285**[FS19], 687 (1987)
244. M. Lüscher and U. Wolff, Nucl. Phys. **B339**, 222 (1991)
245. A.A. Lushnikov, Sov. Phys. JETP **64**, 811 (1986); Phys. Lett. **120A**, 135 (1987)
246. J. Marro, J. Lebowitz and M.H. Kalos, Phys. Rev. Lett. **43**, 282 (1979)
247. P. Martin, *Potts Model and Related Problems in Statistical Mechanics*, World Scientific (Singapore 1990)
248. P.P. Martin and V. Rittenberg, Int. J. Mod. Phys. **A7**, Suppl. 1B, 707 (1992)
249. M.J. Martins, Phys. Rev. Lett. **65**, 2091 (1990); Phys. Lett. **257B**, 317 (1991)
250. M.J. Martins, Santa Barbara preprint UCSBTH-91-45
251. Y. Matsuo and S. Yahikozawa, Phys. Lett. **178B**, 211 (1986)
252. B.M. McCoy and T.T. Wu, Phys. Rev. **162**, 436 (1967)
253. B.M. McCoy and T.T. Wu, Phys. Rev. **176**, 631 (1968)
254. B.M. McCoy and J.H.H. Perk, Phys. Rev. Lett. **44**, 840 (1980)
255. N.D. Mermin and H. Wagner, Phys. Rev. Lett. **17**, 1133 (1966)
256. A. Meurman and A. Rocha-Caridi, Comm. Math. Phys. **107**, 263 (1986)
257. G. Mussardo, Phys. Rep. **218C**, 215 (1992)
258. Th. Niemeijer, Physica **36**, 377 (1967)
259. U. Niederer, Helv. Phys. Acta **45**, 802 (1972); **46**, 191 (1973); **47**, 119, 167 (1974); **51**, 220 (1978)
260. B. Nienhuis, J. Stat. Phys. **34**, 731 (1984); in [G6], Vol. 11, p. 1
261. M.P. Nightingale, Physica **83A**, 561 (1976)
262. M.P. Nightingale and H.W. Blöte, J. Phys. **A16**, L657 (1983)
263. M.P. Nightingale in [G8], ch. VII, p. 287
264. S. Ostlund, Phys. Rev. **B24**, 398 (1981)
265. A.O. Parry, J. Phys. **A25**, L1015 (1992)
266. V. Pasquier, Nucl. Phys. **B285**, 162 (1987)

267. V. Pasquier, J. Phys. **A20**, 5707 (1987); Nucl. Phys. **B295**[FS21], 491 (1988)
268. P.A. Pearce and K.A. Seaton, J. Phys. **A23**, 1191 (1990)
269. P. Pfeuty, Ann. of Phys. **57**, 79 (1970)
270. J.L. Pichard and G. Sarma, J. Phys. **C14**, L617 (1981)
271. P. Piercy and H. Pfnür, Phys. Rev. Lett. **59**, 1124 (1987)
272. K. Penson and M. Kolb, Phys. Rev. **B29**, 2854 (1983)
273. I. Peschel and K.D. Schotte, Z. Phys. **B54**, 305 (1984)
274. I. Peschel, L. Turban and F. Igloi, J. Phys. **A24**, L1229 (1991)
275. A.M. Polyakov, Sov. Phys. JETP Lett. **12**, 381 (1970)
276. R.B. Potts, Proc. Cam. Phil. Soc. **48**, 106 (1952)
277. V. Privman and M.E. Fisher, J. Stat. Phys. **33**, 385 (1983)
278. V. Privman and M.E. Fisher, Phys. Rev. **B30**, 322 (1984)
279. V. Privman, Phys. Rev. **B32**, 6089 (1985)
280. V. Privman and S. Redner, J. Phys. **A18**, L781 (1985)
281. V. Privman, Phys. Rev. **B38**, 9261 (1988)
282. V. Privman in [G8], ch. I, p. 1
283. V. Privman, P.C. Hohenberg and A. Aharony in [G6], Vol. 14, p. 1
284. P.F. Price, C.J. Hamer and D. O'Shaughnessy, Canberra preprint 1992
285. V. Privman, Int. J. Mod. Phys. **C5**, to appear
286. F. Ravanini, Phys. Lett. **274B**, 345 (1992)
287. F. Ravanini, P. Tateo and A. Valleriani, Phys. Lett. **293B**, 361 (1992)
288. J.S. Reeve and A.J. Guttmann, Phys. Rev. Lett. **45**, 1581 (1980)
289. P. Reinicke, J. Phys. **A20**, 4501 and 5325 (1987)
290. P. Reinicke and T. Vescan, J. Phys. **A20**, L653 (1987)
291. P.A. Rikvold, W. Kinzel, J.D. Gunton and K. Kaski, Phys. Rev. **B28**, 2686 (1983)
292. V. Rittenberg in P. Dita and V. Georgescu (Eds) *Conformal Invariance and String Theory*, Academic Press (San Diego 1989), p. 37
293. A. Rocha-Caridi in J. Lepowski, S. Mandelstam and I.M. Singer (Eds) *Vertex Operators in Mathematics and Physics*, MSRI Publications #3, Springer (New York 1985), p. 451
294. H.H. Roomany and H.W. Wyld, Phys. Rev. **D21**, 3341 (1980)
295. A. Ruhe in V.A. Barker (Ed) *Sparse Matrix Techniques*, Springer Lecture Notes in Mathematics, Vol. 572, Springer (Heidelberg 1977), p. 130
296. H. Saleur, J. Phys. **A20**, L1127 (1987)
297. H. Saleur and C. Itzykson, J. Stat. Phys. **48**, 449 (1987)
298. M. Scheunert, *The Theory of Lie Superalgebras*, Springer Lecture Notes in Mathematics, Vol. 716, Springer (Heidelberg 1979)
299. M. Schick, Prog. Surf. Sci. **11**, 245 (1981)
300. M. Schlottmann, Diplomarbeit Bonn-IR-88-41 (unpublished); G. Odor (unpublished)
301. H.J. Schulz and T.A.L. Ziman, Europhys. Lett. **18**, 355 (1992)
302. W. Selke, N.M. Švrakić and P.J. Upton, Z. Phys. **B89**, 231 (1992)
303. W. Selke, in [G6], Vol. 15, p. 1
304. E. Siggia, Phys. Rev. **B16**, 2319 (1977)
305. S. Singh and R.K. Pathria, Phys. Rev. **B31**, 4483 (1983)
306. J. Sólyom and P. Pfeuty, Phys. Rev. **B24**, 218 (1981)
307. C. Sommerfield, Phys. Rev. **176**, 2019 (1968)
308. B. Spain and M.G. Smith, *Functions of Mathematical Physics*, Van Nostrand Reinhold (New-York 1970)
309. H.E. Stanley, Phys. Rev. **176**, 718 (1968)
310. H. Sugawara, Phys. Rev. **170**, 1659 (1968)
311. M. Suzuki, Prog. Theor. Phys. **46**, 1337 (1971)

312. J. Suzuki, T. Nagao and M. Wadati, Int. J. Mod. Phys. **B6**, 1119 (1992)
313. H.N.V. Temperley and E.H. Lieb, Proc. Roy. Soc. (London) **A322**, 251 (1971)
314. K. Thielemans, Int. J. Mod. Phys. **C2**, 787 (1991)
315. C.A. Tracy, J. Phys. **A21**, L603 (1988)
316. L. Turban, J. Phys. **A18**, L325 (1985)
317. L. Turban, Phys. Rev. **B44**, 7051 (1991)
318. L. Turban, J. Phys. **A25**, L127 (1992)
319. K. Uzelac and R. Jullien, J. Phys. **A14**, L151 (1981)
320. J.M. van den Broeck and L.W. Schwartz, SIAM J. Math. Anal. **10**, 658 (1979)
321. E. Verlinde, Nucl. Phys. **B300**, 360 (1988)
322. M.A. Virasoro, Phys. Rev. **D1**, 2933 (1970)
323. F. Wegner in [G5], Vol. 6, p. 8
324. E.J. Weniger, Comp. Phys. Rep. **10**, 189 (1989)
325. J. Wess and J. Bagger, *Supersymmetry and Supergravity*, Princeton University Press (Princeton 1983)
326. R.A. Weston, Phys. Lett. **248B**, 340 (1990)
327. K.G. Wilson, Phys. Rev. **179**, 1499 (1969)
328. W.F. Wolff, P. Hoever and J. Zittartz, Z. Phys. **B42**, 259 (1981)
329. F.Y. Wu, Rev. Mod. Phys. **54**, 235 (1982)
330. F.Y. Wu, Rev. Mod. Phys. **64**, 1099 (1992)
331. T.T. Wu, B.M. McCoy, C.A. Tracy and E. Barouch, Phys. Rev. **B13**, 316 (1976)
332. C.N. Yang and T.D. Lee, Phys. Rev. **87**, 404, 410 (1952)
333. C.N. Yang, Phys. Rev. **85**, 808 (1952)
334. C.N. Yang and C.P. Yang, J. Math. Phys. **10**, 1115 (1969)
335. S.K. Yang, Nucl. Phys. **B285**[FS19], 183 (1987)
336. S.K. Yang and H.B. Zheng, Nucl. Phys. **B285**[FS19], 410 (1987)
337. V.P. Yurov and Al.B. Zamolodchikov, Int. J. Mod. Phys. **A5**, 3221 (1990)
338. V.P. Yurov and Al.B. Zamolodchikov, Int. J. Mod. Phys. **A6**, 4557 (1991)
339. A.B. Zamolodchikov and Al.B. Zamolodchikov, Ann. of Phys. **120**, 253 (1979)
340. A.B. Zamolodchikov and V.A. Fateev, Sov. Phys. JETP, **62**, 215 (1985)
341. A.B. Zamolodchikov, Sov. J. Nucl. Phys. **44**, 529 (1986)
342. A.B. Zamolodchikov, Sov. Phys. JETP Lett. **43**, 730 (1986)
343. A.B. Zamolodchikov, Int. J. Mod. Phys. **A3**, 743 (1988)
344. A.B. Zamolodchikov, Int. J. Mod. Phys. **A4**, 4235 (1989)
345. A.B. Zamolodchikov, Adv. Stud. Pure Math. **19**, 641 (1989)
346. A.B. Zamolodchikov, Int. J. Mod. Phys. **A4**, 2371 (1989)
347. A.B. Zamolodchikov, Nucl. Phys. **B358**, 497, 524 (1991)
348. Al.B. Zamolodchikov, Nucl. Phys. **B342**, 695 (1990)
349. Al.B. Zamolodchikov, Nucl.Phys. **B348**, 619 (1991)
350. Al.B. Zamolodchikov, Phys. Lett. **253B**, 391 (1991)
351. J. Zinn-Justin, Phys. Rev. Lett. **57**, 3296 (1986)
352. J.-B. Zuber, Phys. Lett. **176B**, 127 (1986)

Index

— D —

— E —

— F —

— G —

— H —

— I —

GPSR Compliance

The European Union's (EU) General Product Safety Regulation (GPSR) is a set of rules that requires consumer products to be safe and our obligations to ensure this.

If you have any concerns about our products, you can contact us on ProductSafety@springernature.com

In case Publisher is established outside the EU, the EU authorized representative is:

Springer Nature Customer Service Center GmbH
Europaplatz 3
69115 Heidelberg, Germany

Batch number: 10620881

Printed by Printforce, the Netherlands